Polarization and Correlation Phenomena in Atomic Collisions

A Practical Theory Course

PHYSICS OF ATOMS AND MOLECULES

A Chronological Listing of Volumes in this series appears at the back of this volume.

Polarization and Correlation Phenomena in Atomic Collisions
A Practical Theory Course

Vsevolod V. Balashov
Alexei N. Grum-Grzhimailo

and

Nikolai M. Kabachnik
Moscow State University
Moscow, Russia

Kluwer Academic / Plenum Publishers
New York, Boston, Dordrecht, London, Moscow

Library of Congress Cataloging-in-Publication Data

Balashov, V. V. (Vsevolod Viacheslavovich)
    Polarization and correlation phenomena in atomic collisions : a practical theory course /
Vsevolod V. Balashov, Alexei N. Grum-Grzhimailo, Nikolai M. Kabachnik.
    p. cm. -- (Physics of atoms and molecules)
    Includes bibliographical references and index.

    1. Collisions (Nuclear physics) 2. Pairing correlations (Nuclear physics) 3. Polarization
(Nuclear physics) I. Grum-Grzhimailo, Alexei N. II. Kabachnik, N. M. III. Title. IV.
Series.

QC794.6.C6 B35 2000
539.7'57--dc21
                                                                        99-049492

ISBN 978-1-4419-3328-7

©2010 Kluwer Academic/Plenum Publishers, New York
233 Spring Street, New York, New York 10013

http://www.wkap.nl

10 9 8 7 6 5 4 3 2 1

A C.I.P. record for this book is available from the Library of Congress.

Printed in the United States of America

Preface

This book is based on our training course with the general title "Theoretical Practicum in Atomic and Nuclear Physics," which is used in Moscow State University (MSU) as well as in other universities of our country as one of the components in the education program for students specializing in theoretical atomic, nuclear, and particle physics. The concept of the theoretical practicum and the main aspects of its content are also well known in a number of universities in other countries as a result of special lessons given there either by ourselves or by those who had learned the theoretical practicum during their studies at MSU. Among them are the universities of Rome and Athens, Freiburg and Bielefeld (Germany), Adelaide (Australia), Missouri-Rolla (United States), Cairo, and Ulan Bator.

The course was organized in the academic year 1962–63 at the Faculty of Physics of the Moscow State University [1]. Since that time its content has been regularly modernized taking into account current progress in research in this area [2–4]. The principal goal of the course is to serve as a guide for students who are being trained to perform theoretical calculations, including those which need a long time of intensive individual work, based on the most important methods used in practice. On the other hand, we know that the books in this series are widely used by researchers, theoreticians and experimentalists, in their everyday work as a collection of useful formulas and tables.

The complete program of the theoretical practicum contains such subjects as modern methods in quantum theory of collisions, group theory, solid-state physics, atomic spectroscopy, physics of nuclear reactions, computer simulation of particle propagation through matter, and others. The concepts of density matrix and statistical tensors as well as their application in polarization and correlation studies has always been central to the program. To this end, we started from the famous review by Devons and Goldfarb [5], accepted its style of presentation of the subject, and intensified its pedagogical aspect.

In contrast to all our earlier publications, this book is devoted exclusively to atomic physics. It is known that polarization and correlation studies play a growing role in this branch of modern physics. The main tendencies in atomic physics along this line are the subject of excellent review papers and monographs such as those by Fano and Macek [6], Kessler [7], Blum [8], Andersen et al. [9–12], Danos and Fano [13], and many others. Several recent conferences and workshops have been devoted to polarization and correlation phenomena in atomic collisions, and many particular contributions are briefly described, for example, in Refs. 14–19.

These studies show that the density matrix and tensor operator method is one of the most important and effective instruments in theoretical investigations in the field. This method has been widely used and partly developed in our own work on excitation of atomic autoionizing states in electron–atom collisions, the coincidence $(e, 2e)$ method in autoionization studies, polarization and angular anisotropy of Auger and photoelectrons, inelastic and superelastic scattering in electron–atom collisions, the coincidence $(e, e'\gamma)$ method, photon- and electron-impact excitation of atomic autoionizing and Auger states from laser-excited targets, the theory of cascade processes, polarization effects in antiproton–atom interactions, and other studies. The reader will notice the obvious influence of our research work on the general approach to the problem as well as on the examples chosen to illustrate the basic theoretical methods.

We address this book to a reader with a background in elementary quantum mechanics and general atomic physics at the undergraduate level. All necessary information about the physical meaning of special polarization and correlation phenomena considered in the book is given in the text. It is not our aim to present a wide variety of approaches to describing the dynamics of the processes considered. We see our duty as demonstrating the universality of the density matrix and tensor operator methods. For this reason, throughout the book, we do not link any of them to any special choice among the models and approximations of the dynamic aspects of the processes under consideration as well as the choice of the atomic wave functions. To make the book a real *practicum*, we supplied it with a set of tables that summarize the most important information on the algebra of angular momentum. They overlap considerably with those collected in the well-known book by Varshalovich, Moskalev, and Khersonskii [20].

This book is the first in the theoretical practicum series which is published in English. We are very glad to meet our new readers. Let this book be our small contribution to the international exchange of ideas and experience in the education field.

ACKNOWLEDGMENTS

There is a long list of people whose kind advice and active support have been of great importance for us on the long path of the theoretical practicum to this special edition. First of all, we are grateful to our colleagues at Moscow State University and the Joint Institute of Nuclear Research, Dubna: Grigory Korenman, Yuri Smirnov, Nikolai Yudin, and Vladimir Belyaev, who were the first authors of the series. We thank V. Korotkikh, V. Mileev, V. Popov, V. Senashenko, S. Strakhova, V. Dolinov, and Yu. Krementsova, who continued their work.

We are also indebted to those who are not among the authors of the Theoretical Practicum, but with whom we conducted our research in atomic physics and had numerous discussions on the problems considered in this book. They include S. Baier, E. Berezhko, I. Bodrenko, V. Bubelev, S. Burkov, A. Dorn, I. Golokhov, M. Gorelenkova, U. Hergenhahn, O. Klochkova, V. Kondratiev, I. Kozhevnikov, O. Lee, O. Lhagwa, S. Lipovetski, B. Lohmann, A. Magunov, S. Martin, I. Sazhina, V. Shakirov, V. Sizov, M. Wedowski, K. Ueda, and B. Zhadamba.

We always met a deep interest in our theoretical practicum and in our research work on polarization and correlations in many leading institutes in the physics of electronic and atomic collisions and atomic spectroscopy. We thank Profs. V. Afrosimov, K. Bartschat, K. Becker, U. Becker, D. Berenyi, K. Blum, A. Crowe, S. Datz, H. Ehrhardt, U. Heinzmann, H. Klar, V. Lengyel, H. Lutz, J. Macek, N. Martin, I. McCarthy, W. Mehlhorn, V. Schmidt, B. Sonntag, E. Weigold, and P. Zimmermann for very interesting discussions, their support for our visits to their institutes, and their fruitful collaboration.

Our special thanks go to Prof. Hans Kleinpoppen for his very important support of this edition.

V. Balashov
A. Grum-Grzhimailo
N. Kabachnik

Contents

Density Matrix and Statistical Tensors

1.1. Description of Mixed States by Density Matrix

1.1.1. Pure and Mixed States

The states of physical systems in quantum mechanics can be classified in two categories: *pure states* and *mixed states*. A pure state is characterized by a certain wave function or a state vector $| \psi \rangle$ in an abstract Hilbert space. No vector $| \psi \rangle$ can be related to the mixed state. The necessity of introducing mixed states arises for quantum systems that are not closed. For example, a subsystem that is part of a larger closed system, cannot be characterized by its own wave function, which depends on coordinates of the subsystem, and therefore it is in a mixed state. The system cannot be considered closed even when it is isolated at present if it was not isolated in the past due to interaction with another system. This is a typical situation in the analysis of different characteristics of reaction products, when only part of the products (or part of the characteristics of the product) are observed. Mixed states are described by a *statistical* or *density operator* ρ, which operates in the Hilbert space and in general has the form:

$$\rho = \sum_n W_n | \psi_n \rangle \langle \psi_n | \tag{1.1}$$

where $| \psi_n \rangle$ is a complete set of state vectors and the weight coefficients, W_n, are real positive numbers which, as we will see later, characterize the probability of finding the system in a particular pure state. The density operator is a generalization of the concept of the state vector. Every mixed state is characterized by its density operator.

The explicit form of the wave function that describes a pure state $|\psi\rangle$ depends on the representation ξ: $\psi(\xi) = \langle\xi|\psi\rangle$. Similarly, the mixed state is described by the matrix elements of the density operator in the representation ξ:

$$\langle\xi|\rho|\xi'\rangle = \sum_n W_n\langle\xi|\psi_n\rangle\langle\psi_n|\xi'\rangle = \sum_n W_n\psi_n(\xi)\psi_n^*(\xi') \qquad (1.2)$$

All physical results, such as mean values and distributions of arbitrary observables, are independent of the representation. The matrix of the density operator, $\langle\xi|\rho|\xi'\rangle$, is called the *density matrix*.

If only one coefficient $W_n = 1$ and all others are equal to zero, the system is in a pure state $|\psi_n\rangle$. The density operator for the pure state $|\psi\rangle$ is

$$\rho = |\psi\rangle\langle\psi| \qquad (1.3)$$

and the corresponding density matrix is

$$\langle\xi|\rho|\xi'\rangle = \langle\xi|\psi\rangle\langle\psi|\xi'\rangle = \psi(\xi)\psi^*(\xi') \qquad (1.4)$$

We see that a pure state is a particular case of a mixed state. A description of a system by means of the density operator is the most general form of the quantum-mechanical description of the system.

1.1.2. Main Properties of Density Operator and Density Matrix

In this section we give some important properties of the density operator and density matrix.

- The density matrix is a positively defined Hermitian matrix:

$$\langle\xi|\rho|\xi'\rangle^* = \langle\xi'|\rho|\xi\rangle \qquad (1.5)$$

or, in the operator form:

$$\rho^+ = \rho \qquad (1.6)$$

which follows immediately from definition (1.2).

- Usually the density operator is normalized in such a way that the trace of the density matrix is unity:

$$\mathrm{Tr}\langle\xi|\rho|\xi'\rangle = \sum_\xi \langle\xi|\rho|\xi\rangle = 1 \qquad (1.7)$$

or, in the operator form:

$$\mathrm{Tr}\rho = 1 \qquad (1.8)$$

In this case the weighting coefficients W_n can be interpreted as the probabilities of finding the system in the state $|\psi_n\rangle$. The sum of all probabilities is, naturally, unity:

$$\langle \xi | \rho | \xi \rangle = \sum_n W_n |\psi_n(\xi)|^2 \tag{1.9}$$

$$\sum_\xi \langle \xi | \rho | \xi \rangle = \sum_n W_n = 1 \tag{1.10}$$

Other normalizations of the density operator are also possible. It is often convenient to normalize the density matrix for the final state of the collision process in such a way that the trace of the matrix is equal to the cross section σ of the process considered:

$$\mathrm{Tr}\,\rho = \sigma \tag{1.11}$$

In general, it is more convenient to use the normalization (1.10) when calculating the polarization parameters. The alternative normalization is usually used in calculations of the cross sections. However, in this chapter we always suggest that the density matrix has a unit trace.

We formulate here the properties of the density matrix for the case of an integrable basis $|\xi\rangle$ (discrete basis). If the basis includes a continuum the properties remain valid formally after introducing a corresponding normalization.

- In general, the matrix elements of a density matrix are complex. The N-dimensional complex matrix $\langle n | \rho | n' \rangle$ contains $2N^2$ real parameters. However, from Eqs. (1.5) and (1.7), it follows that in fact the density matrix is characterized only by $N^2 - 1$ independent real parameters.

- The mean value of an arbitrary observable described by the operator F in a state described by the density operator ρ is given by

$$\langle F \rangle = \mathrm{Tr}(\rho F) \tag{1.12}$$

or in matrix notation,

$$\langle F \rangle = \sum_{\alpha\beta} \langle \alpha | \rho | \beta \rangle \langle \beta | F | \alpha \rangle \tag{1.13}$$

Taking eigenvectors $|\varphi_i\rangle$ of the operator F as the basis, one obtains for the mean value

$$\langle F \rangle = \sum_i \langle \varphi_i | \rho | \varphi_i \rangle F_i = \sum_i W(F_i) F_i \tag{1.14}$$

where $F | \varphi_i \rangle = F_i | \varphi_i \rangle$ and the probability of the observable F having the eigenvalue F_i in a state ρ is defined by the weight of the corresponding pure state $| \varphi_i \rangle$ in the state ρ:

$$W(F_i) = \langle \varphi_i | \rho | \varphi_i \rangle \tag{1.15}$$

From the completeness of the set $| \varphi_i \rangle$ and from (1.7), it follows that

$$\sum_i W(F_i) = 1 \tag{1.16}$$

In a general case, Eqs. (1.14) and (1.16) include an integral over a continuum and Eq. (1.15) gives, not only the probabilities of the discrete values F_i, but also the probability density in the continuum of values of the observable F.

If the density matrix is normalized, not by condition (1.8), but by any other condition, then the mean value of the observable F is given by

$$\langle F \rangle = \frac{\text{Tr}(\rho F)}{\text{Tr}\rho} \tag{1.17}$$

- Under a unitary transformation U corresponding to a change in the basis (transformation to another representation), the density operator is transformed by the same law as the operators of observables:

$$F' = UFU^+, \qquad \rho' = U\rho U^+ \tag{1.18}$$

The corresponding transformation for the density matrix is

$$\langle \xi | \rho | \xi' \rangle = \sum_{\alpha\alpha'} \langle \xi | \alpha \rangle \langle \alpha | \rho | \alpha' \rangle \langle \alpha' | \xi' \rangle \tag{1.19}$$

- A particular case of the unitary transformation is the time evolution, which is governed by the *evolution operator* (atomic units $e = \hbar = m_e = 1$ are used throughout):

$$U(t,t_0) = \exp[-iH(t - t_0)] \tag{1.20}$$

Here H is the Hamiltonian of the system, which we assumed to be time independent. In the Schrödinger representation, the density operator $\rho(t)$ at time t is related to the density operator $\rho(t_0)$ at time t_0 by

$$\rho(t) = U(t,t_0)\rho(t_0)U^+(t,t_0) \tag{1.21}$$

Taking the time derivative of both sides of Eqs. (1.20) and (1.21), one finds that the density operator $\rho(t)$ satisfies the *Liouville equation*:

$$i\frac{\partial\rho(t)}{\partial t} = [H, \rho(t)] \tag{1.22}$$

- Operator $U(\infty, -\infty)$ in Eq. (1.21) coincides with the S-operator in the scattering theory. It follows that

$$\rho^f = S\rho^i S^+ \qquad (1.23)$$

where ρ^i is the initial density operator of a system (i.e., before scattering) and ρ^f is the final density operator of the system (i.e., after scattering). Excluding elastic scattering in a forward direction, which is not directly observable and will not be considered in this book, one can write

$$\rho^f = T\rho^i T^+ \qquad (1.24)$$

where T is the *T-operator*, or *transition operator* $T = S - I$, I being the unit operator. In the matrix notation, Eq. (1.24) takes the form

$$\langle \xi \,|\, \rho^f \,|\, \xi' \rangle = \sum_{\alpha\alpha'} \langle \xi \,|\, T \,|\, \alpha \rangle \, \langle \xi' \,|\, T \,|\, \alpha' \rangle^* \, \langle \alpha \,|\, \rho^i \,|\, \alpha' \rangle \qquad (1.25)$$

Here $\langle \xi \,|\, T \,|\, \alpha \rangle$ is a T-matrix element or the amplitude of transition from the state α to the state ξ. Note that in scattering theory the amplitude is defined as proportional to the T-matrix on the energy shell. In this book we do not need to distinguish between the T-matrix element and the amplitude. The transition operator conserves total energy, total linear momentum, and total angular momentum with its projection, while a general normalization constant is usually not relevant in derivations for correlation and polarization characteristics of reactions.

- As all Hermitian matrices the density matrix can always be transformed into diagonal form by some unitary transformation. Let $|\,\psi_n\,\rangle$ be the eigenvectors of the statistical operator ρ, so that the density matrix is diagonal in the representation of the vectors $|\,\psi_n\,\rangle$: $\langle \psi_n \,|\, \rho \,|\, \psi_n \rangle = \rho_n \delta_{nn'}$, where $\delta_{nn'}$ is the Kronecker delta symbol. The fact that the density operator ρ is positively defined means that all eigenvalues ρ_n of the density matrix are not negative:

$$\rho_n \geq 0 \qquad (1.26)$$

Note that ρ_n are equivalent w_n in Eq. (1.1). Combining Eq. (1.26) with the normalization condition (1.10), one obtains

$$\sum_n \rho_n = 1, \qquad 0 \leq \rho_n \leq 1 \qquad (1.27)$$

where the summation is over all eigenvalues of the density matrix. The eigenvectors $|\,\psi_n\,\rangle$ make up a complete set:

$$\sum_n |\,\psi_n\,\rangle \langle \psi_n| = I \qquad (1.28)$$

Introduce the projection operator $\mathcal{P}_n = |\psi_n\rangle\langle\psi_n|$, which is the density operator of the pure state $|\psi_n\rangle$ [see Eq. (1.3)]. After the operator ρ acts on both sides of Eq. (1.28), one obtains an expansion of the density operator of the mixed state in terms of density operators of the pure states:

$$\rho = \sum_n \rho_n |\psi_n\rangle\langle\psi_n| = \sum_n \rho_n \mathcal{P}_n \qquad (1.29)$$

Different terms of the sum (1.29) will not interfere in calculations of any observable [see Eq. (1.12)]. So Eq. (1.29) shows that a mixed state is an *incoherent mixture* of pure states, while the eigenvalues ρ_n are weights of the corresponding pure states in the mixed state. The density matrix describes a pure state if the density matrix has only one nonvanishing eigenvalue. Otherwise it describes an incoherent mixture of pure states. In the first case one says also that we have a *fully coherent* state.

- Using Eqs. (1.29) and (1.27), one obtains

$$\mathrm{Tr}\left(\rho^2\right) = \sum_n \rho_n^2 \leq 1 \qquad (1.30)$$

The equality

$$\mathrm{Tr}\left(\rho^2\right) = 1 \qquad (1.31)$$

is valid if only one of the eigenvalues ρ_n is nonzero (and hence equal to unity): $\rho_n = \delta_{nn_0}$. This is the case for a pure state:

$$\rho = |\psi_{n_0}\rangle\langle\psi_{n_0}| \qquad (1.32)$$

In a mixed state always

$$\mathrm{Tr}\left(\rho^2\right) < 1 \qquad (1.33)$$

The relation (1.31) gives a guideline in deciding whether a state with a given density matrix is pure or mixed. Note that in accordance with Eq. (1.32), an even stronger relation is valid for the pure state:

$$\rho^2 = \rho \qquad (1.34)$$

- Consider two arbitrary states of a system which are characterized by the density operators ρ and τ. Suppose that the system is in one of the states, ρ. Then the probability of finding the system in another state τ is given by the projection of the state ρ onto the state τ:

$$W = \mathrm{Tr}(\rho\tau) \qquad (1.35)$$

or in matrix notation,

$$W = \sum_{\alpha\beta} \langle \alpha \,|\, \rho \,|\, \beta \rangle \langle \beta \,|\, \tau \,|\, \alpha \rangle \tag{1.36}$$

Expression (1.35) is a generalization of the corresponding rule for the pure quantum-mechanical states Ψ and Φ: $W = |\langle \Phi | \Psi \rangle|^2$.

- Suppose that a system consisting of two parts (subsystems), 1 and 2, is in a mixed state described by the density operator ρ. One can express the density operators of each of the subsystems, $\rho^{(1)}$ and $\rho^{(2)}$, in terms of the density operator ρ of the entire system. We require that the mean value of an arbitrary observable, F_1 (a characteristic of the first subsystem) in the state $\rho^{(1)}$ coincide with the mean value of the same observable F_1 calculated for the entire system:

$$\langle F_1 \rangle = \text{Tr}(\rho^{(1)} F_1) = \text{Tr}(\rho F_1) \tag{1.37}$$

It follows that $\rho^{(1)} = \text{Tr}^{(2)} \rho$ (and similarly $\rho^{(2)} = \text{Tr}^{(1)} \rho$) where the super-script 1 or 2 at the trace sign means that the summation is over the variables of the corresponding subsystem. In matrix notation,

$$\left\langle n_1 \left| \rho^{(1)} \right| n_1' \right\rangle = \sum_{n_2} \langle n_1 n_2 \,|\, \rho \,|\, n_1' n_2 \rangle \tag{1.38}$$

where n_1 and n_2 symbolize basis state vectors or variables of the subsystems 1 and 2, respectively. That is, in order to find the density matrix of a subsystem, one must calculate the trace of the density matrix of the whole system over all of the variables except the variables of the subsystem being considered.

In a particular case, the density matrix of the composite system can be written as a product of the density matrices of the subsystems:

$$\langle n_1 n_2 \,|\, \rho \,|\, n_1' n_2' \rangle = \left\langle n_1 \left| \rho^{(1)} \right| n_1' \right\rangle \left\langle n_2 \left| \rho^{(2)} \right| n_2' \right\rangle \tag{1.39}$$

Then the distribution of an arbitrary observable, F_1, of the subsystem 1 is independent of the distribution of an arbitrary observable, F_2, of the subsystem 2, and vice versa. One may say in this case that the subsystems are independent.

1.2.　Spin Density Matrix and Statistical Tensors

1.2.1.　Spin Density Matrix

A particular case of a density matrix is a *spin density matrix*. This is a density matrix of the system in the representation of angular momenta: spin, orbital, and/or total. For example, the spin density matrix describing the spin state of a particle with fixed spin s consists of matrix elements $\langle sm_s \mid \rho \mid sm_s' \rangle \equiv \rho_{m_s m_s'}$, where $m_s, m_s' = -s, \ldots, +s$ are the projections of spin onto the chosen z-axis. It has the dimension $2s + 1$. The spin density matrix of a particle with spin s and orbital angular momentum l in the representation of uncoupled l and s includes matrix elements $\langle lm_l sm_s \mid \rho \mid lm_l' sm_s' \rangle$ with $m_l, m_l' = -l, \ldots, +l; m_s, m_s' = -s, \ldots, +s$. This matrix has the dimension $(2l + 1)(2s + 1)$. Angular momenta in the spin density matrix need not be fixed. Mixtures of states with different values of angular momenta, described by nondiagonal matrix elements, such as $\langle lm \mid \rho \mid l'm' \rangle$, are very important.

In the following paragraphs we deal almost exclusively with the spin density matrices and often write "density matrix" for brevity instead of "spin density matrix."

Consider a system with sharp angular momentum j. The density matrix of this system $\langle jm \mid \rho \mid jm' \rangle$ has the dimension $(2j + 1)$. Given that the density operator ρ is Hermitian and is normalized by the condition (1.7), we find that the number of independent real parameters characterizing the density matrix $\langle jm \mid \rho \mid jm' \rangle$ is $(2j + 1)^2 - 1 = 4j(j + 1)$. Note that this number increases rapidly as a function of j: there are 3 parameters for $j = \frac{1}{2}$, but 8 parameters for $j = 1$ and 24 parameters for $j = 2$. The particular values of the matrix elements depend on the choice of the reference frame. The diagonal elements of the spin density matrix give the probabilities of the system having certain projections of the spin (angular momentum) onto the chosen z-axis. Therefore, they characterize an orientation of the angular momentum in space. If the angular momentum is randomly oriented (there is no preferential direction of its orientation), then all of the projections are equally probable and all magnetic substates are equally populated. In this case the spin density matrix is diagonal in any reference frame and has the form

$$\langle jm \mid \rho \mid jm' \rangle = \delta_{mm'} (2j + 1)^{-1} \tag{1.40}$$

1.2.2. Definition of Statistical Tensors

An equivalent description of a system with angular momenta can be given in terms of *statistical tensors*. Consider first a case when a system is characterized by an angular momentum j (which can take different values). We construct the following combinations of the matrix elements of the density matrix $\langle jm \mid \rho \mid j'm' \rangle$:

$$\rho_{kq}(j, j') = \sum_{mm'} (-1)^{j'-m'} (jm, j'-m' \mid kq) \langle jm \mid \rho \mid j'm' \rangle \tag{1.41}$$

where $(jm, j' - m' \mid kq)$ are the *Clebsch–Gordan coefficients* (see Section A.8 in the Appendix). Equation (1.41) is the definition of the statistical tensor $\rho_{kq}(j, j')$. The reverse transformation follows from the unitarity of the Clebsch–Gordan coefficients (A.88):

$$\langle jm | \rho | j'm' \rangle = \sum_{kq} (-1)^{j'-m'} (jm, j' - m' | kq) \, \rho_{kq}(j, j') \tag{1.42}$$

According to the general properties of the Clebsch–Gordan coefficients (see Section A.8.1), index k is an integral number within the range from $|j - j'|$ to $j + j'$ at fixed j and j'; for each k the value of index q may be $q = -k, \ldots, +k$. Altogether there are $\sum_k (2k + 1) = (2j + 1)(2j' + 1)$ statistical tensors $\rho_{kq}(j, j')$. This coincides with the number of different combinations of indices m and m'. The set of $2k + 1$ quantities $\rho_{kq}(j, j'), q = -k, \ldots, +k$ for a certain k forms an *irreducible tensor* of rank k. This means that $\rho_{kq}(j, j')$ transforms in a corresponding way under rotations of the coordinate frame (see Section A.7). To derive this transformation, we write the statistical tensor (1.41) in the rotated coordinate frame $\tilde{S}$:

$$\widetilde{\rho_{k'q'}}(j, j') = \sum_{\mu\mu'} (-1)^{j'-\mu'} (j\mu, j' - \mu' | k'q') \left\langle \widetilde{j\mu} | \rho | \widetilde{j'\mu'} \right\rangle \tag{1.43}$$

Expressing bra and ket vectors at the right side of Eq. (1.43) in terms of the states $|jm\rangle$ in the initial coordinate frame S by applying the *Wigner D-functions* [see Eq. (A.42)], using Eqs. (1.42), (A.48), and (A.87) one can obtain

$$\widetilde{\rho_{k'q'}}(j, j') = \delta_{kk'} \sum_q \left(D^k_{qq'}(\omega) \right)^* \rho_{kq}(j, j') \quad . \tag{1.44}$$

Equation (1.44) gives the statistical tensors of a system (particle, photon, atom) in the coordinate frame $\tilde{S}$ if the statistical tensors of the system are known in the coordinate frame S. The rotation ω in Eq. (1.44) characterizes the transformation of the frame S into the frame $\tilde{S}$ and is described by the set of three *Euler angles*: $\omega = (\alpha, \beta, \gamma)$ (see Section A.6). Comparing Eq. (1.44) with Eq. (A.59), we see that the set of quantities (1.41) transforms under rotation as a covariant (or conjugate) irreducible tensor of the rank k. The value of q is called the *projection of the statistical tensor*. (Equivalent names for the statistical tensors are used in literature: *multipole moments, state multipoles, tensors of orientation,* etc.)

If a system with angular momentum j is isotropic (i.e., the angular momentum is randomly oriented), it is described by the spin density matrix (1.40). Substituting it into the definition (1.41) of the statistical tensors and using Eqs. (A.85) and (A.86), we obtain

$$\rho_{kq}(j, j) = \frac{1}{\hat{j}} \delta_{k0} \delta_{q0} \tag{1.45}$$

Thus the only nonzero statistical tensor of an *isotropic (unpolarized)* system is $\rho_{00}(j, j)$. We have introduced here a brief notation $\hat{a}\hat{b}\ldots\hat{c} = [(2a + 1)(2b + 1)\cdots(2c + 1)]^{1/2}$ which will be used throughout the book.

If at least one nontrivial $(k > 0)$ statistical tensor of the system is nonzero, the system is called *polarized*. A polarized system characterized by statistical tensors of even rank only is called *aligned*. Finally, if at least one odd rank statistical tensor is nonzero, the system is called *oriented* (or *vector polarized*).

1.2.3. Some General Properties of Statistical Tensors

Owing to the fact that the density matrix $\langle jm \,|\, \rho \,|\, j'm' \rangle$ is Hermitian, the following relation holds for the statistical tensors

$$\rho_{kq}^{*}\left(j, j'\right) = (-1)^{j'-j+q} \rho_{k-q}\left(j', j\right) \tag{1.46}$$

Then, consider restrictions on the statistical tensors due to the normalization condition for the density matrix. Substituting Eq. (1.42) into the normalization condition (1.7) with the use of a summation formula (A.86), one obtains

$$\sum_{j} \hat{j} \rho_{00}\left(j, j\right) = 1 \tag{1.47}$$

or, in the case of normalization to the cross section [see Eq. (1.11)], Eq. (1.47) transforms into

$$\sum_{j} \hat{j} \rho_{00}\left(j, j\right) = \sigma \tag{1.48}$$

where the sum over j runs over all possible values of the angular momentum of the system. Of special importance is the case of a system with sharp angular momentum. For a system with fixed angular momentum j, Eq. (1.47) gives

$$\rho_{00}\left(j, j\right) = \frac{1}{\hat{j}} \tag{1.49}$$

which is also in accordance with Eq. (1.40). The dimensionless polarization and correlation parameters mainly considered in this book do not depend on the normalization conditions. It is convenient to introduce *reduced statistical tensors*,

$$\mathcal{A}_{kq}(j, j) = \frac{\rho_{kq}\left(j, j\right)}{\rho_{00}\left(j, j\right)} \tag{1.50}$$

which are independent of normalization of the density matrix.

As was discussed in Section 1.2.1, the density matrix of a system with sharp angular momentum j $\langle jm \,|\, \rho \,|\, jm' \rangle$ is determined by $4j(j+1)$ real parameters. One can obtain the same number $4j(j+1)$ by counting the independent real parameters that determine the complete set of statistical tensors $\rho_{kq}\left(j, j\right)$:

$k = 0, \ldots, 2j; q = -k, \ldots, +k$ describing the system. As follows from Eq. (1.46), in the case of $j = j'$ components with positive and negative q are related by

$$\rho_{kq}^*(j,j) = (-1)^q \rho_{k-q}(j,j) \tag{1.51}$$

and, therefore, all components of the tensors with $q = 0$ are real for all possible values of the rank k:

$$\rho_{k0}^*(j,j) = \rho_{k0}(j,j) \tag{1.52}$$

Condition (1.51) limits the number of independent real parameters characterizing the tensor $\rho_{kq}(j,j)$ at a given k to $2k+1$. Summation over k from 0 up to $2j$ gives the value $(2j+1)^2$. Since a statistical tensor with zero rank is always fixed by normalization, the number of independent real parameters of the statistical tensors $\rho_{kq}(j,j)$ is $(2j+1)^2 - 1 = 4j(j+1)$. This is precisely the result obtained for the number of independent real parameters of the density matrix. An actual number of the parameters of the spin density matrix or statistical tensors for sharp j can be less than $4j(j+1)$ owing to additional restrictions imposed by symmetry or other particular physical features of the system and by its "history."

The criterion of a pure state (1.31) is transformed, after applying the equality (1.42), the Hermitian property of the statistical tensors (1.46), and the orthogonality of the Clebsch–Gordan coefficients (A.87), into

$$\sum_{jj'} \sum_{k=0}^{j+j'} \sum_{q=-k}^{k} \rho_{kq}(j,j') \rho_{kq}^*(j',j) = 1 \tag{1.53}$$

For a system with fixed angular momentum j, Eq. (1.53) takes the form

$$\sum_{k=0}^{2j} \sum_{q=-k}^{k} |\rho_{kq}(j,j)|^2 = 1 \tag{1.54}$$

According to the definition (1.41), all components of the statistical tensor with nonzero q vanish if the density matrix $\langle jm | \rho | j'm' \rangle$ is diagonal with respect to the projections $(m = m')$:

$$\rho_{kq}(j,j') = \delta_{q0} \rho_{k0}(j,j') \tag{1.55}$$

The inverse statement follows from Eq. (1.42): The density matrix of the state is diagonal with respect to m if the statistical tensors of this state have only components with $q = 0$.

The time evolution of statistical tensors can be found from the general relations (1.20)–(1.22) for the density matrix. Using eigenvectors of the Hamiltonian, $H|\alpha jm\rangle = E_{\alpha j}|\alpha jm\rangle$, as a basis set for the density matrix, one obtains

$$\rho_{kq}(\alpha j, \alpha' j'; t) = \rho_{kq}(\alpha j, \alpha' j'; t = 0) \exp i\omega_{\alpha j, \alpha' j'} t \tag{1.56}$$

where $\omega_{\alpha j,\alpha' j'} = E_{\alpha j} - E_{\alpha' j'}$. Here we used notation α in the state vector for a set of other possible quantum numbers that specify the state.

An important equation relates the statistical tensors of the final state of a system to the statistical tensors of its initial state in terms of the T-matrix elements, or transition amplitudes. Writing Eq. (1.25) in the representation of the total angular momentum j of the system, we have

$$
\langle \beta jm \,|\, \rho^f \,|\, \beta' j'm' \rangle
$$
$$
= \sum_{\substack{\alpha JM \\ \alpha' J'M'}} \langle \beta jm \,|\, T \,|\, \alpha JM \rangle \, \langle \beta' j'm' \,|\, T \,|\, \alpha' J'M' \rangle^* \langle \alpha JM \,|\, \rho^i \,|\, \alpha' J'M' \rangle
$$
$$
= \sum_{\alpha\alpha'} \langle \beta jm \,|\, T \,|\, \alpha jm \rangle \, \langle \beta' j'm' \,|\, T \,|\, \alpha' j'm' \rangle^* \langle \alpha jm \,|\, \rho^i \,|\, \alpha' j'm' \rangle \quad (1.57)
$$

where β denotes all other quantum numbers that specify the final state. In the second equation we used the fact that the transition operator conserves the total angular momentum and its projection (being a scalar operator in the full configuration space). Applying the *Wigner–Eckart theorem* (A.62), the relation between the spin density matrix and statistical tensors (1.42), and the orthogonality of the Clebsch–Gordan coefficients (A.87), we obtain the general relation between the statistical tensors of the initial and final states:

$$
\rho^f_{kq}(\beta j,\beta' j') = \frac{1}{\hat{j}\hat{j}'} \sum_{\alpha\alpha'} \langle \beta j \,\|\, T \,\|\, \alpha j \rangle \, \langle \beta' j' \,\|\, T \,\|\, \alpha' j' \rangle^* \, \rho^i_{kq}(\alpha j,\alpha' j') \quad (1.58)
$$

where the *reduced amplitudes* $\langle \beta j \,\|\, T \,\|\, \alpha j \rangle$ are introduced [see Eq. (A.62)]. A conclusion from Eq. (1.58) is that the tensorial components of statistical tensors of the total angular momentum of a system evolve independently and the system can not gain new tensorial components that were absent in the initial state. This basic rule is a consequence of conservation of the total angular momentum and its projection and is valid until the system can be treated as isolated.

1.2.4. Statistical Tensors of Several Angular Momenta

Consider a system that is composed of two subsystems with angular momenta j_1 and j_2. They can be the orbital angular momentum and spin of a particle, or angular momenta of different sorts of particles (for example, the angular momenta of an ejected electron and a residual ion). It is convenient to use the representation of the total angular momentum for the density matrix of the whole system:

$$
\langle j_1 j_2 JM \,|\, \rho \,|\, j'_1 j'_2 J'M' \rangle = \sum_{m_1 m_2 m'_1 m'_2} (j_1 m_1, j_2 m_2 | JM)\,(j'_1 m'_1, j'_2 m'_2 | J'M')
$$
$$
\times \langle j_1 m_1 j_2 m_2 \,|\, \rho \,|\, j'_1 m'_1 j'_2 m'_2 \rangle \quad (1.59)
$$

Expression (1.59) is a particular case of the transformation (1.19). Obviously, the density matrices on the left and on the right sides of Eq. (1.59) are square matrices of equal dimensions. The density matrix on the right side of Eq. (1.59) has a product form if there is no correlation between angular momenta of the two subsystems,

$$\langle j_1 m_1 j_2 m_2 \,|\, \rho \,|\, j_1' m_1' j_2' m_2' \rangle = \langle j_1 m_1 \,|\, \rho^{(1)} \,|\, j_1' m_1' \rangle \langle j_2 m_2 \,|\, \rho^{(2)} \,|\, j_2' m_2' \rangle \quad (1.60)$$

while there is no factorization in the representation of the total angular momentum J.

To construct the statistical tensors of the composite system in the representation of the total angular momentum, one should use Eq. (1.41):

$$\rho_{kq}\left(j_1 j_2 J, j_1' j_2' J'\right) = \sum_{MM'} (-1)^{J'-M'} \left(JM, J'-M' \,|\, kq\right)$$

$$\times \langle j_1 j_2 JM \,|\, \rho \,|\, j_1' j_2' J' M' \rangle \quad (1.61)$$

The statistical tensor (1.61) shows all the properties described in Section 1.2.3. In the representation of projections of the angular momenta of the subsystems, however, one must introduce new quantities, *double statistical tensors*. These are defined by the relation:

$$\rho_{k_1 q_1 k_2 q_2}\left(j_1 j_2, j_1' j_2'\right) = \sum_{m_1 m_2 m_1' m_2'} (-1)^{j_1' - m_1'} \left(j_1 m_1, j_1' - m_1' \,|\, k_1 q_1\right)(-1)^{j_2' - m_2'}$$

$$\times \left(j_2 m_2, j_2' - m_2' \,|\, k_2 q_2\right) \langle j_1 m_1 j_2 m_2 \,|\, \rho \,|\, j_1' m_1' j_2' m_2' \rangle$$

$$(1.62)$$

The relation between the statistical tensors (1.62) and (1.61) is a transformation between two different coupling schemes: $\mathbf{J} = \mathbf{j}_1 + \mathbf{j}_2$; $\mathbf{J}' = \mathbf{j}_1' + \mathbf{j}_2'$; $\mathbf{k} = \mathbf{J} + \mathbf{J}'$ and $\mathbf{k}_1 = \mathbf{j}_1 + \mathbf{j}_1'$; $\mathbf{k}_2 = \mathbf{j}_2 + \mathbf{j}_2'$; $\mathbf{k} = \mathbf{k}_1 + \mathbf{k}_2$. It has the form

$$\rho_{kq}\left(j_1 j_2 J, j_1' j_2' J'\right) = \sum_{k_1 k_2} \langle (j_1 j_2) J (j_1' j_2') J' : k \,|\, (j_1 j_1') k_1 (j_2 j_2') k_2 : k \rangle$$

$$\times \sum_{q_1 q_2} (k_1 q_1, k_2 q_2 \,|\, kq) \, \rho_{k_1 q_1 k_2 q_2}\left(j_1 j_2, j_1' j_2'\right) \quad (1.63)$$

which can be derived from Eqs. (1.61) and (1.62) and the definition of the Clebsch–Gordan coefficients (A.82). It is usually written in terms of $9j$-*symbols* (A.117):

$$\rho_{kq}\left(j_1 j_2 J, j_1' j_2' J'\right)$$

$$= \sum_{k_1 q_1 k_2 q_2} \hat{k}_1 \hat{k}_2 \hat{J} \hat{J}' \, (k_1 q_1, k_2 q_2 \,|\, kq) \begin{Bmatrix} j_1 & j_2 & J \\ j_1' & j_2' & J' \\ k_1 & k_2 & k \end{Bmatrix} \rho_{k_1 q_1 k_2 q_2}\left(j_1 j_2, j_1' j_2'\right) \quad (1.64)$$

The transformation (1.64) is unitary. The inverse transformation has the form

$$\rho_{k_1 q_1 k_2 q_2}\left(j_1 j_2, j_1' j_2'\right)$$

$$= \sum_{JJ'kq} \hat{k}_1 \hat{k}_2 \hat{J} \hat{J}' \left(k_1 q_1, k_2 q_2 | k q\right) \left\{ \begin{array}{ccc} j_1 & j_2 & J \\ j_1' & j_2' & J' \\ k_1 & k_2 & k \end{array} \right\} \rho_{kq}\left(j_1 j_2 J, j_1' j_2' J'\right) \quad (1.65)$$

If the density matrix $\langle j_1 m_1 j_2 m_2 | \rho | j_1' m_1' j_2' m_2' \rangle$ is of the product form [see Eq. (1.60)], then the double statistical tensor also factorizes:

$$\rho_{k_1 q_1 k_2 q_2}\left(j_1 j_2, j_1' j_2'\right) = \rho_{k_1 q_1}^{(1)}\left(j_1, j_1'\right) \rho_{k_2 q_2}^{(2)}\left(j_2, j_2'\right) \quad (1.66)$$

where the statistical tensors of the subsystems on the right side of Eq. (1.66) are related to the corresponding density matrices of the subsystems $\langle j_1 m_1 | \rho^{(1)} | j_1' m_1' \rangle$ and $\langle j_2 m_2 | \rho^{(2)} | j_2' m_2' \rangle$ by Eq. (1.41). In the particular case when $k_1 = q_1 = 0$ or $k_2 = q_2 = 0$, the product form (1.66) is valid irrespective of the existence of correlations between angular momenta of the subsystems. For example, putting $k_2 = q_2 = 0$ into the definition (1.62), using the property (A.93), the general rule (1.38), and Eq. (1.49), one obtains

$$\rho_{k_1 q_1 0 0}\left(j_1 j_2, j_1' j_2'\right) = \frac{1}{\hat{j}_2} \delta_{j_2 j_2'} \rho_{k_1 q_1}^{(1)}\left(j_1, j_1'\right) = \rho_{00}^{(2)}\left(j_2, j_2'\right) \rho_{k_1 q_1}^{(1)}\left(j_1, j_1'\right) \quad (1.67)$$

The statistical tensors of the subsystem can be found from the statistical tensors of the whole system by the relations

$$\rho_{kq}^{(2)}\left(j_2, j_2'\right) = \sum_{JJ'j_1} (-1)^{J+j_1+j_2'+k} \hat{J}\hat{J}' \left\{ \begin{array}{ccc} j_2 & j_2' & k \\ J' & J & j_1 \end{array} \right\} \rho_{kq}\left(j_1 j_2 J; j_1 j_2' J'\right)$$

$$= \sum_{j_1} \hat{j}_1 \rho_{00kq}\left(j_1 j_2, j_1 j_2'\right) \quad (1.68)$$

and

$$\rho_{kq}^{(1)}\left(j_1, j_1'\right) = \sum_{JJ'j_2} (-1)^{J'+j_1+j_2+k} \hat{J}\hat{J}' \left\{ \begin{array}{ccc} j_1 & j_1' & k \\ J' & J & j_2 \end{array} \right\} \rho_{kq}\left(j_1 j_2 J; j_1' j_2 J'\right)$$

$$= \sum_{j_2} \hat{j}_2 \rho_{kq00}\left(j_1 j_2, j_1' j_2\right) \quad (1.69)$$

where we introduced $6j$-symbols (see Section A.9). Equations (1.68) and (1.69) are equivalent to the calculation of the trace of the spin density matrix over variables of the unobserved system. Their derivation is straightforward. It uses the definition of the statistical tensors (1.41), Eq. (1.38), the summation formula (A.89), and properties of the Clebsch–Gordan coefficients. Additional sums should be introduced on the right sides of Eqs. (1.68) and (1.69) over other quantum numbers of the unobserved subsystem, if they exist.

One can introduce statistical tensors of three and more angular momenta and the corresponding statistical tensors of a higher order, such as triple, and so on. We will not use the higher-order tensors in this book, but the statistical tensors of several angular momenta are very important. In this case, the angular momenta can be coupled in different orders and the relations between the corresponding statistical tensors can be easily written using recoupling coefficients (Sections A.9 and A.10). As an example, we give useful relations between the statistical tensors of three and four angular momenta:

$$\rho_{kq}\left(j_1, (j_2 j_3) j_{23} : J; j_1', (j_2' j_3') j_{23}' : J'\right)$$

$$= (-1)^{j_1+j_2+j_3+j_1'+j_2'+j_3'+J+J'} \sum_{j_{12} j_{12}'} \hat{j}_{12} \hat{j}_{12}' \hat{j}_{23} \hat{j}_{23}' \begin{Bmatrix} j_1 & j_2 & j_{12} \\ j_3 & J & j_{23} \end{Bmatrix}$$

$$\times \begin{Bmatrix} j_1' & j_2' & j_{12}' \\ j_3' & J' & j_{23}' \end{Bmatrix} \rho_{kq}\left((j_1 j_2) j_{12}, j_3 : J; (j_1' j_2') j_{12}', j_3' : J'\right) \tag{1.70}$$

$$\rho_{kq}\left((j_1 j_2) j_{12}, (j_3 j_4) j_{34} : J; (j_1' j_2') j_{12}', (j_3' j_4') j_{34}' : J'\right)$$

$$= \sum_{j_{13} j_{13}' j_{24} j_{24}'} \hat{j}_{12} \hat{j}_{12}' \hat{j}_{13} \hat{j}_{13}' \hat{j}_{24} \hat{j}_{24}' \hat{j}_{34} \hat{j}_{34}' \begin{Bmatrix} j_1 & j_2 & j_{12} \\ j_3 & j_4 & j_{34} \\ j_{13} & j_{24} & J \end{Bmatrix}$$

$$\times \begin{Bmatrix} j_1' & j_2' & j_{12}' \\ j_3' & j_4' & j_{34}' \\ j_{13}' & j_{24}' & J' \end{Bmatrix} \rho_{kq}\left((j_1 j_3) j_{13}, (j_2 j_4) j_{24} : J; (j_1' j_3') j_{13}', (j_2' j_4') j_{24}' : J'\right) \tag{1.71}$$

1.3. Spin Density Matrix and Statistical Tensors for Simple Systems

1.3.1. Spin-Tensor Operators

The concept of spin-tensor operators is very helpful in considering the density matrix for a particle with definite spin. It follows from Eq. (1.42) that the density operator $\rho(j)$ acting in a space of spin (angular momentum) j can be presented as an expansion in terms of some irreducible tensor operators:

$$\rho(j) = \sum_{k=0}^{2j} \sum_{q=-k}^{k} \rho_{kq}(j,j) \, T_{kq}(j) \tag{1.72}$$

with the matrix elements of the operators $T_{kq}(j)$:

$$\langle jm \,|\, T_{kq}(j) \,|\, jm' \rangle = (-1)^{j-m'} \,(jm, j - m' \,|\, kq) \tag{1.73}$$

The operator $T_{kq}(j)$, called the *spin-tensor operator* or simply the *spin tensor*, is a component with the projection q of an irreducible tensor operator of rank k acting in the spin space. As follows from (1.73) and the Wigner–Eckart theorem (A.62), the reduced matrix element of the spin tensor has the simple form

$$\langle j \,\|\, T_k(j) \,\|\, j \rangle = \sqrt{2k+1} \tag{1.74}$$

The spin tensor can be built up in unique way out of components of spin $\mathbf{j}$. For this purpose one can use the irreducible tensor product (see Section A.7) of the operator $\mathbf{j}$, which is an operator of the first rank (as well as all other vector operators):

$$T_{kq}(j) = N_k(j) \left\{ \{ \dots \{\{ \mathbf{j} \otimes \mathbf{j} \}_2 \otimes \mathbf{j} \}_3 \dots \}_{k-1} \otimes \mathbf{j} \right\}_{kq} \tag{1.75}$$

$$N_k(j) = 2^k \left(\frac{(2k+1)!! \,(2j-k)!}{k! \,(2j+k+1)!} \right)^{1/2} \tag{1.76}$$

The normalization constant (1.76) can be found from Eq. (1.74) by successive application of Eqs. (A.63) and (A.107) and using Eq. (A.69). An equivalent method for building the spin-tensor operator is to use the corresponding *solid spherical harmonic* $\mathcal{Y}_{kq}(\mathbf{j})$ (see Section A.5) as an operator function of $\mathbf{j}$. Note that because the components of the spin operator do not commute, in solid harmonics one should replace products such as xy, not by $j_x j_y$, but by the symmetric product $(j_x j_y + j_y j_x)/2$, and so on. With allowance for normalization, a general expression for the spin-tensor operator can be presented as

$$T_{kq}(j) = N_k(j) \left(4\pi \frac{k!}{(2k+1)!!} \right)^{1/2} \mathcal{Y}_{kq}(\mathbf{j}) \tag{1.77}$$

Since the matrix elements (1.73) are real, the Hermitian conjugated spin tensor $T_{kq}^{+}(j)$ reduces to the transposed one $\widetilde{T}_{kq}(j)$:

$$\left\langle jm \,\left|\, T_{kq}^{+}(j) \,\right|\, jm' \right\rangle = \langle jm' \,|\, T_{kq}(j) \,|\, jm \rangle \tag{1.78}$$

Using Eqs. (1.41) and (1.73), the statistical tensor $\rho_{kq}(j,j)$ of a mixed state, described by the density operator ρ, can be written as the mean value of the conjugate spin tensor $T_{kq}^{+}(j)$ in this state:

$$\begin{aligned}
\rho_{kq}(j,j) &= \sum_{mm'} \langle jm \,|\, T_{kq}(j) \,|\, jm' \rangle \langle jm \,|\, \rho \,|\, jm' \rangle \\
&= \sum_{mm'} \langle jm \,|\, \rho \,|\, jm' \rangle \left\langle jm' \,\left|\, T_{kq}^{+}(j) \,\right|\, jm \right\rangle \\
&= \mathrm{Tr}(\rho T_{kq}^{+}(j)) = \langle T_{kq}^{+}(j) \rangle \tag{1.79}
\end{aligned}$$

In the literature, the density operator $\rho(j)$ acting in the space of spin j is used as an expansion in two quite equivalent forms:

$$\rho(j) = \sum_{k=0}^{2j} \sum_{q=-k}^{k} \rho_{kq}(j,j) T_{kq}(j) = \sum_{k=0}^{2j} \sum_{q=-k}^{k} \langle T_{kq}^{+}(j) \rangle T_{kq}(j) \qquad (1.80)$$

The criterion of a pure state (1.54) may be written in the form

$$\sum_{k=0}^{2j} \sum_{q=-k}^{k} |\langle T_{kq}^{+}(j) \rangle|^2 = 1 \qquad (1.81)$$

Equation (1.77) and Table A.4 from Section A.5 allow one to express the spin tensors with the lowest ranks and for an arbitrary spin in terms of the components of the spin (angular momentum) operator. Substituting them into Eq. (1.79), one obtains the statistical tensors in terms of mean values of the corresponding components, for example,

$$\rho_{1q}(j,j) = \langle T_{1q}(j) \rangle^* = \sqrt{\frac{3}{2}} [j(j+1)(2j+1)]^{-1/2} \langle j_q \rangle \qquad (1.82)$$

$$q = 0, \pm 1,$$

$$\rho_{20}(j,j) = \langle T_{20}(j) \rangle^*$$

$$= C(j) \sqrt{\frac{2}{3}} [3\langle j_z^2 \rangle - \langle j^2 \rangle]$$

$$= C(j) \sqrt{\frac{2}{3}} [3\langle j_z^2 \rangle - j(j+1)] \qquad (1.83)$$

$$\rho_{2\pm 1}(j,j) = \langle T_{2\pm 1}(j) \rangle^*$$

$$= \mp C(j) [\langle j_x j_z \rangle + \langle j_z j_x \rangle \mp i (j_y j_z + j_z j_y)] \qquad (1.84)$$

$$\rho_{2\pm 2}(j,j) = \langle T_{2\pm 2}(j) \rangle^* = C(j) [j_x^2 - j_y^2 \mp i (j_x j_y + j_y j_x)] \qquad (1.85)$$

where

$$C(j) = \frac{\sqrt{15}}{4} [j(j+1)(2j-1)(2j+1)(2j+3)]^{-1/2} \qquad (1.86)$$

In Eq. (1.82) we introduced the conventional *circular components* of a vector by the relations

$$j_0 = j_z, \qquad j_{+1} = -\frac{1}{\sqrt{2}}(j_x + ij_y), \qquad j_{-1} = \frac{1}{\sqrt{2}}(j_x - ij_y) \qquad (1.87)$$

Note that the conventional definition of the *spin polarization vector* of a particle with spin j is

$$\mathbf{P} = \langle \mathbf{j} \rangle / j \qquad (1.88)$$

Comparing Eqs. (1.88) and (1.87) with Eq. (1.82), we see that within a factor the statistical tensor of the first rank (complex conjugate) coincides with the spin polarization vector.

It follows from the Wigner–Eckart theorem (A.62) that the mean values of the spin tensors $\langle T_{kq}(j) \rangle$ and, equivalently, the conjugate statistical tensors $\rho^*_{kq}(j,j)$ are proportional to the corresponding components of any quantum characteristic of the system, which has the same tensor rank and nonzero value. For example, the components of the statistical tensors $\rho_{1q}(j,j)$ are proportional to the conjugate circular components of the vector of the dipole magnetic moment (provided the magnetic moment exists in the system).

1.3.2. Density Matrix and Statistical Tensors of Spin-$\frac{1}{2}$ Particles

This case can be considered rather easily without applying the spin-tensor formalism of the preceding section. The density matrix of spin $\frac{1}{2}$ is a 2×2 matrix. An arbitrary unitary 2×2 matrix can be expanded in terms of four independent matrices: the unit matrix I and three *Pauli matrices*, σ_x, σ_y, and σ_z (see Section A.1):

$$\rho = \frac{1}{2}\left(I + P_x\sigma_x + P_y\sigma_y + P_z\sigma_z\right) = \frac{1}{2}\left(I + \mathbf{P}\boldsymbol{\sigma}\right) \qquad (1.89)$$

Here $\mathbf{P}$ is a vector with the components P_x, P_y, and P_z. The normalization constant in Eq. (1.89) is chosen in accordance with the condition $\mathrm{Tr}\,\rho = 1$. Using the explicit form of the Pauli matrices (A.2), one obtains for the density matrix,

$$\left\langle \frac{1}{2}m \,|\rho|\, \frac{1}{2}m' \right\rangle = \frac{1}{2}\left(\begin{array}{cc} 1+P_z & P_x - iP_y \\ P_x + iP_y & 1 - P_z \end{array} \right) \qquad (1.90)$$

Throughout this book we follow the conventional order of numeration of lines and columns in the matrices of angular momenta: from the largest positive value of projection of the angular momentum (first line, first column) to the largest negative one (last line, last column). The first index in the matrix element (and the correspondent bra vector $\langle \frac{1}{2}m |$) indicates the line, while the second index (and the correspondent ket vector $| \frac{1}{2}m' \rangle$) indicates the column.

The physical meaning of the vector $\mathbf{P}$ becomes clear from the following observation: As is known, the spin operator $\mathbf{s}$ of the spin-$\frac{1}{2}$ particle is connected with the Pauli matrices:

$$\mathbf{s} = \frac{1}{2}\boldsymbol{\sigma} \qquad (1.91)$$

According to the general definition (1:88), the spin polarization vector is

$$\frac{\langle \mathbf{s} \rangle}{1/2} = 2\langle \mathbf{s} \rangle = 2\,\mathrm{Tr}\,(\rho\mathbf{s}) = \mathrm{Tr}\,(\rho\boldsymbol{\sigma}) = \mathbf{P} \qquad (1.92)$$

The last equation can be proved using the expansion (1.89) and the explicit form and properties of the Pauli matrices (see Section A.1). Hence, formally introduced vector $\mathbf{P}$ in Eq. (1.89) is the spin polarization vector. Its direction coincides with the direction of the average spin of the particle, while the value $P = |\mathbf{P}|$ is the *degree of spin polarization* of the particle. The natural condition $0 \le P \le 1$ is in accordance with the general requirement (1.30) imposed on the matrix elements of the density matrix. The equality in Eq. (1.30) is achieved only for $P = 1$. Therefore, only spin states with $P = 1$ are pure and vice versa, only in pure states does $P = 1$. The three parameters, P_x, P_y, and P_z, fully characterize the spin state of a particle with $s = \frac{1}{2}$. A particle with $s = \frac{1}{2}$ is polarized if $P \neq 0$ and unpolarized if $P = 0$.

The statistical tensors $\rho_{kq}\left(\frac{1}{2}, \frac{1}{2}\right)$ corresponding to the density matrix (1.90) are easily obtained from the definition (1.41) and values of the Clebsch–Gordan coefficients from Table A.7 on pp. 218–220 (Section A.8). Only tensors with rank 0 and 1 exist for the spin-$\frac{1}{2}$ particle:

$$\rho_{00}\left(\frac{1}{2}, \frac{1}{2}\right) = \frac{1}{\sqrt{2}} \tag{1.93}$$

$$\rho_{10}\left(\frac{1}{2}, \frac{1}{2}\right) = \frac{1}{\sqrt{2}} P_z \tag{1.94}$$

$$\rho_{1\pm1}\left(\frac{1}{2}, \frac{1}{2}\right) = \mp\frac{1}{2}\left(P_x \mp iP_y\right) \tag{1.95}$$

A particle with $s = \frac{1}{2}$ cannot be aligned. The components of the statistical tensor $\rho_{1q}\left(\frac{1}{2}, \frac{1}{2}\right)$ $(q = 0, \pm1)$ coincide to within the common factor $1/\sqrt{2}$ with the conjugate circular components P_q^* of the polarization vector $\mathbf{P}$ [see Eq. (1.87)]. The same result immediately follows from the spin-tensor formalism [Eqs. (1.82) and (1.92)]. Applying (1.54), one can also check the criterion $P = 1$ of a pure state.

1.3.3. Density Matrix and Statistical Tensors of a Free Spinless Particle

Now consider a free spinless particle of mass M moving with the energy E_0 in the direction $\mathbf{n}_0$ characterized by the angles (ϑ, φ). The particle is described by the state vector $|E_0\mathbf{n}_0\rangle$, which can be represented in the coordinate space by the plane-wave function $\phi_{\mathbf{k}_0}(\mathbf{r}) = e^{i\mathbf{k}_0\mathbf{r}}$ with $\mathbf{k}_0 = k_0\mathbf{n}_0$, $k_0 = \sqrt{2ME_0}$. The corresponding angular part of the density operator is defined by expression (1.3):

$$\rho^{\mathbf{n}_0} = |\mathbf{n}_0\rangle\langle\mathbf{n}_0| \tag{1.96}$$

In the representation of orbital angular momentum l and its projection m, the density matrix of the state (1.96) is given by

$$\langle lm|\rho^{\mathbf{n}_0}|l'm'\rangle = \langle lm|\mathbf{n}_0\rangle\langle\mathbf{n}_0|l'm'\rangle \tag{1.97}$$

Taking into account $\langle \mathbf{n}_0 | lm \rangle = Y_{lm}(\vartheta, \varphi)$, see Eq. (A.26), the angular momentum part of the density matrix (1.97) has the form:

$$\langle lm | \rho^{n_0} | l'm' \rangle = Y_{lm}^*(\vartheta, \varphi) \, Y_{l'm'}(\vartheta, \varphi) \qquad (1.98)$$

Substituting Eq. (1.98) into the definition (1.41) and using Eqs. (A.29) and (A.87), one obtains the statistical tensors in the form

$$\rho_{kq}(l, l') = \frac{(-1)^{l'}}{\sqrt{4\pi}} \frac{\hat{l}\hat{l}'}{\hat{k}} (l0, l'0 \mid k0) Y_{kq}^*(\vartheta, \varphi)$$

$$\equiv C_{k0}(l, l') \sqrt{\frac{4\pi}{2k+1}} Y_{kq}^*(\vartheta, \varphi) \qquad (1.99)$$

where we introduce the *radiation parameters* of a spinless particle, $C_{k0}(l, l')$, widely used in practice:

$$C_{k0}(l, l') = \frac{(-1)^{l'}}{4\pi} \hat{l}\hat{l}' \, (l0, l'0 | k0) \qquad (1.100)$$

The radiation parameters are the statistical tensors of a particle moving along the quantization axis z:

$$\rho_{kq}(l, l')\big|_{z \| \mathbf{n}_0} = C_{k0}(l, l') \sqrt{\frac{4\pi}{2k+1}} Y_{k9}^*(0, 0) = \delta_{q0} C_{k0}(l, l') \qquad (1.101)$$

Equation (1.101) shows that in the reference frame with the z-axis along the direction of the particle motion ($\vartheta = \varphi = 0$), only states with zero projection contribute to the density matrix (1.98) due to relation (A.20).

Note that the trace of the density matrix $\langle lm | \rho^{n_0} | lm \rangle$ diverges, which is due to a special case with the density operator (1.96). Hence, the statistical tensors (1.98) do not satisfy the usual normalization condition (1.47).

1.3.4. Density Matrix and Statistical Tensors of a Free Electron

The motion of a free electron may be described by the density matrix and the statistical tensors of the total angular momentum, which include both the spin and the orbital motion. Equation (1.64) provides a connection between the statistical tensors of the electron in different representations:

$$\rho_{kq}\left(l\frac{1}{2}j, l'\frac{1}{2}j'\right) = \sum_{k_l q_l k_s q_s} \hat{k}_l \hat{k}_s \hat{j}\hat{j}' \, (k_l q_l, k_s q_s | kq) \begin{Bmatrix} l & \frac{1}{2} & j \\ l' & \frac{1}{2} & j' \\ k_l & k_s & k \end{Bmatrix}$$

$$\times \rho_{k_l q_l k_s q_s}\left(l\frac{1}{2}, l'\frac{1}{2}\right) \qquad (1.102)$$

Here j is the total angular momentum of the electron: $\mathbf{j} = \mathbf{l} + \mathbf{s}$. The representation of the total angular momentum is especially convenient in the case of nonvanishing relativistic coupling between the orbital angular momentum and spin of the electron (the *spin–orbit interaction*).

If the orbital angular momentum and spin of the electron are not correlated, the double statistical tensor is factorized into the orbital and spin parts:

$$\rho_{k_l q_l k_s q_s} \left(l\frac{1}{2}, l'\frac{1}{2} \right) = \rho_{k_l q_l} \left(l, l' \right) \rho_{k_s q_s} \left(\frac{1}{2}, \frac{1}{2} \right) \tag{1.103}$$

The statistical tensors $\rho_{k_s q_s} \left(\frac{1}{2}, \frac{1}{2} \right)$ and $\rho_{k_l q_l} \left(l, l' \right)$ were considered in Sections 1.3.2 and 1.3.3, respectively.

For the spin-unpolarized electron beam ($k_s = q_s = 0$), one finds with the use of Eqs. (1.102), (1.67), (1.99), and (A.122):

$$\rho_{kq} \left(l\frac{1}{2} j, l'\frac{1}{2} j' \right) = C_{k0}(lj, l'j') \sqrt{\frac{4\pi}{2k+1}} Y_{kq}^*(\vartheta, \varphi) \tag{1.104}$$

where

$$C_{k0}(lj, l'j') = \frac{(-1)^{j'+1/2}}{4\pi} \hat{l}\hat{l'}\hat{j}\hat{j'} \left(10, l'0 | k0 \right) \left\{ \begin{matrix} j & l & \frac{1}{2} \\ l' & j' & k \end{matrix} \right\} \tag{1.105}$$

are the *radiation parameters of the electron*.

In the next section we omit for brevity the quantum number of fixed spin $s = \frac{1}{2}$ in the arguments of the statistical tensors of the total angular momentum j, as in Eq. (1.105), for example $\rho_{kq} \left(lj, l'j' \right) \equiv \rho_{kq} \left(l\frac{1}{2}j, l'\frac{1}{2}j' \right)$.

It is implied here and throughout the book that the order of coupling of the orbital angular momentum and spin is always $\mathbf{l} + \frac{1}{2} = \mathbf{j}$. A transgression of a fixed order of coupling $l + \frac{1}{2} = j$, or, alternatively, $\frac{1}{2} + l = j$ in calculations often leads to mistakes. This pertains equally to any two angular momenta.

1.3.5. Density Matrix and Statistical Tensors of a Photon

1.3.5.1. Representation of Helicity; Stokes Parameters

The density matrix and statistical tensors of a photon need special consideration. The main reason is that the photon is a particle without mass, which cannot be at rest. At the same time, although the photon has intrinsic spin 1, owing to its transverse character, the projection of the photon spin on its linear momentum, the *helicity*, can take only two values: $\lambda = \pm 1$, which is unusual for an ordinary particle with spin 1.

Consider a photon with a fixed wave vector $\mathbf{k}$. The helicity part of the density matrix, which describes the photon polarization, is a 2×2 matrix that can be parameterized by three real parameters, similar to the case of a particle with spin $\frac{1}{2}$. Most often this matrix is parameterized in the form

$$\langle \mathbf{k}\lambda \,|\, \rho \,|\, \mathbf{k}\lambda' \rangle = \frac{1}{2} \begin{pmatrix} 1+P_3 & -P_1+iP_2 \\ -P_1-iP_2 & 1-P_3 \end{pmatrix} \qquad (1.106)$$

or equivalently,

$$\langle \mathbf{k}\lambda \,|\, \rho \,|\, \mathbf{k}\lambda' \rangle = \frac{1}{2} \left[\delta_{\lambda\lambda'}(1+\lambda P_3) + (1-\delta_{\lambda\lambda'})(-P_1+i\lambda P_2) \right] \qquad (1.107)$$

The order of lines and columns in the matrix (1.106) is the same as for the usual angular momentum: the first line (column) corresponds to $\lambda = +1$ and the second line (column) corresponds to $\lambda = -1$. The quantities P_1, P_2, and P_3 are called *Stokes parameters* of the photon. The definition of the Stokes parameters may be different in different texts.

The Stokes parameters characterize the polarization properties of the photon beam [21]. Consider this in more detail. Photons with the helicity $\lambda = +1(-1)$ correspond in classical electrodynamics to electromagnetic waves with right (left) hand rotation of the electric and magnetic field vectors around the direction of $\mathbf{k}$. Therefore, photons with $\lambda = +1$ are called *right circularly polarized photons*, and photons with $\lambda = -1$ are called *left circularly polarized photons*. Note that for historical reasons, in optics a terminology is usually used that is opposite to that used in quantum electrodynamics. The diagonal matrix elements in Eq. (1.106) give the weights of the purely circularly polarized states:

$$W(\lambda = \pm 1) = \frac{1}{2}(1 \pm P_3) \qquad (1.108)$$

and consequently, the Stokes parameter P_3 is the *degree of circular polarization*, P_c, of the photon beam:

$$P_c \equiv \frac{W(+1) - W(-1)}{W(+1) + W(-1)} = P_3 \qquad (1.109)$$

Right and left circularly polarized photon states can serve as a basis for describing a state with arbitrary polarization. We choose the z-axis along the wave vector $\mathbf{k}$ and present the circularly polarized state in terms of states linear polarized along the x- and y-axes:

$$|\mathbf{k}, \lambda = \pm 1\rangle = \mp \frac{1}{\sqrt{2}}(|\mathbf{k}, \mathbf{e}_x\rangle \pm i |\mathbf{k}, \mathbf{e}_y\rangle) \qquad (1.110)$$

The state of a photon *linearly polarized* along some direction φ in the xy-plane can be then written in the form

$$\begin{aligned} |\mathbf{k}, \mathbf{e}_\varphi\rangle &= \cos\varphi \,|\mathbf{k}, \mathbf{e}_x\rangle + \sin\varphi \,|\mathbf{k}, \mathbf{e}_y\rangle \\ &= -\frac{1}{\sqrt{2}}\left(e^{-i\varphi}|\mathbf{k}, \lambda = +1\rangle - e^{i\varphi}|\mathbf{k}, \lambda = -1\rangle \right) \end{aligned} \qquad (1.111)$$

The corresponding density matrix is

$$\langle \mathbf{k}\lambda \,|\, \rho_\varphi \,|\, \mathbf{k}\lambda' \rangle = \langle \mathbf{k}\lambda \,|\, \mathbf{k}, \mathbf{e}_\varphi \rangle \langle \mathbf{k}, \mathbf{e}_\varphi \,|\, \mathbf{k}\lambda' \rangle = \frac{1}{2}\lambda\lambda' e^{-i(\lambda-\lambda')\varphi} \qquad (1.112)$$

The probability that the state described by the density matrix (1.106) is contained in the state described by the density matrix (1.112) is given by the general rule (1.35) with the result:

$$W(\mathbf{e}_\varphi) = \frac{1}{2}\left(1 + P_1\cos 2\varphi + P_2\sin 2\varphi\right) \qquad (1.113)$$

It follows from Eq. (1.113) that the Stokes parameter P_1 characterizes the degree of linear polarization relative to the x ($\varphi = 0°$) and y ($\varphi = 90°$) axes

$$\frac{W(0°) - W(90°)}{W(0°) + W(90°)} = P_1 \qquad (1.114)$$

while the Stokes parameter P_2 characterizes the degree of linear polarization relative to the axes $\varphi = 45°$ and $\varphi = 135°$

$$\frac{W(45°) - W(135°)}{W(45°) + W(135°)} = P_2 \qquad (1.115)$$

Hence, in the case of arbitrary P_1, P_2, and P_3, the photon is circularly and linearly polarized. The general requirement (1.30) for the density matrix (1.106) is satisfied by the relation

$$P_1^2 + P_2^2 + P_3^2 \leq 1 \qquad (1.116)$$

The equality $\sum_{i=1}^3 P_i^2 = 1$ indicates that the polarization state of the photon is pure. The parameter

$$P_l = \sqrt{P_1^2 + P_2^2} \qquad (1.117)$$

is called the *degree of linear polarization*. Condition (1.116) imposes the restriction $P_c^2 + P_l^2 \leq 1$. If both P_c and P_l are nonzero, while the latter inequality turns to equality, the photon is called *elliptically polarized*.

It is convenient to introduce the angle φ_0 by the relations

$$\cos 2\varphi_0 = \frac{P_1}{P_l}, \quad \sin 2\varphi_0 = \frac{P_2}{P_l}. \qquad (1.118)$$

where P_l is the degree of linear polarization (1.117). With the definition (1.118), Eq. (1.113) takes the form

$$W(\mathbf{e}_\varphi) = \frac{1}{2}\left(1 + P_l\cos 2(\varphi - \varphi_0)\right) \qquad (1.119)$$

Let us assume that one measures the intensity of the photon beam passing through an ideal polarization filter that is transparent only for photons linearly polarized along the axis of the polarimeter. The angle φ_0 describes the direction of the axis when the intensity of the transmitted photon beam is the largest, while the lowest intensity is observed when the axis of the polarimeter is perpendicular to this direction. The angle φ_0 indicates the *principal axis* of the *polarization ellipse* of the photon beam. In the coordinate frame with the x-axis directed along this principal axis ($\varphi_0 = 0$), one finds from Eqs. (1.114), (1.115), and (1.119) that the Stokes parameter P_1 takes the largest value $|P_1| = P_l$, while the Stokes parameter P_2 vanishes. This demonstrates that the Stokes parameters P_1 and P_2 for a given photon beam depend on the choice of x-axis (with z-axis fixed along the beam). At the same time, the degree of linear polarization P_l is invariant. Therefore, it is often convenient in an analysis of different experimental situations to use the degree of linear polarization P_l and the direction of the principal axis of the polarization ellipse φ_0 instead of the Stokes parameters P_1 and P_2.

1.3.5.2. *Representation of Multipoles and Statistical Tensors*

To obtain the statistical tensors of the photon, we consider the representation of the total angular momentum. This is accomplished by use of the multipole expansion of the angular part of the state $|\mathbf{k}, \lambda\rangle$ of a photon with the wave vector $\mathbf{k} = \{k, \vartheta, \varphi\}$ and helicity λ:

$$|\{\vartheta\varphi\}, \lambda\rangle = \sum_{p=0,1} \sum_{LM} \langle pLM \,|\, \vartheta\varphi, \lambda\rangle \,|pLM\rangle \qquad (1.120)$$

Wave functions $|pLM\rangle$ describe the states of the photon with angular momentum L and its projection M on the quantization axis z. Summation in Eq. (1.120) starts from $L = 1$ (an absence of photons with $L = 0$ is a result of their transverse polarization). The quantum number p specifies the type of photon: either electric ($p = 0$) or magnetic ($p = 1$). The parity of the photon is defined by the expression

$$\pi = (-1)^{L+p} \qquad (1.121)$$

Photons with $p = 0$ and different L are called electric 2^L-pole (dipole, quadrupole, octupole, etc.) or EL photons. Similarly, photons with $p = 1$ are called magnetic 2^L-pole or ML photons. The transformation coefficients $\langle pLM \,|\, \vartheta\varphi, \lambda\rangle$ in Eq. (1.120) are known [5]:

$$\langle pLM \,|\, \vartheta\varphi, \lambda\rangle = \sqrt{\frac{2L+1}{8\pi}}\, \lambda^p D^L_{M\lambda}(\varphi, \vartheta, 0) \qquad (1.122)$$

Table 1.1. Statistical Tensors of the Dipole Photon in the Two Coordinate Frames (see Figure 1.1)

$\rho^{\gamma}_{kq}(P_1, P_2, P_3)$	S	S'
ρ^{γ}_{00}	$1/\sqrt{3}$	$1/\sqrt{3}$
ρ^{γ}_{10}	$P_3/\sqrt{2}$	0
$\rho^{\gamma}_{1\pm1}$	0	$\mp P_3/2$
ρ^{γ}_{20}	$1/\sqrt{6}$	$-(1+3P_1)/2\sqrt{6}$
$\rho^{\gamma}_{2\pm1}$	0	$iP_2/2$
$\rho^{\gamma}_{2\pm2}$	$-(P_1 \mp iP_2)/2$	$(1-P_1)/4$

In a particular coordinate frame with the quantization axis z along the wave vector $\mathbf{k}$ ($\vartheta = 0, \varphi = 0$) (we call this the *photon frame*) this equation reduces to

$$\langle pLM \,|\, 00, \lambda \rangle = \delta_{M\lambda} M^p \sqrt{\frac{2L+1}{8\pi}} \qquad (1.123)$$

Using this formula one can transform the 2×2 density matrix in the helicity representation (1.106) into a standard form of the density matrix of angular momentum of the dimension $(2L+1) \times (2L+1)$ corresponding to the multipolarity L of the photon. In the majority of studies of photon–atom interactions, especially in studies of angular distributions and polarizations, dipole transitions ($E1$ transitions) are considered. Substituting $p = p' = 0$, $L = L' = 1$ in Eq. (1.123) and using (1.106), we obtain for the normalized density matrix of the dipole photon

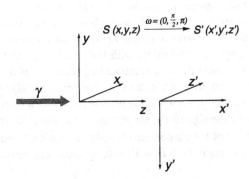

FIGURE 1.1. Two coordinate systems used to describe the polarization state of photons: S (photon frame) and S'.

in the photon frame

$$\langle p=0, L=1, M \,|\, \rho \,|\, p'=0, L'=1, M' \rangle = \frac{1}{2} \begin{pmatrix} 1+P_3 & 0 & -P_1+iP_2 \\ 0 & 0 & 0 \\ -P_1-iP_2 & 0 & 1-P_3 \end{pmatrix}$$

(1.124)

The nonzero elements of this matrix coincide with the corresponding elements of the density matrix in the helicity representation, Eq. (1.106). The difference is in the extra line and column in the matrix (1.124), which correspond to the zero projection M.

Generally, using Eqs. (1.106) and (1.123), one obtains the angular momentum density matrix and, hence, corresponding statistical tensors of the photon of any multipolarity. In the photon frame only statistical tensors with projections $0, \pm 2$ turn out to be nonzero. They may be presented in the form

$$\rho_{k0} \left(pL, p'L' \right) = (-1)^{L'-1} \frac{\hat{L}\hat{L}'}{16\pi} \left(L1, L'-1 \,|\, k0 \right) \left[1 + (-1)^f + P_3 \left(1 - (-1)^f \right) \right]$$

(1.125)

$$\rho_{k, \pm 2} \left(pL, p'L' \right) = (-1)^{L'+p'} \frac{\hat{L}\hat{L}'}{16\pi} \left(L1, L'1 \,|\, k2 \right) (\pm 1)^f \left(P_1 \mp iP_2 \right)$$

(1.126)

where $f = L + p + L' + p' - k$. Similar to Eqs. (1.100) and (1.105), the statistical tensors (1.125) and (1.126) are the *radiation parameters of the photon*. We shall denote them as $C_{kq}(pL, p'L')$.

The statistical tensors of the dipole photons $\rho_{kq}^{\gamma} (P_1, P_2, P_3)$, which depend on the Stokes parameters, can be obtained either from the general expressions (1.125) and (1.126) for arbitrary multipolarity (after the corresponding normalization), or from the density matrix (1.124). They are shown in column S of Table 1.1. Here and later we indicate the statistical tensors of the dipole photon by the superscript γ. Arguments P_1, P_2, and P_3 will often be suppressed for brevity. We remind the reader that the reference frame S is the photon frame with $z \parallel \mathbf{k}$.

For a dipole photon, only tensors with the rank $k \leq 2$ exist. In the photon frame, the statistical tensors $\rho_{1, \pm 1}^{\gamma}$ are zero and ρ_{20}^{γ} always has the same constant value regardless of the polarization state of the photon. The only statistical tensor of the first rank, ρ_{10}^{γ}, is nonzero when the photon possesses a circular component ($P_3 \neq 0$). In changing the sign of P_3, we change at the same time the sign of ρ_{10}^{γ}. The linear polarization of the photon affects only the statistical tensors $\rho_{2, \pm 2}^{\gamma}$. For an unpolarized photon beam, $P_1 = P_2 = P_3 = 0$, two statistical tensors are nonzero:

$$\rho_{00}^{\gamma} = \frac{1}{\sqrt{3}}, \qquad \rho_{20}^{\gamma} = \frac{1}{\sqrt{6}}$$

(1.127)

The fact that ρ_{20}^{γ} is nonzero for an unpolarized photon reflects a particular feature of the photon, which is related to the existence of the characteristic direction along the wave vector **k**.

To find the statistical tensors in another frame, we must use the standard expression (1.44) for transformation of the statistical tensors. For example, a coordinate frame S' is often used with the x'-axis along the photon propagation and the z'-axis along the x-axis of the photon frame (Figure 1.1). The statistical tensors of the dipole photon in this important case are shown in the last column of Table 1.1. (Sometimes the axis x' of the system S' is chosen against the photon momentum, while y' is parallel to y. In this case, the statistical tensors with odd projections in the column S' of Table 1.1 change sign. One more possible choice of the coordinate system is discussed in Section 4.2.1.2.)

1.4. Symmetry Restrictions on the Density Matrix and Statistical Tensors

If a physical system is unaltered by some transformation of the coordinate frame, for example, by rotation or reflection in some plane, the transformation is called the *symmetry transformation* of the system. Under symmetry transformations, the observables of this system do not change. The statistical tensors or their linear combinations are quantities that can be observed. Therefore, all of the statistical tensors in a transformed coordinate frame must coincide with the statistical tensors in the initial coordinate frame if the transformation is a symmetry transformation:

$$\rho_{kq}\left(j,j'\right) = \rho_{kq}^{\text{tran}}\left(j,j'\right) \qquad \text{(all } k \text{ and } q) \qquad (1.128)$$

The requirement (1.128) gives additional relations between the statistical tensors and brings in additional restrictions on the possible values of k and q. We are interested in symmetries connected with rotations, reflections, and inversion of the coordinate system. The transformations of the statistical tensors under rotations of the coordinate system are given by Eq. (1.44). Under inversion, the statistical tensors acquire a phase factor $\eta = \pi\pi'$, where π and π' are internal parities of states with j and j', respectively. If at least one of the states does not have definite parity, the transformation, which includes inversion, in general is not the transformation of symmetry. The reflection in a plane is a product of two transformations: rotation through an angle $180°$ about the axis perpendicular to the plane and the inversion. Rotations and inversion are commutative.

1.4.1. Isotropic System

For an isotropic system, an arbitrary rotation is the symmetry transformation:

$$\rho_{kq}\left(j,j'\right) = \rho_{kq}^{\text{tran}}\left(j,j'\right) = \sum_{q'}\left(D_{q'q}^{k}\left(\alpha,\beta,\gamma\right)\right)^{*}\rho_{kq'}\left(j,j'\right) \tag{1.129}$$

where α, β, and γ are arbitrary angles. The identity is possible only in the case $D_{q'q}^{k}\left(\alpha,\beta,\gamma\right) = \delta_{qq'}$, which implies that $k = 0$ (hence $q = q' = 0$). As a result, an isotropic system possesses only statistical tensors with rank zero $\rho_{00}\left(j,j'\right)$. From the properties of the Clebsch–Gordan coefficient in Eq. (1.42), only the diagonal matrix elements of the density matrix $\langle jm \mid \rho \mid j'm'\rangle$ with $j = j'$ and $m = m'$ survive for the isotropic system. The density matrix is diagonal in an arbitrary coordinate frame; all magnetic substates are equally populated and the density matrix is given by Eq. (1.40). There is no coherence between states with different j and different m. To create the coherence one must disturb the isotropy of the system.

1.4.2. A System with Axial Symmetry

We choose the z-axis of the coordinate frame along the symmetry axis. A rotation about the z-axis is then the symmetry transformation. Using Eq. (A.54), one obtains

$$\rho_{kq}\left(j,j'\right) = \rho_{kq}^{\text{tran}}\left(j,j'\right)$$
$$= \sum_{q'}\left(D_{q'q}^{k}\left(\alpha,0,0\right)\right)^{*}\rho_{kq'}\left(j,j'\right) = e^{-iq\alpha}\rho_{kq}\left(j,j'\right) \tag{1.130}$$

where α is the angle of rotation. The identity (1.130) is valid for arbitrary α only when $q = 0$. As a result, a system with axial symmetry possesses only statistical tensors with zero projection, $\rho_{k0}\left(j,j'\right)$, in the coordinate frame with the z-axis parallel to the axis of symmetry. As follows from Eq. (1.42), in this frame the density matrix $\langle jm \mid \rho \mid j'm'\rangle$ contains only elements diagonal in m: $m = m'$. There is no coherence between states with different projections. This statement, however, is valid in the chosen frame only. The statistical tensors $\rho_{k0}\left(j,j'\right)$ are either real or imaginary, depending on the parity of the difference, $j' - j$ [see Eq. (1.46)]. In particular, the tensors $\rho_{k0}\left(j,j\right)$ are real. The number of independent real parameters characterizing the density matrix of the system with a sharp j is reduced from $4j(j+1)$ to $2j$.

1.4.3. A System with Plane Symmetry

First consider a case with the quantization axis z in the plane of symmetry, which we choose as the xz-plane. The reflection in the xz-plane is equivalent to rotation about the y-axis through an angle of $180°$ and inversion. This gives [see Eq. (A.57)]

$$\rho_{kq}\left(j,j'\right) = \rho_{kq}^{\text{tran}}\left(j,j'\right) = \eta \sum_{q'} \left(D_{q'q}^{k}\left(0,\pi,0\right)\right)^{*} \rho_{kq'}\left(j,j'\right)$$

$$= \eta(-1)^{k+q} \rho_{k-q}\left(j,j'\right) \qquad (1.131)$$

where η is the product of the internal parities of the states j and j'. This relation shows that the statistical tensors with odd ranks k and $q = 0$ vanish when $\eta = +1$ and the statistical tensors with even ranks k and $q = 0$ vanish when $\eta = -1$. With allowance for the relation (1.46) imposed by the Hermitian character of the density matrix, Eq. (1.131) is equivalent to

$$\rho_{kq}\left(j,j'\right) = (-1)^{k+j-j'} \eta \rho_{kq}^{*}\left(j',j\right) \qquad (1.132)$$

For a sharp j, the plane symmetry of the system leads to a reduction in the number of independent real parameters of the density matrix from $4j(j+1)$ to $2j(j+1)$ (j — integer), or to $2j(j+1) - \frac{1}{2}$ (j — half integer). Usually a system with a sharp j (for example, an atomic state) has a certain parity. Then Eq. (1.132) reduces to a simpler relation:

$$\rho_{kq}\left(j,j\right) = (-1)^{k} \rho_{kq}^{*}\left(j,j\right) \qquad (1.133)$$

which shows that all even statistical tensors are real, whereas odd tensors are purely imaginary. Combining Eq. (1.133) with the relation (1.51), which follows from the hermiticity of the density matrix, we obtain:

$$\rho_{kq}\left(j,j\right) = (-1)^{k+q} \rho_{k-q}\left(j,j\right) \qquad (1.134)$$

Symmetry relations between the elements of the density matrix follow from Eqs. (1.131) and (1.42) and the properties of Clebsch–Gordan coefficients:

$$\langle jm|\rho|j'm'\rangle = \eta \langle j-m|\rho|j'-m'\rangle \qquad (1.135)$$

When the z-axis is chosen perpendicular to the plane of symmetry, the transformation of symmetry is the rotation through $180°$ around the z-axis and inversion. This gives

$$\rho_{kq}\left(j,j'\right) = \rho_{kq}^{\text{tran}}\left(j,j'\right)$$

$$= \eta \sum_{q'} \left(D_{q'q}^{k}\left(\pi,0,0\right)\right)^{*} \rho_{kq'}\left(j,j'\right) = (-1)^{q} \eta \rho_{kq}\left(j,j'\right) \qquad (1.136)$$

The statistical tensors (1.136) become zero for odd q, when $\eta = +1$, while for $\eta = -1$ the tensors with even q become zero. Equation (1.136) leads to the symmetry relation between elements of the density matrix:

$$\langle jm|\rho|j'm'\rangle = (-1)^{m-m'} \eta \langle jm|\rho|j'm'\rangle \qquad (1.137)$$

Table 1.2. Independent Parameters Characterizing the Polarization State of a System with the Angular Momentum j and Different Symmetries[a]

j	No symmetry	Reflection in xz-plane	Rotation around z-axis	Reflection in xz-plane and rotation around z-axis	Isotropy
$\frac{1}{2}$	$N = 3$: $\rho_{10}, \rho_{11}(2)$	$N = 1$: $i\rho_{11}$	$N = 1$: ρ_{10}	$N = 0$	$N = 0$
1	$N = 8$: $\rho_{10}, \rho_{11}(2)$ $\rho_{20}, \rho_{21}(2), \rho_{22}(2)$	$N = 4$: $i\rho_{11}$ $\rho_{20}, \rho_{21}, \rho_{22}$	$N = 2$: ρ_{10} ρ_{20}	$N = 1$: ρ_{20}	$N = 0$
$\frac{3}{2}$	$N = 15$: $\rho_{10}, \rho_{11}(2)$ $\rho_{20}, \rho_{21}(2), \rho_{22}(2)$ $\rho_{30}, \rho_{31}(2),$ $\rho_{32}(2), \rho_{33}(2)$	$N = 7$: $i\rho_{11}$ $\rho_{20}, \rho_{21}, \rho_{22}$ $i\rho_{31}, i\rho_{32}, i\rho_{33}$	$N = 3$: ρ_{10} ρ_{20} ρ_{30}	$N = 1$: ρ_{20}	$N = 0$
2	$N = 24$: $\rho_{10}, \rho_{11}(2)$ $\rho_{20}, \rho_{21}(2), \rho_{22}(2)$ $\rho_{30}, \rho_{31}(2),$ $\rho_{32}(2), \rho_{33}(2)$ $\rho_{40}, \rho_{41}(2), \rho_{42}(2),$ $\rho_{43}(2), \rho_{44}(2)$	$N = 12$: $i\rho_{11}$ $\rho_{20}, \rho_{21}, \rho_{22}$ $i\rho_{31}, i\rho_{32}, i\rho_{33}$ $\rho_{40}, \rho_{41},$ $\rho_{42}, \rho_{43}, \rho_{44}$	$N = 4$: ρ_{10} ρ_{20} ρ_{30} ρ_{40}	$N = 2$: ρ_{20} ρ_{40}	$N = 0$

[a] The general number N of real, independent, nontrivial parameters is also shown. The number 2 in parentheses at some statistical tensors in the second column indicates that real and imaginary parts are both nonvanishing. Otherwise the parameters are real.

1.4.4. A System with a Plane of Symmetry and an Axis of Symmetry in This Plane

In this case we choose the z-axis along the symmetry axis and consider the xz-plane as the symmetry plane. We can obtain the restrictions imposed by the symmetry of the system on the statistical tensors by directly combining the results of Sections 1.4.2 and 1.4.3. This shows that only statistical tensors with zero projection are nonzero. They are real and obey the relation [see Eq. (1.131)]

$$\rho_{k0}(j, j') = (-1)^k \eta \rho_{k0}(j', j) \qquad (1.138)$$

Therefore, for states with definite parity, only statistical tensors with an even rank k and $q = 0$ can exist if $\eta = +1$, whereas for $\eta = -1$, only tensors with odd k and $q = 0$ are possible. Owing to the symmetry, for a sharp j, the number of real independent parameters of the density matrix is reduced as a result of the symmetry to j (j — integer), or to $j - \frac{1}{2}$ (j — half integer). The density matrix in the chosen coordinate system is diagonal with respect to the projection of the angular momenta, and for these diagonals over m elements, the following relation is valid:

$$\langle jm \,|\, \rho \,|\, j'm \rangle = \eta \, \langle j - m \,|\, \rho \,|\, j' - m \rangle \qquad (1.139)$$

For a state with definite angular momentum and parity, the populations of magnetic sublevels with different signs of the projection m are equal.

1.4.5. A System with the Symmetry Axis Perpendicular to the Plane of Symmetry

We choose the z-axis along the axis of symmetry. The symmetry restrictions can be found by combining the results of Section 1.4.2 and Eq. (1.136). It follows that only those statistical tensors can exist that have $q = 0$ and obey the relation

$$\rho_{k0}\,(j, j') = \eta \, \rho_{k0}\,(j, j') \qquad (1.140)$$

Therefore, this type of symmetry can be realized only for $\eta = +1$; otherwise all statistical tensors will be zero. This means that such a physical system must have definite parity. No other restrictions follow from Eq. (1.140) and hence, for a single state with definite parity, adding the plane of symmetry to the axis of symmetry, which is normal to this plane, will not reduce the number of independent parameters. In other words, from the viewpoint of symmetry, a system with definite parity and a symmetry axis is equivalent to a system with a symmetry axis and a plane of symmetry perpendicular to this axis.

Concluding this section, we collect in Table 1.2 a list of possible independent statistical tensors of systems with the angular momenta $j \leq 2$.

1.5.   Efficiency Matrix and Efficiency Tensors

1.5.1.   Concepts of Efficiency Operator and Efficiency Matrix

We assume that as a result of some process, a state of the system is formed that is described by the density operator ρ. The physical characteristics of this system are measured by a detector (or a number of detectors). Let us assume that the detector is "tuned" to the state, which is characterized by the density operator ε. That is, the registration occurs only if the detector is affected by the system in the state ε. The state ε is an inherent property of the detector and is called the *efficiency operator* of the detector. The probability of the detector registering the state ε while observing the system in the state ρ is then given by Eq. (1.35):

$$W = \mathrm{Tr}\,(\rho \varepsilon) \qquad (1.141)$$

The probability (1.141) defines the counting rate of the detector. Equation (1.141) can be explained as follows. Suppose that the detector reacts if the system is in a pure state $|\psi\rangle$, but does not react if the system is in any other pure state orthogonal to $|\psi\rangle$. In accordance with (1.15), the probability of detecting the state $|\psi\rangle$ in the state ρ is given by $W = \langle\psi|\rho|\psi\rangle$. Now we generalize this statement to the case when the detector registers a pure state $|\psi_1\rangle$ with a probability ε_1, another pure state $|\psi_2\rangle$ with a probability ε_2, and so on, where $|\psi_k\rangle$ forms some basis set. Obviously, the total probability of registration by this detector is now defined by $W = \sum_k \varepsilon_k \langle\psi_k|\rho|\psi_k\rangle$. Introducing the efficiency operator by the relation

$$\varepsilon = \sum_k \varepsilon_k |\psi_k\rangle\langle\psi_k| \tag{1.142}$$

one immediately obtains (1.141). Equation (1.142) coincides formally with (1.29), and the operator ε, which is the efficiency operator, shows all the mathematical properties of the density operator ρ.

The *efficiency matrix* of the detector is a matrix of the efficiency operator in a particular representation. As follows from above, the efficiency matrix $\langle\xi|\varepsilon|\xi'\rangle$ is mathematically equivalent to the density matrix $\langle\xi|\rho|\xi'\rangle$. The normalization of the efficiency matrix will be discussed later.

Throughout this book we consider ideal detectors which are sensitive to the particle energy, to the direction of the particle motion, and to the spin state of the particle, and the efficiency of detecting each of these quantities is independent of the others. This means that the efficiency operator is a product of the three operators:

$$\varepsilon = \varepsilon^E \, \varepsilon^d \, \varepsilon^s \tag{1.143}$$

where the superscripts E, d, and s indicate that the corresponding factors relate to the energy, direction (space), and spin parts of the efficiency operator, respectively. The first factor in Eq. (1.143) does not influence the angular distributions or polarization. For the detector registering particles moving in a well-defined direction $\mathbf{n}_0$, the space part of its efficiency operator is given by

$$\varepsilon^d = |\mathbf{n}_0\rangle\langle\mathbf{n}_0| \tag{1.144}$$

[see Eq. (1.96)]. In reality, the detector registers particles within a finite solid angle. The finite angular resolution of real detectors can be accounted for by the methods described, for example, in Ref. 22, where the corresponding efficiency matrices are analyzed. Here the finite solid angle seen by a detector can be further allowed for by integrating over the corresponding solid angle. The spin part of the efficiency operator ε^s in the representation of the particle spin s and its projection μ gives the *efficiency matrix of spin*, $\langle s\mu|\varepsilon|s\mu'\rangle$, which is formally equivalent to the spin density matrix, $\langle sm_s|\rho|sm'_s\rangle$, from Section 1.2.1.

Let us discuss the normalization of the efficiency matrix of spin. The spin density matrix of the observed particle with spin s is normalized according to Eq. (1.7) to obtain the total probability 1. We assume that the particle is observed by a spin-insensitive detector; that is, all particles are registered with equal probabilities by the detector, regardless of their spin state. Hence, the efficiency matrix corresponds to the spin density matrix of the isotropic system, Eq. (1.40): $\langle s\mu \,|\, \varepsilon \,|\, s\mu'\rangle = C\delta_{\mu\mu'}$, where C is a normalization constant. Using (1.141) one obtains

$$W = \mathrm{Tr}\,(\rho\varepsilon) = \sum_{\mu\mu'} \langle s\mu \,|\, \rho \,|\, s\mu'\rangle \langle s\mu' \,|\, \varepsilon \,|\, s\mu \rangle = C\sum_{\mu} \langle s\mu \,|\, \rho \,|\, s\mu \rangle = C \quad (1.145)$$

The requirement that the particle be detected with certainty ($W = 1$) gives $C = 1$ and, hence $\mathrm{Tr}\,(\rho\varepsilon) = 1$. This results in the efficiency matrix for the spin-insensitive detector:

$$\langle s\mu \,|\, \varepsilon \,|\, s\mu'\rangle = \delta_{\mu\mu'} \quad (1.146)$$

This gives the normalization condition $\mathrm{Tr}\,\varepsilon^s = 2s + 1$, which differs from the normalization of the density operator, Eq. (1.8).

In reality, the normalization of the efficiency matrix is often not important for the analysis of the angular distributions and spin polarization in the scattering process, since corresponding parameters are usually obtained from relative measurements.

1.5.2. Efficiency Tensors

Consider the efficiency matrix in the representation of the total angular momentum j of the observed system and its projection m: $\langle \alpha jm \,|\, \varepsilon \,|\, \alpha' j'm'\rangle$, where a set of other quantum numbers, which specify the representation, is denoted by α. We introduce the *efficiency tensors* $\varepsilon_{kq}(\alpha j, \alpha' j')$, in the same way as has been done for the statistical tensors [Eq. (1.41)]:

$$\varepsilon_{kq}(\alpha j, \alpha' j') = \sum_{mm'} (-1)^{j'-m'} (jm, j'-m'\,|\,kq) \langle \alpha jm \,|\, \varepsilon \,|\, \alpha' j'm'\rangle \quad (1.147)$$

The reverse transformation is similar to Eq. (1.42). The efficiency tensors possess all the properties of the statistical tensors except the normalization condition, as discussed above. Normalization in the spirit of Eq. (1.146) leads to the relation

$$\varepsilon_{00}(\alpha j, \alpha' j') = \hat{j}\delta_{jj'} \quad (1.148)$$

which can be compared with Eq. (1.49). The efficiency tensor of the total angular momentum J can be presented in terms of the efficiency tensors of the decoupled angular momenta of subsystems, just as has been done for the statistical tensors [see Eq. (1.64)]. Since the efficiency tensor for a number of independent detectors is the product of the efficiency tensors of individual detectors, one can write

$$\varepsilon_{kq}\left(j_1 j_2 J, j_1' j_2' J'\right) = \sum_{k_1 q_1 k_2 q_2} \hat{k}_1 \hat{k}_2 \hat{J} \hat{J}' \left(k_1 q_1, k_2 q_2 \mid kq\right) \begin{Bmatrix} j_1 & j_2 & J \\ j_1' & j_2' & J' \\ k_1 & k_2 & k \end{Bmatrix}$$

$$\times \varepsilon_{k_1 q_1}\left(j_1, j_1'\right) \varepsilon_{k_2 q_2}\left(j_2, j_2'\right) \tag{1.149}$$

Rewriting Eq. (1.141) in the representation of the total angular momentum of the system, one obtains

$$W = \text{Tr}\left(\rho\varepsilon\right) = \sum_{\substack{\alpha\alpha' jj' \\ mm'}} \left\langle \alpha jm \mid \rho \mid \alpha' j'm' \right\rangle \left\langle \alpha' j'm' \mid \varepsilon \mid \alpha jm \right\rangle \tag{1.150}$$

Using definitions (1.41) and (1.147), the Hermitian character of the efficiency matrix, and Eq. (A.87) we obtain

$$W = \sum_{\substack{\alpha\alpha' jj' \\ kq}} \rho_{kq}\left(\alpha j, \alpha' j'\right) \varepsilon_{kq}^*\left(\alpha j, \alpha' j'\right) \tag{1.151}$$

The statistical tensors $\rho_{kq}\left(\alpha j, \alpha' j'\right)$ and the efficiency tensors $\varepsilon_{kq}^*\left(\alpha j, \alpha' j'\right)$ in Eq. (1.151) are taken in one and the same coordinate system.

Equation (1.151) can serve as a starting point for solving most of the problems relevant to the angular distributions, angular correlations, and polarization of the reaction products. The form of a scalar product of two tensors reflects the general isotropy of the space, i.e., the invariance of the laws of nature with respect to rotations.

The characteristics of the detector and in particular its efficiency matrix are given primarily in its own coordinate system. At the same time, the density matrix of the physical system is more naturally presented in a coordinate system related to the process being investigated. Therefore, the form (1.151) is usually more convenient for calculating probabilities than (1.150) because it allows one to easily transform from one system to another.

Equation (1.151) can be written in terms of higher-order statistical tensors in the case of several angular momenta. For example, it is straightforward to show by using Eqs. (1.151), (1.64) (and its counterpart for the double efficiency tensor), (A.135) and (A.88) that in the case of two angular momenta,

$$W = \sum_{\substack{\beta\beta' j_1 j_1' j_2 j_2' \\ k_1 q_1 k_2 q_2}} \rho_{k_1 q_1 k_2 q_2}\left(\beta j_1 j_2, \beta' j_1' j_2'\right) \varepsilon_{k_1 q_1 k_2 q_2}^*\left(\beta j_1 j_2, \beta' j_1' j_2'\right) \tag{1.152}$$

Another problem arises when the polarization state of a subsystem 2 is to be found if another subsystem 1 of the whole system is detected in a polarization state $\varepsilon_{k_1 q_1}^{(1)} (\alpha_1 j_1, \alpha_1' j_1')$. The problem can be solved by making a projection of the density matrix ρ of the whole system onto the corresponding efficiency matrix $\varepsilon^{(1)}$ of subsystem 1:

$$\left\langle \alpha_2 j_2 m_2 \left| \rho^{(2)} \right| \alpha_2' j_2' m_2' \right\rangle = \sum_{\substack{j_1 m_1 j_1' m_1' \\ \alpha_1 \alpha_1'}} \left\langle \alpha_1 j_1 m_1 \alpha_2 j_2 m_2 \left| \rho \right| \alpha_1' j_1' m_1' \alpha_2' j_2' m_2' \right\rangle$$
$$\times \left\langle \alpha_1' j_1' m_1' \left| \varepsilon^{(1)} \right| \alpha_1 j_1 m_1 \right\rangle \qquad (1.153)$$

where the desired density matrix of subsystem 2 is at the left of Eq. (1.153). Coupling j_1 and j_2 into the total angular momentum J of the system, turning to the statistical and efficiency tensors, and performing summations over projections, one obtains

$$\rho_{k_2 q_2}^{(2)} (\alpha_2 j_2, \alpha_2' j_2') = \sum_{JJ'kq} \sum_{\substack{j_1 j_1' k_1 q_1 \\ \alpha_1 \alpha_1'}} \hat{J}\hat{J}'\hat{k}_1 \hat{k}_2 (k_1 q_1, k_2 q_2 | kq) \begin{Bmatrix} j_1 & j_2 & J \\ j_1' & j_2' & J' \\ k_1 & k_2 & k \end{Bmatrix}$$
$$\times \rho_{kq} (\alpha_1 j_1 \alpha_2 j_2 : J; \alpha_1' j_1' \alpha_2' j_2' : J') \, \varepsilon_{k_1 q_1}^{(1)*} (\alpha_1 j_1, \alpha_1' j_1') \qquad (1.154)$$

The result for subsystem 1 takes a similar form:

$$\rho_{k_1 q_1}^{(1)} (\alpha_1 j_1, \alpha_1' j_1') = \sum_{JJ'kq} \sum_{\substack{j_2 j_2' k_2 q_2 \\ \alpha_2 \alpha_2'}} \hat{J}\hat{J}'\hat{k}_1 \hat{k}_2 (k_1 q_1, k_2 q_2 | kq) \begin{Bmatrix} j_1 & j_2 & J \\ j_1' & j_2' & J' \\ k_1 & k_2 & k \end{Bmatrix}$$
$$\times \rho_{kq} (\alpha_1 j_1 \alpha_2 j_2 : J; \alpha_1' j_1' \alpha_2' j_2' : J') \, \varepsilon_{k_2 q_2}^{(2)*} (\alpha_2 j_2, \alpha_2' j_2') \qquad (1.155)$$

Equations (1.154) and (1.155) are of a simpler form, with the use of the double statistical tensors:

$$\rho_{k_2 q_2}^{(2)} (\alpha_2 j_2, \alpha_2' j_2') = \sum_{\substack{\alpha_1 \alpha_1' j_1 j_1' \\ k_1 q_1}} \rho_{k_1 q_1 k_2 q_2} (\alpha_1 j_1 \alpha_2 j_2, \alpha_1' j_1' \alpha_2' j_2')$$
$$\times \varepsilon_{k_1 q_1}^{(1)*} (\alpha_1 j_1, \alpha_1' j_1') \qquad (1.156)$$

$$\rho_{k_1 q_1}^{(1)} (\alpha_1 j_1, \alpha_1' j_1') = \sum_{\substack{\alpha_2 \alpha_2' j_2 j_2' \\ k_2 q_2}} \rho_{k_1 q_1 k_2 q_2} (\alpha_1 j_1 \alpha_2 j_2, \alpha_1' j_1' \alpha_2' j_2')$$
$$\times \varepsilon_{k_2 q_2}^{(2)*} (\alpha_2 j_2, \alpha_2' j_2') \qquad (1.157)$$

The case of a detector insensitive to polarization is of special importance. Consider this case using as an example a system with a sharp angular momentum j and a definite set of additional quantum numbers α. The efficiency matrix and the efficiency tensor of the detector follow from Eqs. (1.146) and (1.147):

$$\langle \alpha j m \,|\, \varepsilon \,|\, \alpha j m' \rangle = \delta_{mm'} \tag{1.158}$$

$$\varepsilon_{kq}(\alpha j, \alpha j) = \hat{j} \, \delta_{k0} \, \delta_{q0} \tag{1.159}$$

Assume that a system consists of the two subsystems, 1 and 2. Substituting Eq. (1.159) into Eq. (1.154), one obtains the statistical tensors of subsystem 2 under the condition that subsystem 1 is observed by the detector insensitive to polarization:

$$\rho^{(2)}_{k_2 q_2}(\alpha_2 j_2, \alpha_2' j_2') = \delta_{k_2 k} \delta_{q_2 q} \sum_{JJ' \alpha_1 j_1} (-1)^{J+j_1+j_2'+k} \hat{f}\hat{f}' \begin{Bmatrix} j_2 & j_2' & k \\ J' & J & j_1 \end{Bmatrix}$$

$$\times \rho_{kq}(\alpha_1 j_1 \alpha_2 j_2 J; \alpha_1 j_1 \alpha_2' j_2' J') \tag{1.160}$$

(The result for subsystem 1 when the detector of subsystem 2 is insensitive to its polarization has a similar form.) Comparing Eq. (1.160) with Eq. (1.68), we see that the case when the detector of one of the subsystems is insensitive to its polarization is equivalent, from the viewpoint of the observed polarization characteristics of the partner, to the case when the first subsystem is not observed at all.

In the next section we will find efficiency tensors describing the ideal detector system in typical experiments on angular correlations and polarization of the reaction products. Later they will serve as building blocks for constructing formulas for different kinds of angular correlations.

1.5.3. Detection of Electrons

A general method for finding the efficiency tensors of the total angular momentum of an entire system is to express them by a sequential application of the general coupling rule (1.149) in combination with the efficiency tensors of individual detectors. The latter are known up to normalization constants from Section 1.3.

Consider a case when an electron and an atomic system A (atom, ion, molecule, etc.), which we call hereafter an "atom," in the final state

$$A(\alpha_f J_f) + e \tag{1.161}$$

of some atomic reaction are detected simultaneously. Here J_f is the total angular momentum of the atom, and α_f is a set of quantum numbers that specify the final state of the atom. On the other hand, instead of an electron, another particle with spin $\frac{1}{2}$, for example, a proton, can be considered in Eq. (1.161) without changing anything in the formulas of this section.

Let us find the efficiency tensor for the system (1.161) in the representation of its total angular momentum. Using Eq. (1.149) for decoupling the angular momenta of the ion and the electron, we obtain

$$\varepsilon_{kq}\left(\alpha_f J_f, lj : J; \alpha'_f J'_f, l'j' : J'\right) = \sum_{k_f q_f k_j q_j} \hat{J}\hat{J}'\hat{k}_f\hat{k}_j \left(k_f q_f, k_j q_j | kq\right)$$

$$\times \begin{Bmatrix} J_f & j & J \\ J'_f & j' & J' \\ k_f & k_j & k \end{Bmatrix} \varepsilon_{k_f q_f}\left(\alpha_f J_f, \alpha'_f J'_f\right) \varepsilon_{k_j q_j}\left(lj, l'j'\right) \qquad (1.162)$$

Here l and j are the orbital and total angular momenta of the electron, respectively and J is the total angular momentum of the system (1.161) $\mathbf{J} = \mathbf{J}_f + \mathbf{j}$. The efficiency tensor of the electron detector, $\varepsilon_{k_j q_j}\left(lj, l'j'\right)$, which detects electrons moving in the direction (ϑ, φ) is given by Eqs. (1.104) and (1.105) except for the normalization constant. It can be further decoupled into the orbital and spin parts, similar to Eqs. (1.102) and (1.103):

$$\varepsilon_{kq}\left(lj, l'j'\right) = \sum_{k_l q_l k_s q_s} \hat{J}\hat{J}'\hat{k}_s\hat{k}_l \left(k_l q_l, k_s q_s | k_j q_j\right) \begin{Bmatrix} l & \frac{1}{2} & j \\ l' & \frac{1}{2} & j' \\ k_l & k_s & k_j \end{Bmatrix}$$

$$\times \varepsilon_{k_l q_l}\left(l, l'\right) \varepsilon_{k_s q_s}\left(\frac{1}{2}, \frac{1}{2}\right) \qquad (1.163)$$

where we factored the decoupled tensor into the orbital $\varepsilon_{k_l q_l}\left(l, l'\right)$ and spin $\varepsilon_{k_s q_s}\left(\frac{1}{2}, \frac{1}{2}\right)$ parts using the ansatz of independence of the corresponding measurements by the "ideal" detector, as stated in Section 1.5.1. All efficiency tensors in Eqs. (1.162) and (1.163) are given in a common coordinate frame, which we choose as being the laboratory frame.

The spin part of the efficiency tensor, $\varepsilon_{k_s q_s}\left(\frac{1}{2}, \frac{1}{2}\right)$, can be expressed in terms of the three components of the polarization vector $\mathbf{P}$ in the same way as for the statistical tensors (1.93)–(1.95) with allowance for normalization (1.148):

$$\varepsilon_{00}^{\text{det}}\left(\frac{1}{2}, \frac{1}{2}\right) = \sqrt{2}$$

$$\varepsilon_{10}^{\text{det}}\left(\frac{1}{2}, \frac{1}{2}\right) = \sqrt{2}P_z$$

$$\varepsilon_{1\pm1}^{\text{det}}\left(\frac{1}{2}, \frac{1}{2}\right) = \mp(P_x \mp iP_y) \qquad (1.164)$$

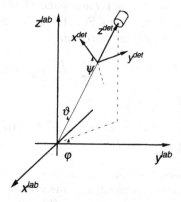

FIGURE 1.2. Laboratory and detector coordinate systems.

Here the superscript "det" indicates that the efficiency tensors of spin are taken in a coordinate system, S^{det}, related to the detector of the electrons. Our standard choice of the detector coordinate system is shown in Figure 1.2. The z^{det}-axis is directed to the detector, while the axes x^{det} and y^{det} are conveniently chosen in the plane perpendicular to the detected beam. For example, when the detector is a polarimeter, which selects electrons transversely polarized along some direction, we can choose the x^{det}-axis along this direction. In this case one obtains from Eq. (1.164)

$$\varepsilon_{00}^{det}\left(\frac{1}{2},\frac{1}{2}\right) = \sqrt{2}, \quad \varepsilon_{10}^{det}\left(\frac{1}{2},\frac{1}{2}\right) = 0, \quad \varepsilon_{1\pm1}^{det}\left(\frac{1}{2},\frac{1}{2}\right) = \mp1 \qquad (1.165)$$

To transform the spin part of the efficiency tensor into the laboratory coordinate system, one can use a transformation reverse to Eq. (1.44):

$$\varepsilon_{kq}\left(\frac{1}{2},\frac{1}{2}\right) = \sum_{q'} D_{qq'}^{k}(\varphi,\vartheta,\psi)\,\varepsilon_{kq'}^{det}\left(\frac{1}{2},\frac{1}{2}\right) \qquad (1.166)$$

where the angles (φ,ϑ,ψ) shown in Figure 1.2 are Euler angles characterizing the rotation from the laboratory coordinate system, S^{lab}, to the detector coordinate system S^{det}.

The space part of the efficiency tensor in the laboratory system, $\varepsilon_{k_l q_l}(l,l')$, is given by Eqs. (1.99) and (1.100) except for the common normalization factor. Collecting Eqs. (1.162), (1.163), (1.166), (1.99), and (1.100), one obtains the general form of the efficiency tensors (1.162) in the laboratory frame for the case when both products in Eq. (1.161) are detected:

$$
\varepsilon_{kq}\left(\alpha_f J_f, l j : J; \alpha'_f J'_f, l' j' : J'\right) = \frac{1}{4\pi}(-1)^{l'}\,\hat{l}\hat{l'}\,\hat{j}\hat{j'}\,\hat{J}\hat{J'} \sum_{k_f k_j k_l k_s} \hat{k}_f \hat{k}_j \hat{k}_l \hat{k}_s
$$

$$
\times (l0, l'0 | k_l 0) \begin{Bmatrix} J_f & j & J \\ J'_f & j' & J' \\ k_f & k_j & k \end{Bmatrix} \begin{Bmatrix} l & \frac{1}{2} & j \\ l' & \frac{1}{2} & j' \\ k_l & k_s & k_j \end{Bmatrix}
$$

$$
\times \sum_{\substack{q_f q_j q_l \\ q_s q'_s}} (k_f q_f, k_j q_j | k q)\,(k_l q_l, k_s q_s | k_j q_j)
$$

$$
\times \varepsilon_{k_f q_f}\left(\alpha_f J_f, \alpha'_f J'_f\right) \varepsilon^{\text{det}}_{k_s q'_s}\left(\frac{1}{2},\frac{1}{2}\right) \sqrt{\frac{4\pi}{2k_l+1}}\, Y^*_{k_l q_l}(\vartheta, \varphi)
$$

$$
\times D^{k_s}_{q_s q'_s}(\varphi, \vartheta, \psi) \tag{1.167}
$$

Here the efficiency tensors of the detector of atom A, $\varepsilon_{k_f q_f}\left(\alpha_f J_f, \alpha'_f J'_f\right)$, are still in the laboratory frame. They can be expressed, if needed, in terms of the efficiency tensors of the atomic detector in its own frame similar to Eq. (1.166) with the corresponding set of angles $(\varphi_A, \vartheta_A, \psi_A)$. This procedure leads to an expression that includes angular and spin correlations of the reaction products (1.161).

Here we turn to a simpler and more widespread case when the polarization state and the direction of movement of the ion are not detected. Then we should integrate over all angles of atom recoil and sum over polarization states of the atom. This is equivalent to a situation when the atom is detected with the unit probability independent of its state. Such a situation is described by the isotropic part $k_f = q_f = 0$ of the efficiency tensor of the atom detector. So, if the atom is left in the sharp eigenstate and no further measurements are made for it, then taking into account Eq. (1.148), we obtain

$$
\varepsilon_{k_f q_f}\left(\alpha_f J_f, \alpha'_f J'_f\right) = \hat{J}_f\, \delta_{k_f 0}\, \delta_{q_f 0}\, \delta_{J_f J'_f}\, \delta_{\alpha_f \alpha'_f} \tag{1.168}
$$

It follows from Eqs. (1.168) and (1.162) that in this case the efficiency tensor for the detector of the whole system (1.161) is proportional to the efficiency tensor for the electron detector. Substituting Eq. (1.168) into Eq. (1.167) gives

$$
\varepsilon_{kq}\left(\alpha_f J_f, l j : J; \alpha'_f J'_f, l' j' : J'\right) = \delta_{J_f J'_f}\, \delta_{\alpha_f \alpha'_f} \frac{1}{4\pi}(-1)^{l'}\,\hat{l}\hat{l'}\,\hat{j}\hat{j'}\,\hat{J}\hat{J'}
$$

$$
\times \sum_{k_l k_s}(-1)^{j'+J+J_f+k}\,\hat{k}_l \hat{k}_s\,(l0, l'0|k_l 0) \begin{Bmatrix} J & j & J_f \\ j' & J' & k \end{Bmatrix} \begin{Bmatrix} l & \frac{1}{2} & j \\ l' & \frac{1}{2} & j' \\ k_l & k_s & k \end{Bmatrix}
$$

$$
\times \sum_{q_l q_s q'_s}(k_l q_l, k_s q_s | k q)\, \varepsilon^{\text{det}}_{k_s q'_s}\left(\frac{1}{2},\frac{1}{2}\right) \sqrt{\frac{4\pi}{2k_l+1}}
$$

$$\times Y^*_{k_l q_l}(\vartheta, \varphi) D^{k_s}_{q_s q'_s}(\varphi, \vartheta, \psi) \tag{1.169}$$

To obtain Eq. (1.169), we used Eq. (A.122). Further simplification follows when the electron detector is insensitive to the polarization

$$\varepsilon^{\text{det}}_{k_s q'_s}\left(\frac{1}{2}, \frac{1}{2}\right) = \sqrt{2}\,\delta_{k_s 0}\,\delta_{q'_s 0} \tag{1.170}$$

Then Eq. (1.169) reduces to

$$\varepsilon_{kq}\left(\alpha_f J_f, lj : J; \alpha'_f J'_f, l'j' : J'\right) = \delta_{J_f J'_f}\,\delta_{\alpha_f \alpha'_f}\,\frac{1}{4\pi}(-1)^{J+J_f+k-1/2}\,\hat{l}\hat{l}'\,\hat{j}\hat{j}'\,\hat{J}\hat{J}'$$

$$\times (l0, l'0|k0) \left\{\begin{matrix} J & j & J_f \\ j' & J' & k \end{matrix}\right\} \left\{\begin{matrix} l & j & \frac{1}{2} \\ j' & l' & k \end{matrix}\right\} \sqrt{\frac{4\pi}{2k+1}}\,Y^*_{kq}(\vartheta, \varphi) \tag{1.171}$$

1.5.4. Detection of Photons

Now we consider the efficiency tensors of a detector for the system

$$A(\alpha_f J_f) + \gamma \tag{1.172}$$

which corresponds to emission of a photon γ in some process. The derivation is very similar to the one in Section 1.5.3. We use the multipole representation for the efficiency tensors of the detector of photons, $\varepsilon^{\text{det}}_{kq}(pL, p'L')$, in the coordinate system S^{det} (Figure 1.2), which coincides with statistical tensors (1.125) and (1.126) within the normalization. The efficiency tensors for the system (1.172) in the representation of the total angular momentum J are obtained by decoupling the angular momenta of the atom A and the radiation, J_f and L, respectively, and by using a transformation similar to Eq. (1.166), with the result:

$$\varepsilon_{kq}\left(\alpha_f J_f, pL : J; \alpha'_f J'_f, p'L' : J'\right) = \sum_{k_f k_L} \hat{J}\hat{J}'\hat{k}_f\hat{k}_L \left\{\begin{matrix} J_f & L & J \\ J'_f & L' & J' \\ k_f & k_L & k \end{matrix}\right\}$$

$$\times \sum_{\substack{q_f q_L \\ q'_L = 0 \pm 2}} (k_f q_f, k_L q_L|kq)\,\varepsilon_{k_f q_f}\left(\alpha_f J_f, \alpha'_f J'_f\right)\,\varepsilon^{\text{det}}_{k_L q'_L}(pL, p'L')$$

$$\times D^{k_L}_{q_L q'_L}(\varphi, \vartheta, \psi) \tag{1.173}$$

One obtains a simple relation when the atom in the final state $A(\alpha_f J_f)$ is not detected. Substitution of Eq. (1.168) into Eq. (1.173) gives

$$\varepsilon_{kq}\left(\alpha_f J_f, pL : J; \alpha'_f J'_f, p'L' : J'\right) = \delta_{J_f J'_f}\,\delta_{\alpha_f \alpha'_f}(-1)^{L'+J+J_f+k}\,\hat{J}\hat{J}'$$

$$\times \left\{\begin{matrix} J' & L' & J_f \\ L & J & k \end{matrix}\right\} \sum_{q'=0\pm 2} \varepsilon^{\text{det}}_{kq'}(pL, p'L')\,D^k_{qq'}(\varphi, \vartheta, \psi) \tag{1.174}$$

For a detector that is insensitive to photon polarization, one must put $P_1 = P_2 = P_3 = 0$. Then only terms with $q' = 0$ contribute to the sum in Eq. (1.174), as follows from Eqs. (1.125) and (1.126), and the efficiency tensors (1.174) take the form

$$\varepsilon_{kq}\left(\alpha_f J_f, pL : J; \alpha'_f J'_f, p'L' : J'\right) = \delta_{J_f J'_f} \delta_{\alpha_f \alpha'_f} (-1)^{L'+J+J_f+k} \hat{f}\hat{f}' \begin{Bmatrix} J' & L' & J_f \\ L & J & k \end{Bmatrix}$$

$$\times \varepsilon_{k0}^{\text{det}}\left(pL, p'L'\right) \sqrt{\frac{4\pi}{2k+1}} Y^*_{kq}(\vartheta, \varphi) \quad (1.175)$$

For dipole photons ($p = p' = 0$, $L = L' = 1$), the efficiency tensors are practically of the form as in column S in Table 1.1

$$\varepsilon_{00}^{\text{det}} = \frac{2}{\sqrt{3}}, \quad \varepsilon_{10}^{\text{det}} = \sqrt{2}P_3^{\text{det}}, \quad \varepsilon_{20}^{\text{det}} = \sqrt{\frac{2}{3}}$$

$$\varepsilon_{2\pm2}^{\text{det}} = -(P_1^{\text{det}} \mp iP_2^{\text{det}}), \quad \varepsilon_{1\pm1}^{\text{det}} = \varepsilon_{2\pm1}^{\text{det}} = 0 \quad (1.176)$$

but normalized according to the condition $\text{Tr}\,\varepsilon = 2$, which follows, in turn, from the more fundamental equality $\text{Tr}(\rho\varepsilon) = 1$.

1.5.5. Detection of Two Electrons in Coincidence

In principle, the examples of the efficiency tensors given above are sufficient for the analysis of a large variety of atomic reactions when one particle is detected in the final state, or even when several reaction products are detected in coincidence, but only when the reaction proceeds in few independent steps, each of which results in ejection of an electron or photon. At the same time, investigation of angular correlation and polarization of the products of reactions in a general case requires knowledge of the efficiency tensors for a number of angle-resolved and/or spin-sensitive detectors. Direct ionization of the type $(\gamma, 2e)$ or $(e, 2e)$, when the two outgoing electrons are detected in coincidence are examples of such processes.

Consider two independent detectors of electrons 1 and 2 located at angles (ϑ_1, φ_1) and (ϑ_2, φ_2) with respect to the laboratory frame, the point of ejection of the electrons being taken as the origin of the coordinate frame. Let angles ψ_1 and ψ_2, together with the two pairs of angles (ϑ_1, φ_1) and (ϑ_2, φ_2), characterize orientation of the internal coordinate systems of the detectors relative to the laboratory frame. Let $\varepsilon_{k_{s1} q_{s1}}^{\text{det}}(1/2, 1/2)$ and $\varepsilon_{k_{s2} q_{s2}}^{\text{det}}(1/2, 1/2)$ stand for their spin efficiency tensors in the corresponding internal coordinate frames. The efficiency tensors of each of the detectors in the representation of the total angular momentum of the electron are given by combining equations (1.102), (1.103) and (1.99), (1.100). In the laboratory frame they read as

$$
\varepsilon_{k_1 q_1}\left(l_1 j_1, l_1' j_1'\right) = \sum_{k_{l1} q_{l1} k_{s1} q_{s1} q_{s1}'} \hat{k}_{s1} \hat{j}_1 \hat{j}_1' \left(k_{l1} q_{l1}, k_{s1} q_{s1} \,|\, k_1 q_1\right)
\left\{
\begin{array}{ccc}
l_1 & 1/2 & j_1 \\
l_1' & 1/2 & j_1' \\
k_{l1} & k_{s1} & k_1
\end{array}
\right\}
$$

$$
\times \frac{(-1)^{l_1'}}{\sqrt{4\pi}} \hat{l}_1 \hat{l}_1' \left(l_1 0, l_1' 0 \,|\, k_{l1} 0\right) Y_{k_{l1} q_{l1}}^* (\vartheta_1, \varphi_1)\, \varepsilon_{k_{s1} q_{s1}}^{\det} (1/2, 1/2)\, D_{q_{s1} q_{s1}'}^{k_{s1}} (\varphi_1, \vartheta_1, \psi_1)
$$

$$\tag{1.177}$$

(corresponding equation for $\varepsilon_{k_2 q_2}\left(l_2 j_2, l_2' j_2'\right)$ has the similar form). Taking into account that detectors 1 and 2 are independent of each other, the efficiency tensor of the whole detector system can be calculated according to Eq. (1.149):

$$
\varepsilon_{kq}\left(l_1 j_1, l_2 j_2 : j; l_1' j_1', l_2' j_2' : j'\right)
$$

$$
= \sum_{k_1 q_1 k_2 q_2} \hat{k}_1 \hat{k}_2 \hat{j} \hat{j}' \left(k_1 q_1, k_2 q_2 \,|\, k q\right)
\left\{
\begin{array}{ccc}
j_1 & j_2 & j \\
j_1' & j_2' & j' \\
k_1 & k_2 & k
\end{array}
\right\}
\varepsilon_{k_1 q_1}\left(l_1 j_1, l_1' j_1'\right) \varepsilon_{k_2 q_2}\left(l_2 j_2, l_2' j_2'\right)
$$

$$
= \frac{1}{4\pi}(-1)^{l_1' + l_2'} \hat{l}_1 \hat{l}_1' \hat{l}_2 \hat{l}_2' \hat{j}_1 \hat{j}_1' \hat{j}_2 \hat{j}_2' \hat{j} \hat{j}' \sum_{\substack{k_{l1} k_{l2} k_{s1} k_{s2} \\ k_1 k_2}} \hat{k}_1 \hat{k}_2 \hat{k}_{s1} \hat{k}_{s2}
$$

$$
\times \left(l_1 0, l_1' 0 \,|\, k_{l1} 0\right) \left(l_2 0, l_2' 0 \,|\, k_{l2} 0\right)
\left\{
\begin{array}{ccc}
j_1 & j_2 & j \\
j_1' & j_2' & j' \\
k_1 & k_2 & k
\end{array}
\right\}
\left\{
\begin{array}{ccc}
l_1 & 1/2 & j_1 \\
l_1' & 1/2 & j_1' \\
k_{l1} & k_{s1} & k_1
\end{array}
\right\}
\left\{
\begin{array}{ccc}
l_2 & 1/2 & j_2 \\
l_2' & 1/2 & j_2' \\
k_{l2} & k_{s2} & k_2
\end{array}
\right\}
$$

$$
\times \sum_{\substack{q_1 q_2 q_{l1} q_{l2} \\ q_{s1} q_{s1}' q_{s2} q_{s2}'}} \left(k_1 q_1, k_2 q_2 \,|\, k q\right) \left(k_{l1} q_{l1}, k_{s1} q_{s1} \,|\, k_1 q_1\right) \left(k_{l2} q_{l2}, k_{s2} q_{s2} \,|\, k_2 q_2\right)
$$

$$
\times \varepsilon_{k_{s1} q_{s1}}^{\det} (1/2, 1/2)\, \varepsilon_{k_{s2} q_{s2}}^{\det} (1/2, 1/2)\, Y_{k_{l1} q_{l1}}^* (\vartheta_1, \varphi_1)\, Y_{k_{l2} q_{l2}}^* (\vartheta_2, \varphi_2)
$$

$$
\times D_{q_{s1} q_{s1}'}^{k_{s1}} (\varphi_1, \vartheta_1, \psi_1)\, D_{q_{s2} q_{s2}'}^{k_{s2}} (\varphi_2, \vartheta_2, \psi_2)
$$

$$\tag{1.178}$$

This equation includes both angular and spin correlations of the outgoing electrons. A simpler expression may be written for the efficiency tensor when both detectors are insensitive to polarization of the electrons. After substituting Eq. (1.170) into Eq. (1.178) for both electrons and straightforward transformations one obtains

$$
\varepsilon_{kq}\left(l_1 j_1, l_2 j_2 : j; l_1' j_1', l_2' j_2' : j'\right)
$$

$$
= \frac{1}{4\pi}(-1)^{j_1' + j_2' + 1} \hat{l}_1 \hat{l}_1' \hat{l}_2 \hat{l}_2' \hat{j}_1 \hat{j}_1' \hat{j}_2 \hat{j}_2' \hat{j} \hat{j}' \sum_{k_1 k_2} \left(l_1 0, l_1' 0 \,|\, k_1 0\right) \left(l_2 0, l_2' 0 \,|\, k_2 0\right)
$$

$$\times \begin{Bmatrix} j_1 & j_2 & j \\ j'_1 & j'_2 & j' \\ k_1 & k_2 & k \end{Bmatrix} \begin{Bmatrix} l_1 & j_1 & 1/2 \\ j'_1 & l'_1 & k_1 \end{Bmatrix} \begin{Bmatrix} l_2 & j_2 & 1/2 \\ j'_2 & l'_2 & k_2 \end{Bmatrix} \{Y_{k_1}(\vartheta_1, \varphi_1) \otimes Y_{k_2}(\vartheta_2, \varphi_2)\}^*_{kq}$$

(1.179)

where $\{Y_{k_1}(\vartheta_1, \varphi_1) \otimes Y_{k_2}(\vartheta_2, \varphi_2)\}_{kq}$ is the tensor product of spherical harmonics, which is called *bipolar spherical harmonic* (see Section A.4).

Always in this section we derive the efficiency tensors for the case when two outgoing electrons are detected, further generalization for a larger number of particles is straightforward.

2

Production of Polarized States

2.1. *Polarization of Compound Systems*

In this and the following sections we consider applications of the density matrix and statistical tensor formalism to various atomic processes. We start from the simplest atomic reaction in which as a result of the interaction of two particles in the initial state only one particle (a "compound" system) is formed in the final state. A typical example of such a process is photoexcitation of a discrete atomic level. A compound system may also be formed in a scattering process as a sufficiently long-living intermediate state (resonance). We assume that the compound system has a certain energy and a total angular momentum. Our goal is to determine the polarization state of the compound system and to find out how it is related to the characteristics of the initial colliding partners.

2.1.1. *Density Matrix and Statistical Tensors of Photoexcited Atomic States*

We consider an excitation of a discrete atomic state by photoabsorption:

$$\gamma + A(\alpha_0 J_0) \longrightarrow A^*(\alpha J) \tag{2.1}$$

Both the initial (ground) and the excited atomic states are characterized by sharp total angular momenta, J_0 and J, respectively. All other quantum numbers that are necessary for specifying the states are denoted by α_0 and α. Generalization to the case when the initial and/or the final atomic states are mixtures of states with different values of the total angular momenta is straightforward. Absorption of a photon, as well as its emission, are considered rigorously in quantum electrodynamics, where the concept of an S-matrix has been developed for the processes of the interaction of electromagnetic field with charged particles. Here we use the results of this consideration without discussing the basic aspects. Our aim is to obtain the density matrix and statistical tensors of the excited state and to express them in terms of the parameters characterizing the initial atomic state and the photon beam.

We know that the initial ρ^i and final ρ^f density operators are connected by the relation (1.24)

$$\rho^f = T\rho^i T^+ \tag{2.2}$$

where the transition operator T describes photoabsorption. Matrix elements of the density matrix of the final state can therefore be presented as [see Eq. (1.57)]

$$
\begin{aligned}
\langle \alpha JM \,|\, \rho^f \,|\, \alpha JM' \rangle = \sum_{M_0 M_0' \lambda \lambda'} & \langle \alpha JM \,|\, T \,|\, \alpha_0 J_0 M_0, \mathbf{k}_i \lambda \rangle \\
& \times \langle \alpha JM' \,|\, T \,|\, \alpha_0 J_0 M_0', \mathbf{k}_i \lambda' \rangle^* \\
& \times \langle \alpha_0 J_0 M_0, \mathbf{k}_i \lambda \,|\, \rho^i \,|\, \alpha_0 J_0 M_0', \mathbf{k}_i \lambda' \rangle
\end{aligned} \tag{2.3}
$$

where we use the helicity representation as a basis for the wave function of the incident photon with the wave vector $\mathbf{k}_i$. The initial state density operator can be presented as a direct product of the density operators of the photon and initial atom:

$$\rho^i = \rho^\gamma \rho^A \tag{2.4}$$

Thus the density matrix of the final state of the excited atom is

$$
\begin{aligned}
\langle \alpha JM \,|\, \rho^f \,|\, \alpha JM' \rangle = \sum_{M_0 M_0' \lambda \lambda'} & \langle \alpha JM \,|\, T \,|\, \alpha_0 J_0 M_0, \mathbf{k}_i \lambda \rangle \\
& \times \langle \alpha JM' \,|\, T \,|\, \alpha_0 J_0 M_0', \mathbf{k}_i \lambda' \rangle^* \\
& \times \langle \alpha_0 J_0 M_0 \,|\, \rho^A \,|\, \alpha_0 J_0 M_0' \rangle \langle \mathbf{k}_i \lambda \,|\, \rho^\gamma \,|\, \mathbf{k}_i \lambda' \rangle
\end{aligned} \tag{2.5}
$$

In principle, this equation solves our problem. However, it can be simplified if we expand the photon wave functions in multipoles (1.120) and use the long-wave approximation for calculating the matrix elements of the transition operator T. The *long-wave approximation* is valid when the wavelength of the photon is much larger than the size of the target atom. This is a typical situation in atomic physics.

In a long-wave approximation, we express the transition matrix elements $\langle \alpha JM \,|\, T \,|\, \alpha_0 J_0 M_0, \mathbf{k}_i \lambda \rangle$ in terms of matrix elements of multipole operators. In atomic physics in the majority of experiments on photoabsorption and photoexcitation, electric dipole (E1) transitions dominate. The contribution of higher electric and magnetic multipoles is negligible, as a rule. In the *dipole approximation* and assuming that the photon beam is directed along the z-axis of the coordinate system, the transition matrix element in Eq. (2.5) is proportional to the *dipole amplitude*

$$\langle \alpha JM \,|\, T \,|\, \alpha_0 J_0 M_0, \mathbf{k}_i \lambda \rangle \sim c \, \langle \alpha JM \,|\, D_\lambda \,|\, \alpha_0 J_0 M_0 \rangle \tag{2.6}$$

Here D_λ are the circular components [see Eq. (1.87)] of the vector of the dipole momentum of the atom:

$$\mathbf{D} = \sum_{n=1}^{N} \mathbf{r}_n \qquad (2.7)$$

where the summation is taken over all atomic electrons. Thus the density matrix of the excited state (2.5) can be presented as

$$\langle \alpha JM \, | \, \rho^f \, | \, \alpha JM' \rangle = c \sum_{M_0 M_0' \lambda \lambda'} \langle \alpha JM \, | \, D_\lambda \, | \, \alpha_0 J_0 M_0 \rangle$$

$$\times \langle \alpha JM' \, | \, D_{\lambda'} \, | \, \alpha_0 J_0 M_0' \rangle^* \langle \alpha_0 J_0 M_0 \, | \, \rho^A \, | \, \alpha_0 J_0 M_0' \rangle \langle \lambda \, | \, \rho^\gamma \, | \, \lambda' \rangle \quad (2.8)$$

where the photon density matrix in the dipole approximation is given by Eq. (1.124) [see also Eq. (1.106)]. The coefficient c in Eq. (2.8) is calculated in accordance with the normalization condition chosen for ρ^A, ρ^γ and ρ^f. We will follow the condition (1.8). Taking a particular case when the initial atomic state and the photon beam are unpolarized and starting from Eq. (2.8), we get

$$\mathrm{Tr}\,\rho^f = \frac{c}{2(2J_0+1)} \sum_{MM_0\lambda} |\langle \alpha JM \, | \, D_\lambda \, | \, \alpha_0 J_0 M_0 \rangle|^2 = \frac{c}{3(2J_0+1)} |\langle \alpha J \| D \| \alpha_0 J_0 \rangle|^2$$

$$(2.9)$$

where $\langle \alpha J \| D \| \alpha_0 J_0 \rangle$ are reduced matrix elements of the dipole operator defined according to the Wigner–Eckart theorem (A.62)[*]:

$$\langle \alpha JM \, | \, D_\lambda \, | \, \alpha_0 J_0 M_0 \rangle = \frac{1}{\hat{J}} \, (J_0 M_0, 1\lambda \, | \, JM) \, \langle \alpha J \| D \| \alpha_0 J_0 \rangle \qquad (2.10)$$

The condition $\mathrm{Tr}\,\rho^f = 1$ means that

$$c = \frac{3(2J_0+1)}{|\langle \alpha J \| D \| \alpha_0 J_0 \rangle|^2} \qquad (2.11)$$

Putting Eq. (2.11) into Eq. (2.8) and using Eq. (1.41) to relocate the spin-density matrix to the corresponding statistical tensors one obtains the statistical tensors of the final excited atom

$$\rho_{kq}(\alpha J) = 3(2J_0+1) \sum_{\substack{k_0 q_0 \\ k_\gamma q_\gamma}} \hat{k}_0 \hat{k}_\gamma \, (k_0 q_0, k_\gamma q_\gamma | kq) \begin{Bmatrix} J_0 & 1 & J \\ J_0 & 1 & J \\ k_0 & k_\gamma & k \end{Bmatrix}$$

$$\times \rho^\gamma_{k_\gamma q_\gamma} \rho_{k_0 q_0}(\alpha_0 J_0) \qquad (2.12)$$

---

[*]Standard methods of quantum electrodynamics give the total photoexcitation cross section in the process (2.1) in terms of the reduced matrix elements $\langle \alpha J \| D \| \alpha_0 J_0 \rangle$ in the form $\sigma = \frac{\pi k_l}{3(2J_0+1)} |\langle \alpha J \| D \| \alpha_0 J_0 \rangle|^2$, where it is implied that the photon beam and the target atoms are unpolarized.

Table 2.1. Statistical Tensors $\rho_{kq}(\alpha J)$ of States Excited by a Photon from an Unpolarized Target with the Total Angular Momentum J_0 [in the photon reference frame, $z \parallel \mathbf{k}_i$, see Eq. (2.15)]

P_1		arbitrary	arbitrary	+1
P_2		arbitrary	arbitrary	0
P_3		arbitrary	+1	arbitrary
J_0	J	ρ_{20}	ρ_{10}	ρ_{22}
0	1	$1/\sqrt{6}$	$1/\sqrt{2}$	$-1/2$
1	1	$-1/2\sqrt{6}$	$1/2\sqrt{2}$	$1/4$
1	2	$\sqrt{7}/10\sqrt{2}$	$3/2\sqrt{10}$	$-\sqrt{21}/20$
2	1	$1/10\sqrt{6}$	$-1/2\sqrt{2}$	$-1/20$
2	2	$-\sqrt{7}/10\sqrt{2}$	$1/2\sqrt{10}$	$\sqrt{21}/20$
2	3	$\sqrt{3}/5\sqrt{7}$	$1/\sqrt{7}$	$-3/5\sqrt{14}$
3	2	$1/5\sqrt{14}$	$-1/\sqrt{10}$	$-\sqrt{3}/10\sqrt{7}$
3	3	$-\sqrt{3}/4\sqrt{7}$	$1/4\sqrt{7}$	$3/4\sqrt{14}$
3	4	$\sqrt{11}/12\sqrt{7}$	$\sqrt{5}/4\sqrt{3}$	$-\sqrt{11}/4\sqrt{42}$
4	3	$1/4\sqrt{21}$	$-3/4\sqrt{7}$	$-1/4\sqrt{14}$
4	4	$-\sqrt{77}/60$	$1/4\sqrt{15}$	$\sqrt{77}/20\sqrt{6}$
4	5	$\sqrt{13}/5\sqrt{66}$	$3/\sqrt{110}$	$-\sqrt{13}/10\sqrt{11}$
1/2	1/2	0	$1/\sqrt{2}$	0
1/2	3/2	$1/4$	$\sqrt{5}/4$	$-\sqrt{6}/8$
3/2	1/2	0	$-1/2\sqrt{2}$	0
3/2	3/2	$-1/5$	$1/2\sqrt{5}$	$\sqrt{3}/5\sqrt{2}$
3/2	5/2	$\sqrt{7}/10\sqrt{3}$	$\sqrt{7}/2\sqrt{10}$	$-\sqrt{7}/10\sqrt{2}$
5/2	3/2	$1/20$	$-3/4\sqrt{5}$	$-\sqrt{3}/20\sqrt{2}$
5/2	5/2	$-4/5\sqrt{21}$	$1/\sqrt{70}$	$2\sqrt{2}/5\sqrt{7}$
5/2	7/2	$\sqrt{3}/4\sqrt{14}$	$3\sqrt{3}/4\sqrt{14}$	$-3/8\sqrt{7}$
7/2	5/2	$1/4\sqrt{21}$	$-\sqrt{5}/2\sqrt{14}$	$-1/4\sqrt{14}$
7/2	7/2	$-1/\sqrt{42}$	$-1/2\sqrt{42}$	$1/2\sqrt{7}$
7/2	9/2	$\sqrt{11}/20\sqrt{3}$	$\sqrt{11}/2\sqrt{30}$	$-\sqrt{11}/20\sqrt{2}$

Here $\rho_{k_0 q_0}(\alpha_0 J_0)$ is the statistical tensor of the atom before photoabsorption, and $\rho^\gamma_{k_\gamma q_\gamma}$ is the statistical tensor of the dipole photon. Note that hereafter we use the short notation for statistical tensors with two identical arguments:

$$\rho_{kq}(\alpha J) \equiv \rho_{kq}(\alpha J, \alpha J) \qquad (2.13)$$

This straightforward but somewhat tedious way to the resulting expression (2.12) may be replaced by a more elegant derivation if we use the properties of the statistical tensors we already know. Indeed, statistical tensors of the final atomic state are connected with those of the initial state by the following equation, which is in our case equivalent to Eq. (1.58):

$$\rho_{kq}(\alpha J) \sim \hat{J}^{-2} \langle \alpha J \parallel D \parallel \alpha_0 J_0 \rangle \, \rho^i_{kq}(J) \, \langle \alpha J \parallel D \parallel \alpha_0 J_0 \rangle^* \qquad (2.14)$$

The statistical tensor $\rho_{kq}^i(J)$ characterizes the initial state of the system photon + atom. It can be presented in terms of separate statistical tensors of the photon $\rho_{k_\gamma q_\gamma}^\gamma$ and the atom $\rho_{k_0 q_0}(\alpha_0 J_0)$ using Eqs. (1.64) and (1.66). After normalization according to the condition (1.8), the result coincides with Eq. (2.12).

The statistical tensors (2.12) as well as $\rho_{k_\gamma q_\gamma}^\gamma$ and $\rho_{k_0 q_0}(\alpha_0 J_0)$ are determined in the photon frame (z-axis along the photon beam). In order to transform $\rho_{kq}(\alpha J)$ to any other reference frame, one has to use Eq. (1.44).

Equation (2.12) describes the cross section and polarization state of an atom excited by photoabsorption provided the polarization properties of the initial state and the photon beam are known. Consider as an example an unpolarized target ($k_0 = q_0 = 0$). Then Eq. (2.12) reduces to the form

$$\rho_{kq}(\alpha J) = 3(-1)^{J+J_0+k+1} \begin{Bmatrix} J & 1 & J_0 \\ 1 & J & k \end{Bmatrix} \rho_{kq}^\gamma \qquad (2.15)$$

Equation (2.15) shows that the excited atom is characterized only by those statistical tensors that are present in the description of the photon beam. Thus, if the photon beam is unpolarized or linearly polarized, the excited atom can only be aligned, not oriented. The orientation of the excited atom can be produced by a circularly polarized beam only. Note that the values of the polarization parameters of the excited atom are proportional to the corresponding values of the photon beam. Table 2.1 contains the values of statistical tensors of the excited atom for some of the dipole transitions. The tensor $\rho_{20}(\alpha J)$ is independent of the polarization state of the photon. The tensors $\rho_{22}(\alpha J) = \rho_{2-2}^*(\alpha J)$ are connected with the linear polarization of the photon, while the tensor $\rho_{10}(\alpha J)$ is connected with its circular polarization. The tensors $\rho_{10}(\alpha J)$ corresponding to left and right circularly polarized photons are of opposite sign. The tensors with projections ± 1 are always zero: $\rho_{2\pm 1}(\alpha J) = \rho_{1\pm 1}(\alpha J) = 0$.

Let us now discuss photoexcitation of a polarized target. Equation (2.12) allows one to determine some general qualitative features of polarization of the excited state: The rank of the statistical tensor of the excited state is restricted by the triangle inequality

$$|k_0 - k_\gamma| \le k \le k_0 + k_\gamma \qquad (2.16)$$

where in the dipole approximation $k_\gamma \le 2$; and

$$k + k_0 + k_\gamma = \text{even} \qquad (2.17)$$

[otherwise the $9j$-symbol in (2.12) vanishes owing to the property (A.121)].

Consider several simple examples to illustrate the consequences of the above properties. If the target atom is aligned (k_0 even), then by absorption of linearly polarized light (k_γ even), the atom can only be aligned. Absorption of circularly polarized light (k_γ odd and even) creates both alignment and orientation of the excited state. Here the orientation is simply proportional to the degree of circular polarization of the photons. If the target is oriented (k_0 odd and even), then absorption of linearly polarized light (k_γ even) leads to the alignment and orientation of the excited state; orientation of the target influences only odd-rank statistical tensors of the excited state. Photoexcitation of the oriented target by circularly polarized light also leads to the alignment and orientation of the final state, but here the final orientation is determined by both the degree of circular polarization and the degree of orientation of the target. Moreover, the alignment of the final state has a component that depends on the product of the initial orientation of the atom and the circular polarization of the beam. We will see from the following discussion that this term is responsible for circular dichroism in the angular distribution of the autoionization electrons and the fluorescence produced from oriented targets.

In practice, one often has to calculate $\rho_{kq}(\alpha J)$ when statistical tensors of the initial atomic state and the photon are given in different coordinate frames, e.g., the laboratory frame for the target atom and the photon frame for the photon. Let R be a rotation that brings the laboratory frame into coincidence with the photon frame. Then taking into account Eq. (1.44), we obtain from Eq. (2.12)

$$
\rho_{kq}(\alpha J) = 3(2J_0+1) \sum_{\substack{k_0 q_0 \\ k_\gamma q_\gamma}} \hat{k}_0 \hat{k}_\gamma \left(k_0 q_0, k_\gamma q_\gamma \mid kq\right) \begin{Bmatrix} J_0 & 1 & J \\ J_0 & 1 & J \\ k_0 & k_\gamma & k \end{Bmatrix}
$$
$$
\times \rho_{k_0 q_0}^{\text{lab}}(\alpha_0 J_0) \sum_{q'_\gamma} D_{q'_\gamma q_\gamma}^{k_\gamma *}(R) \rho_{k_\gamma q'_\gamma}^{\gamma} \tag{2.18}
$$

where $\rho_{k_\gamma q_\gamma}^{\gamma}$ are given in the photon frame and $\rho_{k_0 q_0}^{\text{lab}}(\alpha_0 J_0)$ are given in the laboratory frame. Applying Eq. (2.18) several times, one can obtain the statistical tensors of states excited sequentially by several photon beams with arbitrary polarizations and mutual directions.

2.1.2. Density Matrix and Statistical Tensors of Compound Systems Formed in Collisions

Now we consider the formation of a compound system C in a collision of two subsystems (atom and electron, ion and electron, etc.), each having a definite internal total angular momentum (spin):

$$
A(\alpha_A J_A) + x(\alpha_x j_x) \rightarrow C(\alpha_C J_C) \tag{2.19}
$$

We denote by J_A, j_x, and J_C the total angular momenta (spins) of the colliding particles and the compound system, respectively, while α_A, α_x, and α_C denote all other quantum numbers characterizing the particles, including parity, parameters of the internal orbital momentum, and internal spin coupling. We assume that the collision occurs along the direction $\mathbf{n}$ defined by the polar and azimuthal angles ϑ, φ.

The density matrix ρ^i of the entire system $A + x$ in the initial state is a direct product of the spin density matrices ρ^A and ρ^x of the colliding particles and the density matrix

$$\rho^{\mathbf{n}} = |\mathbf{n}\rangle\langle\mathbf{n}| \tag{2.20}$$

describing their relative motion:

$$\rho^i = \rho^A \rho^x \rho^{\mathbf{n}} \tag{2.21}$$

The latter can be taken in the $|lm\rangle$ representation [see Eq. (1.98)]:

$$\langle lm | \rho^{\mathbf{n}} | l'm' \rangle = Y^*_{lm}(\vartheta, \varphi) Y_{l'm'}(\vartheta, \varphi) \tag{2.22}$$

The orbital angular momentum l together with the spins J_A and j_x of the colliding subsystems form the total angular momentum of the system J. Two coupling schemes are used in practice:

$$\text{(a)} \qquad \mathbf{J}_A + \mathbf{j}_x = \mathbf{S}; \quad \mathbf{S} + \mathbf{l} = \mathbf{J} \tag{2.23}$$

Here $\mathbf{J}_A$ and $\mathbf{j}_x$ form the total *channel spin* of the system in the initial state $\mathbf{S}$. Then the orbital angular momentum $\mathbf{l}$ of the relative motion is added to give the total angular momentum $\mathbf{J}$. Another coupling scheme looks as follows:

$$\text{(b)} \qquad \mathbf{l} + \mathbf{j}_x = \mathbf{J}_x; \quad \mathbf{J}_A + \mathbf{J}_x = \mathbf{J} \tag{2.24}$$

Here one introduces the total angular momentum J_x (orbital momentum + spin) of the subsystem x. Together with the spin J_A of the subsystem A, it gives the total angular momentum J of the whole system. Both coupling schemes are equivalent.

The density operators of the initial and final states of the process are connected by Eq. (1.24):

$$\rho^C = T\rho^i T^+ \tag{2.25}$$

Taking into account Eq. (2.21), we can write the matrix element of the density matrix in the representation of the spin projections:

$$\langle \alpha_C J_C M_C | \rho^C | \alpha_C J_C M'_C \rangle = \sum_{\substack{m_x m'_x m m' \\ M_A M'_A}} \langle \alpha_C J_C M_C | T | \alpha_A J_A M_A, \alpha_x j_x m_x, lm \rangle$$

$$\times \langle \alpha_C J_C M'_C | T | \alpha_A J_A M'_A, \alpha_x j_x m'_x, l'm' \rangle^* \langle \alpha_A J_A M_A | \rho^A | \alpha_A J_A M'_A \rangle$$

$$\times \langle \alpha_x j_x m_x | \rho^x | \alpha_x j_x m'_x \rangle \langle lm | \rho^{\mathbf{n}} | l'm' \rangle \tag{2.26}$$

The statistical tensors of the compound state C are simply expressed in terms of the matrix elements (2.26) in a standard way:

$$\rho_{kq}^C(\alpha_C J_C) = \sum_{M_C M_C'} (-1)^{J_C - M_C'} \left(J_C M_C, J_C - M_C' \,|\, kq\right)$$

$$\times \left\langle \alpha_C J_C M_C \,|\, \rho^C \,|\, \alpha_C J_C M_C' \right\rangle \tag{2.27}$$

A more compact expression can be obtained by coupling the angular momenta according to the coupling scheme (2.23) with the help of the Clebsch–Gordan coefficients. Then using the Wigner–Eckart theorem (A.62) for the matrix elements of the transition operator and the definition of statistical tensors (1.41) for the particles A and x, and performing summations over all projections of the angular momenta, it is possible to express the statistical tensors of the compound system C in terms of the statistical tensors of the colliding particles. However, as in the preceding subsection, we will do this in a simpler way, using the known properties of statistical tensors.

Equation (1.58) relates the statistical tensors of the compound system C and the colliding system in the initial state:

$$\rho_{kq}^C(\alpha_C J_C) = \hat{J}_C^{-2} \sum_{\alpha \alpha'} \langle \alpha_C J_C \,\|\, T \,\|\, \alpha J_C \rangle \, \rho_{kq}^i(\alpha J_C, \alpha' J_C')$$

$$\times \langle \alpha_C J_C \,\|\, T \,\|\, \alpha' J_C' \rangle^* \tag{2.28}$$

Here α and α' denote a set of quantum numbers of the initial state, including those related to the internal parameters of subsystems A and x. The particular sets α, α' depend on the coupling scheme of angular momenta for the colliding system.

Let us choose the spin channel representation (2.23):

$$|\alpha J_C\rangle \equiv |\alpha_A J_A, \alpha_x j_x, Sl : J_C\rangle; \qquad |\alpha' J_C\rangle \equiv |\alpha_A J_A, \alpha_x j_x, S'l' : J_C\rangle \tag{2.29}$$

Summation over α, α' in Eq. (2.28) then means summation over $Sl, S'l'$. Note that Eq. (2.28) accounts for conservation of the total angular momentum of the system: $J = J_C$.

Let $\rho_{k_A q_A}^A(\alpha_A J_A)$ and $\rho_{k_x q_x}^x(\alpha_x j_x)$ denote the statistical tensors of spins of subsystems A and x. We introduce the statistical tensor of the channel spin S according to Eq. (1.64):

$$\rho_{k_S q_S}^S(S, S') = \sum_{k_A q_A k_x q_x} \hat{k}_A \hat{k}_x \hat{S} \hat{S}' (k_A q_A, k_x q_x \,|\, k_S q_S)$$

$$\times \begin{Bmatrix} J_A & j_x & S \\ J_A & j_x & S' \\ k_A & k_x & k_S \end{Bmatrix} \rho_{k_A q_A}^A(\alpha_A J_A) \rho_{k_x q_x}^x(\alpha_x j_x) \tag{2.30}$$

Let us introduce further the statistical tensor $\rho^l_{k_l q_l}(l,l')$ of the orbital motion of the colliding subsystems. In the coordinate frame with the quantization axis chosen along the vector $\mathbf{n}$, the latter is given by Eq. (1.101). Then, taking into account conservation of the total angular momentum of the system in the transition (2.19), we relate the statistical tensors $\rho^C_{kq}(\alpha_C J_C)$ of the compound system (2.28) to those of the channel spin and of the orbital angular momenta using again Eq. (1.64) for the two angular momenta S and l:

$$\rho^C_{kq}(\alpha_C J_C) = c \sum_{\alpha \alpha' k_S q_S k_l q_l} \langle \alpha_C J_C \| T \| \alpha J_C \rangle \langle \alpha_C J_C \| T \| \alpha' J_C \rangle^*$$

$$\times \hat{k}_S \hat{k}_l (k_S q_S k_l 0 \,|\, kq) \begin{Bmatrix} S & l & J_C \\ S' & l' & J_C \\ k_S & k_l & k \end{Bmatrix} \rho^S_{k_S q_S}(S,S') \rho^l_{k_l q_l}(l,l') \qquad (2.31)$$

Substituting Eqs. (2.30), (1.101), and (1.100) into (2.31), we express the statistical tensors of the compound system in terms of polarization parameters of angular momenta of the colliding particles A and x:

$$\rho^C_{kq}(\alpha_C J_C) = \frac{1}{4\pi} \sum_{\substack{S l S' l' k_S q_S k_l \\ k_A q_A k_x q_x}} (-1)^{l'} \hat{k}_S \hat{k}_l \hat{k}_A \hat{k}_x \hat{S} \hat{S'} \hat{l} \hat{l'} (l0, l'0 \,|\, k_l 0)(k_S q_S, k_l 0 \,|\, kq)$$

$$\times (k_A q_A, k_x q_x \,|\, k_S q_S) \begin{Bmatrix} S & l & J_C \\ S' & l' & J_C \\ k_S & k_l & k \end{Bmatrix} \begin{Bmatrix} J_A & j_x & S \\ J_A & j_x & S' \\ k_A & k_x & k_S \end{Bmatrix} \rho^A_{k_A q_A}(\alpha_A J_A) \rho^x_{k_x q_x}(\alpha_x j_x)$$

$$\times \langle \alpha_C J_C \| T \| \alpha_A J_A, \alpha_x j_x, Sl : J_C \rangle \langle \alpha_C J_C \| T \| \alpha_A J_A, \alpha_x j_x, S'l' : J_C \rangle^* \qquad (2.32)$$

From this equation it follows that the projections of the tensors are related because

$$q = q_S = q_A + q_x \qquad (2.33)$$

Now consider possible values of rank k of the statistical tensors of the compound system formed. Its alignment ($k = $ even) has its origin in the initial polarization of the colliding subsystems and also can be produced by the collision itself. The maximal rank k_{max} of the alignment parameter is $2J_C$, but limitations on the dynamic character brought into Eq. (2.32) by the matrix elements $\langle \alpha_C J_C \| T \| \alpha J_C \rangle$ can reduce this maximal number considerably.

To investigate the problem of orientation of the system C ($k = $ odd), we use the following interchange relation for the product of three factors in (2.31):

$$\begin{Bmatrix} S & l & J_C \\ S' & l' & J_C \\ k_S & k_l & k \end{Bmatrix} \rho^S_{k_S q_S}(S,S') \rho^l_{k_l 0}(l,l')$$

$$= (-1)^{k+k_A+k_x} \begin{Bmatrix} S' & l' & J_C \\ S & l & J_C \\ k_S & k_l & k \end{Bmatrix} \rho^S_{k_S q_S}(S',S)\rho^l_{k_l 0}(l',l) \tag{2.34}$$

which is valid, taking into account that for the states $|J_A\rangle$ and $|J_C\rangle$ of definite parity, the rank k_l of the statistical tensors of the orbital motion is always even. Then we rewrite Eq. (2.31) in an equivalent form:

$$\rho^C_{kq}(\alpha_C J_C) = \frac{1}{2} \sum_{\alpha\alpha' k_S q_S k_l} \hat{k}_S \hat{k}_l (k_S q_S k_l 0 \mid kq) \begin{Bmatrix} S & l & J_C \\ S' & l' & J_C \\ k_S & k_l & k \end{Bmatrix} \rho^S_{k_S q_S}(S,S')\rho^l_{k_l}(l,l')$$

$$\times \Big[\langle \alpha_C J_C \| T \| \alpha J_C \rangle \langle \alpha_C J_C \| T \| \alpha' J_C \rangle^*$$

$$+ (-1)^{k+k_A+k_x} \Big[\langle \alpha_C J_C \| T \| \alpha J_C \rangle^* \langle \alpha_C J_C \| T \| \alpha' J_C \rangle \Big]$$

$$= \sum_{k+k_A+k_x=\text{even}} \sum_{q_S} \hat{k}_S \hat{k}_l (k_S q_S k_l 0 \mid kq) \sum_{\alpha\alpha'} \begin{Bmatrix} S & l & J_C \\ S' & l' & J_C \\ k_S & k_l & k \end{Bmatrix} \rho^S_{k_S q_S}(S,S')\rho^l_{k_l}(l,l')$$

$$\times \mathrm{Re}\Big(\langle \alpha_C J_C \| T \| \alpha J_C \rangle \langle \alpha_C J_C \| T \| \alpha' J_C \rangle^* \Big)$$

$$+ \sum_{k+k_A+k_x=\text{odd}} \sum_{q_S} \hat{k}_S \hat{k}_l (k_S q_S k_l 0 \mid kq) \sum_{\alpha\alpha'} \begin{Bmatrix} S & l & J_C \\ S' & l' & J_C \\ k_S & k_l & k \end{Bmatrix} \rho^S_{k_S q_S}(S,S')\rho^l_{k_l}(l,l')$$

$$\times \mathrm{Im}\Big(\langle \alpha_C J_C \| T \| \alpha J_C \rangle \langle \alpha_C J_C \| T \| \alpha' J_C \rangle^* \Big) \tag{2.35}$$

Note that, starting from Eq. (1.58) [and, even earlier, from Eq. (1.24)], we were not interested in the normalization of the statistical tensor $\rho^C_{kq}(\alpha_C J_C)$ which is not relevant for discussing any polarization and correlation characteristics of the process under consideration.

Equation (2.35) shows that the compound system can be oriented ($k = \text{odd}$) not only when the colliding partners, or at least one of them are oriented (this is the *transfer mechanism* of producing orientation of the compound system). The system C can be oriented also by collision of nonoriented but aligned partners (both k_A and k_x are even). However, this can happen only if there is a phase difference between the interfering amplitudes $\langle \alpha_C J_C \| T \| \alpha J_C \rangle$ and $\langle \alpha_C J_C \| T \| \alpha' J_C \rangle$. In the latter case, the value $q = 0$ is excluded because $(k_s 0, k_l 0 \mid k0) = 0$ for $k_A + k_x + k = \text{odd}$. This means that the polarization vector of the compound system, given generally by three components, $\rho^C_{1q}(\alpha_C J_C)$, with $q = 0 \pm 1$, has here no component $\rho^C_{10}(\alpha_C J_C)$ along the z-axis and lies in the plane normal to the collision direction $\mathbf{n}$.

If a compound system is formed in the s-wave channel only (e.g., the case of an isolated s-wave resonance), then no direction is fixed by the orbital movement of the colliding subsystems. In this case, the orientation of the compound system can be produced via transfer of the initial spin orientations of the colliding partners only.

2.2. Polarization and Angular Distribution of Scattering Products

In this section we discuss examples of a more complicated case where in the final state of a two-particle collision there are also two particles different, in general, from the initial ones. This is a very frequent case in atomic collision physics. It includes all cases of elastic scattering, inelastic scattering with excitation of the target and/or the projectile, and many others.

2.2.1. Density Matrix and Polarization Parameters of Atoms Excited by Electron Impact and in Ion–Atom Collisions

We begin with a case of electron-impact excitation of an atom A from an initial state with a total angular momentum J_0 to a final state A^* with a total angular momentum J:

$$A(\alpha_0 J_0) + e(\mathbf{p}_0) \rightarrow A^*(\alpha J) + e_{sc}(\mathbf{p}) \tag{2.36}$$

Situations with mixtures of states with different total angular momenta in the initial and final atomic states will be considered later in this section.

Let ρ_A^i and ρ_e^i denote the angular momentum density matrices of the target atom and the incoming electron beam. Then the density matrix of the entire system in the initial state is their direct product:

$$\rho^i = \rho_A^i \rho_e^i \tag{2.37}$$

The final state density operator is related to the initial state one by Eq. (1.24):

$$\rho^f = T(\mathbf{p}_0, \mathbf{p}) \, \rho^i \, T^+(\mathbf{p}_0, \mathbf{p}) = T(\mathbf{p}_0, \mathbf{p}) \, \rho_A^i \rho_e^i \, T^+(\mathbf{p}_0, \mathbf{p}) \tag{2.38}$$

where $T(\mathbf{p}_0, \mathbf{p})$ is the transition operator describing the scattering of electrons with initial linear momentum $\mathbf{p}_0$ to the final state with momentum $\mathbf{p}$.

2.2.1.1. Representation of Spin Projections

The matrix element of the spin density matrix (2.38) in the representation of the spins of individual atoms and electrons takes the form

$$
\left\langle \alpha JM, \frac{1}{2}\mu \,\middle|\, \rho^f(\mathbf{p}_0, \mathbf{p}) \,\middle|\, \alpha JM', \frac{1}{2}\mu' \right\rangle
$$

$$
= \sum_{M_0 M_0' \mu_0 \mu_0'} T_{M_0 \mu_0 \to M\mu}(\mathbf{p}_0, \mathbf{p}) \left\langle \alpha_0 J_0 M_0 \,\middle|\, \rho_A^i \,\middle|\, \alpha_0 J_0 M_0' \right\rangle
$$

$$
\times \left\langle \frac{1}{2}\mu_0 \,\middle|\, \rho_e^i \,\middle|\, \frac{1}{2}\mu_0' \right\rangle T^*_{M_0' \mu_0' \to M'\mu'}(\mathbf{p}_0, \mathbf{p}) \tag{2.39}
$$

where

$$
T_{M_0 \mu_0 \to M\mu}(\mathbf{p}_0, \mathbf{p}) \equiv \left\langle \alpha JM, \mathbf{p}\mu \,\middle|\, T \,\middle|\, \alpha_0 J_0 M_0, \mathbf{p}_0 \mu_0 \right\rangle \tag{2.40}
$$

are matrix elements of the transition operator or amplitudes of the transition from the atomic state $\alpha_0 J_0 M_0$ to the state αJM at electron scattering, with μ_0 and μ being spin projections of the incoming and scattered electrons.

We are interested in the spin density matrix of the excited atom only. According to the prescription of Section 1.1.2, in order to obtain it, one should calculate the trace of the total density matrix over the variables of the unobserved part of the system. Suppose first that the detector of electrons is not sensitive to their spins; thus the spin of the scattered electrons is not observed. In this case, the matrix element of the density matrix of the final atomic state can be written as

$$
\left\langle \alpha JM \,\middle|\, \rho_A^f(\mathbf{p}_0, \mathbf{p}) \,\middle|\, \alpha JM' \right\rangle = \sum_{\mu} \left\langle \alpha JM, \frac{1}{2}\mu \,\middle|\, \rho^f(\mathbf{p}_0, \mathbf{p}) \,\middle|\, \alpha JM', \frac{1}{2}\mu \right\rangle
$$

$$
= \sum_{M_0 M_0' \mu \mu_0 \mu_0'} T_{M_0 \mu_0 \to M\mu}(\mathbf{p}_0, \mathbf{p}) \left\langle \alpha_0 J_0 M_0 \,\middle|\, \rho_A^i \,\middle|\, \alpha_0 J_0 M_0' \right\rangle
$$

$$
\times \left\langle \frac{1}{2}\mu_0 \,\middle|\, \rho_e^i \,\middle|\, \frac{1}{2}\mu_0' \right\rangle T^*_{M_0' \mu_0' \to M'\mu}(\mathbf{p}_0, \mathbf{p}) \tag{2.41}
$$

Equation (2.41) leads to some particular formulas corresponding to various special conditions concerning polarization of the target atom and the incoming beam. In the case of an unpolarized incoming electron beam, its density matrix is

$$
\left\langle \frac{1}{2}\mu_0 \,\middle|\, \rho_e^i \,\middle|\, \frac{1}{2}\mu_0' \right\rangle = \frac{1}{2}\delta_{\mu_0\mu_0'} \tag{2.42}
$$

If the target atom is also unpolarized,

$$
\left\langle \alpha_0 J_0 M_0 \,\middle|\, \rho_A^i \,\middle|\, \alpha_0 J_0 M_0' \right\rangle = \frac{1}{2J_0+1}\delta_{M_0 M_0'} \tag{2.43}
$$

then the density matrix of the excited atom can be written as

$$\left\langle \alpha JM \left| \rho_A^f(\mathbf{p}_0, \mathbf{p}) \right| \alpha JM' \right\rangle = \sum_{M_0\mu_0\mu} T_{M_0\mu_0 \to M\mu}(\mathbf{p}_0, \mathbf{p}) T^*_{M_0\mu_0 \to M'\mu}(\mathbf{p}_0, \mathbf{p}) \quad (2.44)$$

Normalizing the density matrix by the standard condition (1.8) gives

$$\left\langle \alpha JM \left| \rho_A^f(\mathbf{p}_0, \mathbf{p}) \right| \alpha JM' \right\rangle = \frac{\sum_{M_0\mu_0\mu} T_{M_0\mu_0 \to M\mu}(\mathbf{p}_0, \mathbf{p}) T^*_{M_0\mu_0 \to M'\mu}(\mathbf{p}_0, \mathbf{p})}{\sum_{MM_0\mu_0\mu} |T_{M_0\mu_0 \to M\mu}(\mathbf{p}_0, \mathbf{p})|^2}$$

$$(2.45)$$

Using this equation and the definition (1.41), one obtains statistical tensors of the final atomic state for the fixed kinematics of the scattering process:

$$\rho_{kq}^f(\alpha J; \mathbf{p}_0, \mathbf{p}) = \sum_{MM'} (-1)^{J-M'} (JM, J-M'|kq)$$

$$\times \left\langle \alpha JM \left| \rho_A^f(\mathbf{p}_0, \mathbf{p}) \right| \alpha JM' \right\rangle \quad (2.46)$$

(Additional arguments in the notation of the statistical tensors show that we consider the tensors differential with respect to the scattering angle.) This set of statistical tensors contains complete information on polarization properties of the excited atom produced by electron impact with electrons scattered in a certain direction. It can be used, for example, in a description of the angular distribution and polarization of the subsequent radiation in two-step excitation–deexcitation processes such as $(e, e'\gamma)$ when the emitted radiation is detected in coincidence with the scattered electron.

In order to treat noncoincidence experiments where the scattered electron is not detected, one should average the density matrix (2.44) over angles $(\vartheta_{sc}, \varphi_{sc})$ of the scattered electron and introduce, according to Eq. (2.46), the *integral statistical tensors* (or *integral state multipoles*)

$$\int \rho_{kq}^f(\alpha J; \mathbf{p}_0, \mathbf{p}) \, d\Omega_{sc} = \sum_{MM'} (-1)^{J-M'} (JMJ' - M'|kq)$$

$$\times \sum_{M_0\mu_0\mu} \int T_{M_0\mu_0 \to M\mu}(\mathbf{p}_0, \mathbf{p}) \, T^*_{M_0\mu_0 \to M'\mu}(\mathbf{p}_0, \mathbf{p}) d\Omega_{sc} \quad (2.47)$$

After standard normalization, according to Eq. (1.49) the integral statistical tensors of the excited atom take the form:

$$\rho_{kq}^f(\alpha J; \mathbf{p}_0) = \sum_{MM'} (-1)^{J-M'} (JM, J-M'|kq)$$

$$\times \frac{\sum_{M_0\mu_0\mu} \int T_{M_0\mu_0 \to M\mu}(\mathbf{p}_0, \mathbf{p}) \, T^*_{M_0\mu_0 \to M'\mu}(\mathbf{p}_0, \mathbf{p}) d\Omega_{sc}}{\sum_{MM_0\mu_0\mu} \int |T_{M_0\mu_0 \to M\mu}(\mathbf{p}_0, \mathbf{p})|^2 d\Omega_{sc}} \quad (2.48)$$

As shown in Section 1.4, symmetry arguments considerably reduce the number of independent parameters of the density matrix and statistical tensors of the atomic system. In our case of an unpolarized target atom and an unpolarized electron beam, when the detector of scattered electrons is not sensitive to their polarization there is a symmetry with respect to the reflection in the scattering plane. Choose the xz-plane as the scattering plane. According to (1.134) this leads to the following relation:

$$\rho_{kq}^{f}(\alpha J; \mathbf{p}_0, \mathbf{p}) = (-1)^{k+q} \rho_{k-q}^{f}(\alpha J; \mathbf{p}_0, \mathbf{p}) \tag{2.49}$$

which reduces the number of real independent parameters of the final atomic state to one (instead of three) when $J = 1/2$, to four (instead of eight) when $J = 1$, and so on (see Table 1.2 on p. 30). [The same restrictions on the statistical tensors (2.48) can be derived from the symmetry properties of the amplitudes $T_{M_0\mu_0 \to M\mu}(\mathbf{p}_0, \mathbf{p})$.] The average value $\langle \mathbf{J} \rangle$ of the vector of angular momentum of the atom in the final state is proportional to the first-rank statistical tensor. According to Table 1.2, the only nonzero component of the first-rank tensor is ρ_{11} and it is purely imaginary. This means that $\langle \mathbf{J} \rangle$ has only a y-component [see Eqs. (1.82) and (1.87)]. Thus in the conditions under consideration, the average vector of the angular momentum of the excited atom is always directed normally to the scattering plane. This can be easily understood from a simple symmetry consideration. In the case considered, the only axial vector (angular momentum) that can be created from the available vectors is the vector product $\mathbf{p}_0 \times \mathbf{p}$, which is directed perpendicular to the scattering plane.

If the scattered electrons are not detected, the final atomic state is of a higher symmetry: It should be symmetric with respect to rotations about the direction of the incoming beam. Taking into account that the atomic states under consideration have definite parities and choosing the $\mathbf{p}_0$ direction as the quantization axis, we see from symmetry considerations (see Section 1.4) that only diagonal elements of the density matrix and only statistical tensors with an even rank and zero projection do not vanish

$$\left\langle \alpha J M \left| \rho_A^f(\mathbf{p}_0) \right| \alpha J M' \right\rangle = \delta_{MM'} \left\langle \alpha J M \left| \rho_A^f(\mathbf{p}_0) \right| \alpha J M \right\rangle \tag{2.50}$$

$$\rho_{kq}^f(\alpha J, \mathbf{p}_0) = \delta_{q0}\, \rho_{k0}^f(\alpha J, \mathbf{p}_0)|_{k=\text{even}} \tag{2.51}$$

This means, in particular, that if the scattered electrons are not observed, then the excited atom in the state $|J = 1/2\rangle$ turns out to be isotropic, while the atom in the states $|J = 1\rangle$ and $|J = \frac{3}{2}\rangle$ is characterized by a single alignment parameter, $\rho_{20}^f(\alpha J)$. The excited atom can never be oriented in this case. Generally, the number of parameters characterizing the polarization of the final atom is much smaller if the scattered electrons are not observed, than in coincidence experiments (see Section 1.4 and Table 1.2).

In a more general case when the detector of scattered electrons is sensitive to the spin polarization, one should use Eq. (2.39) and the convolution procedure (1.153) to find the density matrix and thereafter the statistical tensors of the excited atom. Instead of Eq. (2.41), one then has

$$\left\langle \alpha JM \left| \rho_A^f(\mathbf{p}_0,\mathbf{p}) \right| \alpha JM' \right\rangle = \sum_{M_0 M_0' \mu_0 \mu_0' \mu \mu'} T_{M_0 \mu_0 \to M\mu}(\mathbf{p}_0,\mathbf{p}) \, T^*_{M_0' \mu_0' \to M'\mu'}(\mathbf{p}_0,\mathbf{p})$$

$$\times \left\langle \alpha_0 J_0 M_0 \left| \rho_A^i \right| \alpha_0 J_0 M_0' \right\rangle \left\langle \tfrac{1}{2}\mu_0 \left| \rho_e^i \right| \tfrac{1}{2}\mu_0' \right\rangle$$

$$\times \left\langle \tfrac{1}{2}\mu' \left| \varepsilon_e \right| \tfrac{1}{2}\mu \right\rangle \tag{2.52}$$

where the efficiency matrix of the spin of the scattered electrons is introduced.

The reduced statistical tensors can be expressed in terms of the scattering amplitudes from definition (1.50). For example, from Eqs. (2.44) and (2.46), the reduced statistical tensors of an atom, when the scattered electron moves with the linear momentum $\mathbf{p}$ and its spin is not detected are of the form

$$\mathcal{A}_{kq}(\alpha J; \mathbf{p}_0, \mathbf{p}) = \rho_{kq}^f(\alpha J; \mathbf{p}_0, \mathbf{p}) / \rho_{00}^f(\alpha J; \mathbf{p}_0, \mathbf{p})$$

$$= \hat{J} \sum_{MM'} (-1)^{J-M'} \left(JM, J-M' | kq \right)$$

$$\times \frac{\sum_{M_0 \mu_0 \mu} T_{M_0 \mu_0 \to M\mu}(\mathbf{p}_0,\mathbf{p}) T^*_{M_0 \mu_0 \to M'\mu}(\mathbf{p}_0,\mathbf{p})}{\sum_{MM_0\mu_0\mu} |T_{M_0\mu_0 \to M\mu}(\mathbf{p}_0,\mathbf{p})|^2} \tag{2.53}$$

The integral reduced statistical tensors

$$\mathcal{A}_{kq}(\alpha J; \mathbf{p}_0) = \frac{\rho_{kq}^f(\alpha J; \mathbf{p}_0)}{\rho_{00}^f(\alpha J; \mathbf{p}_0)} \tag{2.54}$$

can be expressed, according to the definition of the statistical tensor (1.41) and symmetry properties (2.50) and (2.51), in terms of probabilities of excitation $\sigma(\alpha JM)$ for individual magnetic substates αJM (the quantization axis is along the incident beam):

$$\mathcal{A}_{k0}(\alpha J; \mathbf{p}_0) = \frac{1}{\sigma(\alpha J)} \sum_{M=-J}^{J} (-1)^{J-M} (JM, J-M|k0) \, \sigma(\alpha JM) \tag{2.55}$$

where

$$\sigma(\alpha J) = \sum_{M=-J}^{J} \sigma(\alpha JM) \tag{2.56}$$

A frequent case corresponds to the second-rank tensor $\mathcal{A}_{20}(\alpha J; p_0)$, which takes the form

$$\mathcal{A}_{20}(\alpha J; p_0) = \left[\frac{5}{(2J+3)J(J+1)(2J-1)} \right]^{1/2}$$
$$\times \frac{1}{\sigma(\alpha J)} \sum_M [3M^2 - J(J+1)] \sigma(\alpha JM) \qquad (2.57)$$

The component of the second-rank tensor with zero projection (2.57) is often called an *alignment parameter* or simply *alignment* of the excited atom.

Excitation of atoms in ion–atom collisions can be considered in a similar way and within the same formalism as given above. Formally, expressions like (2.53) and (2.48) are valid also for ion–atom inelastic scattering. The difference between the electron-impact and the ion-impact cases occurs in practical calculations of the dependence of the scattering amplitudes on kinematic variables of the scattering process. In a description of ion–atom collisions, the impact parameter **b** instead of scattering angles $(\vartheta_{sc}, \varphi_{sc})$ is often used as a kinematic variable. Thus the integration over Ω_{sc} is transformed into integration over **b** in calculating the integrated cross sections and polarization parameters of the excited atom for noncoincidence experiments when the scattered ions are not observed.

2.2.1.2. Representation of the Total Angular Momentum

Consider the statistical tensor of the final system $A^*(\alpha J) + e_{sc}$ in Eq. (2.36) in the representation of the total angular momentum, which can often be more convenient than the representation of the spin projections. This can be done according to Eq. (1.58) using the statistical tensors of the initial state and the reduced excitation amplitudes:

$$\rho_{\tilde{k}\tilde{q}}\left(\alpha J, lj : J_t; \alpha' J', l'j' : J_t'\right)$$
$$= \frac{1}{\hat{J}_t \hat{J}_t'} \sum_{\substack{l_0 l_0' j_0 j_0' \\ \alpha_0 \alpha_0' J_0 J_0'}} \rho_{\tilde{k}\tilde{q}}\left(\alpha_0 J_0, l_0 j_0 : J_t; \alpha_0' J_0', l_0' j_0' : J_t'\right)$$
$$\times \left\langle \alpha J, lj : J_t \| T \| \alpha_0 J_0, l_0 j_0 : J_t \right\rangle$$
$$\times \left\langle \alpha' J', l'j' : J_t' \| T \| \alpha_0' J_0', l_0' j_0' : J_t' \right\rangle^* \qquad (2.58)$$

where $l_0 j_0$ and lj are the orbital and total angular momenta of the incoming and outgoing electrons, respectively, and J_t is the total angular momentum of the final system. Equation (2.58) takes into account possible mixing of several initial and final atomic levels with different total angular momenta (for example, mixing of fine-structure levels). Decoupling the angular momenta in the statistical tensor of the initial state according to Eqs. (1.64) and (1.66) and using the statistical tensors (1.102) and (1.103) for the initial electron yields

$$
\rho_{\tilde{k}\tilde{q}}\left(\alpha J, l j : J_t; \alpha' J', l' j' : J_t'\right) = \sum_{\substack{l_0 l_0' j_0 j_0' \\ \alpha_0 \alpha_0' J_0 J_0'}} \sum_{\substack{k_0 q_0 k_{j_0} q_{j_0} \\ k_{l_0} q_{l_0} k_{s_0} q_{s_0}}} \hat{\jmath}_0 \hat{\jmath}_0' \hat{k}_0 \hat{k}_{j_0} \hat{k}_{l_0} \hat{k}_{s_0}
$$

$$
\times \left(k_0 q_0, k_{j_0} q_{j_0} \mid \tilde{k}\tilde{q}\right) \left(k_{l_0} q_{l_0}, k_{s_0} q_{s_0} \mid k_{j_0} q_{j_0}\right)
\begin{Bmatrix} J_0 & j_0 & J_t \\ J_0' & j_0' & J_t' \\ k_0 & k_{j_0} & \tilde{k} \end{Bmatrix}
$$

$$
\times \begin{Bmatrix} l_0 & \frac{1}{2} & j_0 \\ l_0' & \frac{1}{2} & j_0' \\ k_{l_0} & k_{s_0} & k_{j_0} \end{Bmatrix}
\rho_{k_0 q_0}\left(\alpha_0 J_0, \alpha_0' J_0'\right) \rho_{k_{l_0} q_{l_0}}\left(l_0, l_0'\right) \rho_{k_{s_0} q_{s_0}}\left(\tfrac{1}{2}\right)
$$

$$
\times \left\langle \alpha J, l j : J_t \parallel T \parallel \alpha_0 J_0, l_0 j_0 : J_t \right\rangle \left\langle \alpha' J', l' j' : J_t' \parallel T \parallel \alpha_0' J_0', l_0' j_0' : J_t' \right\rangle^* \quad (2.59)
$$

To obtain the statistical tensor of the excited atom, we now use the convolution procedure (1.155) and Eq. (1.163) for the efficiency tensor of the electron detector. This gives

$$
\rho_{kq}\left(\alpha J, \alpha' J'\right) = c \sum_{\substack{l_0 l_0' j_0 j_0' \\ \alpha_0 \alpha_0' J_0 J_0'}} \sum_{\substack{l l' j j' \\ J_t J_t' \tilde{k}\tilde{q}}} \sum_{\substack{k_0 q_0 k_{j_0} q_{j_0} \\ k_{l_0} q_{l_0} k_{s_0} q_{s_0}}} \sum_{\substack{k_l q_l k_s q_s \\ k_j q_j}} \hat{\jmath}_t \hat{\jmath}_t' \hat{\jmath}_0 \hat{\jmath}_0' \hat{k}_0 \hat{k}_{j_0} \hat{k}_{l_0} \hat{k}_{s_0} \hat{k} \hat{k}_j \hat{\jmath} \hat{\jmath}' \hat{k}_s \hat{k}_l
$$

$$
\times \left(k_0 q_0, k_{j_0} q_{j_0} \mid \tilde{k}\tilde{q}\right) \left(k_{l_0} q_{l_0}, k_{s_0} q_{s_0} \mid k_{j_0} q_{j_0}\right) \left(k q, k_j q_j \mid \tilde{k}\tilde{q}\right)
$$

$$
\times \left(k_l q_l, k_s q_s \mid k_j q_j\right)
\begin{Bmatrix} J_0 & j_0 & J_t \\ J_0' & j_0' & J_t' \\ k_0 & k_{j_0} & \tilde{k} \end{Bmatrix}
\begin{Bmatrix} l_0 & \frac{1}{2} & j_0 \\ l_0' & \frac{1}{2} & j_0' \\ k_{l_0} & k_{s_0} & k_{j_0} \end{Bmatrix}
\begin{Bmatrix} J & J' & k \\ j & j' & k_j \\ J_t & J_t' & \tilde{k} \end{Bmatrix}
\begin{Bmatrix} l & \frac{1}{2} & j \\ l' & \frac{1}{2} & j' \\ k_l & k_s & k_j \end{Bmatrix}
$$

$$
\times \rho_{k_0 q_0}\left(\alpha_0 J_0, \alpha_0' J_0'\right) \rho_{k_{l_0} q_{l_0}}\left(l_0, l_0'\right) \rho_{k_{s_0} q_{s_0}}\left(\tfrac{1}{2}\right) \varepsilon_{k_l q_l}^*\left(l, l'\right) \varepsilon_{k_s q_s}^*\left(\tfrac{1}{2}\right)
$$

$$
\times \left\langle \alpha J, l j : J_t \parallel T \parallel \alpha_0 J_0, l_0 j_0 : J_t \right\rangle \left\langle \alpha' J', l' j' : J_t' \parallel T \parallel \alpha_0' J_0', l_0' j_0' : J_t' \right\rangle^*
$$

$$
(2.60)
$$

Equation (2.60) is the most general expression for the statistical tensor of the excited atom in conditions when the scattered electron is detected at a certain angle by a spin-sensitive detector. Statistical and efficiency tensors of the electron orbital momenta, $\rho_{k_{l_0} q_{l_0}}\left(l_0, l_0'\right)$ and $\varepsilon_{k_l q_l}\left(l, l'\right)$, are given by Eqs. (1.99) and (1.100). Now one can easily transform different factors in Eq. (2.60) into various coordinate systems. For example, one can write the efficiency tensor of a spin-sensitive detector in the detector frame using Eq. (1.44).

Consider an important particular case of Eq. (2.60). Suppose that the detector of the scattered electrons is not sensitive to spin polarization [$k_s = q_s = 0$, see Eq. (1.164)]; the target atom is not polarized ($k_0 = q_0 = 0$); initial atomic levels $\alpha_0 J_0$ are populated according to their statistical weight: $\rho_{k_0 q_0}(\alpha_0 J_0, \alpha_0' J_0') = \delta_{k_0 0}\, \delta_{q_0 0}\, \delta_{J_0 J_0'} \frac{1}{\hat{J}_0}(2J_0 + 1)$.

Then Eq. (2.60) takes the form (remember that we omit irrelevant constant factors):

$$
\rho_{kq}(\alpha J, \alpha' J') = \sum_{\substack{l_0 l_0' j_0 j_0' \\ k_{l_0} k_{s_0}}} \sum_{\substack{l'' jj' \\ k_l q_l}} \sum_{\substack{J_t J_t' \tilde{k}\tilde{q} \\ J_0}} \hat{J}_t \hat{J}_t' \hat{j}_0 \hat{j}_0' \hat{k}_{l_0} \hat{k}_{s_0} \hat{k} \hat{j} \hat{j}' \hat{k}_l \hat{l}_0 \hat{l}_0' \hat{l} \hat{l}'
$$

$$
\times (-1)^{J_0 + \tilde{k} + j_0' + J_t + l_0' + j' + 1/2}
$$

$$
\times (k_{l_0} 0, k_{s_0} \tilde{q} | \tilde{k} \tilde{q})\,(kq, k_l q_l | \tilde{k} \tilde{q})\,(l_0 0, l_0' 0 | k_{l_0} 0)\,(l 0, l' 0 | k_l 0)
$$

$$
\times \begin{Bmatrix} J_t' & j_0' & J_0 \\ j_0 & J_t & \tilde{k} \end{Bmatrix} \begin{Bmatrix} l' & j' & \frac{1}{2} \\ j & l & k_l \end{Bmatrix} \begin{Bmatrix} l_0 & \frac{1}{2} & j_0 \\ l_0' & \frac{1}{2} & j_0' \\ k_{l_0} & k_{s_0} & \tilde{k} \end{Bmatrix} \begin{Bmatrix} J & J' & k \\ j & j' & k_l \\ J_t & J_t' & \tilde{k} \end{Bmatrix}
$$

$$
\times \rho_{k_{s_0} \tilde{q}}\left(\frac{1}{2}\right) Y_{k_l q_l}^*(\vartheta_{sc}, \varphi_{sc})
$$

$$
\times \langle \alpha J, l j : J_t \| T \| \alpha_0 J_0, l_0 j_0 : J_t \rangle
$$

$$
\times \langle \alpha' J', l' j' : J_t' \| T \| \alpha_0 J_0, l_0' j_0' : J_t' \rangle^* \tag{2.61}
$$

Here we have chosen the quantization axis z along the incoming beam, and angles $\vartheta_{sc}, \varphi_{sc}$ characterize the direction of scattered electrons. Also we fixed a set of quantum numbers α_0 which is practically always the case. For fixed parities of initial and final atomic states, the orbital momenta l_0 and l_0' should have the same parity, as well as l and l', since the transition operator conserves parity of the whole system. Therefore, k_{l_0} and k_l are even, owing to the property of the Clebsch–Gordan coefficients with zero projections (A.94). It follows that the angular dependence of the statistical tensors (2.61), including the angular distribution of scattered electrons, which is determined by $\rho_{00}(\alpha J, \alpha' J')$, is described by the spherical harmonics $Y_{k_l q_l}(\vartheta_{sc}, \varphi_{sc})$ of even ranks. This statement also holds true for a polarized target.

For an unpolarized initial electron beam ($k_{s_0} = \tilde{q} = 0$), further simplification follows:

$$
\rho_{kq}(\alpha J, \alpha' J') = \sum_{\substack{l_0 l_0' j_0 j_0' \\ J_0}} \sum_{\substack{J_t J_t' \tilde{k} k_l \\ l'' jj'}} \hat{J}_t \hat{J}_t' \hat{j}_0 \hat{j}_0' \hat{k} \hat{\tilde{k}} \hat{j} \hat{j}' \hat{l}_0 \hat{l}_0' \hat{l} \hat{l}'
$$

$$
\times (-1)^{J_0 + J_t + j' + \tilde{k} + k}\,(kq, \tilde{k} 0 | k_l q)\,(l_0 0, l_0' 0 | \tilde{k} 0)\,(l 0, l' 0 | k_l 0)
$$

$$
\times \begin{Bmatrix} J_t' & j_0' & J_0 \\ j_0 & J_t & \tilde{k} \end{Bmatrix} \begin{Bmatrix} l' & j' & \frac{1}{2} \\ j & l & k_l \end{Bmatrix} \begin{Bmatrix} l_0' & j_0' & \frac{1}{2} \\ j_0 & l_0 & \tilde{k} \end{Bmatrix} \begin{Bmatrix} J & J' & k \\ j & j' & k_l \\ J_t & J_t' & \tilde{k} \end{Bmatrix} Y_{k_l q}(\vartheta_{sc}, \varphi_{sc})
$$

$$\times \langle \alpha J, l j : J_t \| T \| \alpha_0 J_0, l_0 j_0 : J_t \rangle$$

$$\times \langle \alpha' J', l' j' : J_t' \| T \| \alpha_0 J_0, l_0' j_0' : J_t' \rangle^* \tag{2.62}$$

In this case the experimental conditions possess plane symmetry with respect to the scattering plane. Using the symmetry properties of Clebsch–Gordan coefficients, $6j$-symbols, $9j$-symbols, and spherical harmonics, one can check the corresponding symmetry relation (1.133) for the statistical tensors (2.62).

Finally, after integration over the scattering angles, one obtains the integral statistical tensors of the excited atom:

$$\rho_{kq}\left(\alpha J, \alpha' J'\right) = \delta_{q0} \sum_{\substack{l_0 l_0' j_0 j_0' \\ J_0}} \sum_{J_t J_t' l j} \hat{J}_t \hat{J}_t' \hat{j}_0 \hat{j}_0' \hat{l}_0 \hat{l}_0' (-1)^{J_0 + J_t + J_t' - 1/2 + J + j}$$

$$\times \left(l_0 0, l_0' 0 | k 0\right) \begin{Bmatrix} J_t' & j_0' & J_0 \\ j_0 & J_t & k \end{Bmatrix} \begin{Bmatrix} l_0' & j_0' & \frac{1}{2} \\ j_0 & l_0 & k \end{Bmatrix} \begin{Bmatrix} J' & J & k \\ J_t & J_t' & j \end{Bmatrix}$$

$$\times \langle \alpha J, l j : J_t \| T \| \alpha_0 J_0, l_0 j_0 : J_t \rangle$$

$$\times \langle \alpha' J', l j : J_t' \| T \| \alpha_0 J_0, l_0' j_0' : J_t' \rangle^* \tag{2.63}$$

The statistical tensors (2.63) describe an axially symmetric experiment (there is only one direction in the problem: the direction of the incoming beam). In full accordance with the general consideration in Section 1.4, only even statistical tensors ($k = $ even) with zero projection exist ($q = 0$) in this case [see discussion after Eq. (2.61)].

2.2.1.3. Plane-Wave Born Approximation

The plane-wave Born approximation without exchange (PWBA) is the simplest model for the dynamics of charged particle–atom scattering. The domain of its applicability is restricted, as a rule, to high energies of the incident beam and small scattering angles. The transition amplitude (2.40) in the PWBA takes the form

$$T_{M_0 \mu_0 \to M \mu}\left(\mathbf{p}_0, \mathbf{p}\right) \equiv T_{M_0 \mu_0 \to M \mu}\left(\mathbf{Q}\right)$$

$$= -\frac{1}{Q^2} \left\langle \alpha J M \left| \sum_n \exp\left(i \mathbf{Q} \mathbf{r}_n\right) \right| \alpha_0 J_0 M_0 \right\rangle \delta_{\mu_0 \mu} \tag{2.64}$$

where $\mathbf{r}_n$ is a space coordinate of the nth atomic electron, summation is taken over all atomic electrons, and

$$\mathbf{Q} = \mathbf{p}_0 - \mathbf{p} \tag{2.65}$$

is the momentum transfer. Integration in the matrix element (2.64) is over atomic configuration space, including spatial and spin variables. Expanding the exponent in Eq. (2.64) according to Eq. (A.33) and taking the quantization axes $z \parallel \mathbf{Q}$, one finds [see Eq. (A.20)] with the use of the Wigner–Eckart theorem (A.62) that the projection of the total angular momentum of atomic state on the direction of the momentum transfer vector $\mathbf{Q}$ is conserved during the collision. Owing to this selection rule and also owing to the conservation of the spin state of the scattering electron, it is possible in many cases to find the density matrix and the statistical tensors of the excited atom without any further dynamic calculations.

To obtain statistical tensors in the PWBA, it is much simpler to start with the representation of projections than to use partial wave expansions and the reduced amplitudes. Let us consider sharp values of J_0 and J and find the statistical tensors (2.46) of the excited atom. Generalization to the case of mixtures of the total angular momenta of the initial and final atomic states is simple and will be briefly considered at the end of this section. We take the expression (2.44) for the density matrix of the excited atom. Using Eq. (2.64), expansions (A.33) and (A.29), the summation formula (A.91) together with the symmetry properties of spherical harmonics and Clebsch–Gordan coefficients, we obtain

$$
\rho_{kq}(\alpha J; \mathbf{Q}) = \frac{1}{Q^4} \sum_{\lambda\lambda' k_0 q_0 t\tau} \hat{k}\hat{\lambda}\hat{\lambda}' (\lambda\, 0, \lambda'\, 0 | t\, 0)
\begin{Bmatrix} J & J & k \\ \lambda' & \lambda & t \\ J_0 & J_0 & k_0 \end{Bmatrix}
$$
$$
\times i^{-\lambda-\lambda'} M_\lambda^B (J_0 \to J) M_{\lambda'}^{B*} (J_0 \to J)
$$
$$
\times (t\,\tau, k\,q | k_0\, q_0)\, \rho_{k_0 q_0}(\alpha_0 J_0)\, Y_{t\tau}(\vartheta_Q, \varphi_Q) \qquad (2.66)
$$

where

$$
M_\lambda^B (J_0 \to J) = \left\langle \alpha J \left\| \sum_n j_\lambda(Q r_n) Y_\lambda(\vartheta_n, \varphi_n) \right\| \alpha_0 J_0 \right\rangle \qquad (2.67)
$$

is the reduced multipole amplitude for excitation and $j_\lambda(x)$ is the spherical Bessel function. The amplitude (2.67) is real. Equation (2.66) is valid in an arbitrary coordinate frame and angles ϑ_Q, φ_Q characterize the direction of the momentum transfer in this frame. Note that Eq. (2.66) is independent of the polarization state of the incident electron beam and of the spin sensitivity of the electron detector. This is because we neglected the exchange scattering in the model under consideration. Owing to parity conservation, the values of λ and λ' in the reduced amplitudes should have the same parity. Therefore, t in Eq. (2.66) is even [see Eq. (A.94)].

One can show using the symmetry properties of Clebsch–Gordan and $9j$-coefficients that only terms with $k_0 + k =$ even are nonzero in Eq. (2.66). This means, for example, that for an aligned target, the excited atomic state will also be aligned, but not oriented. In the PWBA, the orientation of the final atomic state can only be transferred from the oriented initial state. It cannot arise as a result of the scattering process. This is a consequence of the PWBA and it is not valid in the general case when the process possesses not the axial but only the plane symmetry with respect to the scattering plane (see discussion in Section 1.4).

Assume now that a single multipole λ contributes to the excitation. In fact, this case is widely met in practice. For example, it is always the case for $J_0 = 0$ and $J_0 = 1/2$ (or $J = 0$ and $J = 1/2$). By fixing λ we obtain from Eq. (2.66):

$$
\rho_{kq}(\alpha J; \mathbf{Q}) = \frac{1}{Q^4} \sum_{k_0 q_0 t \tau} (-1)^\lambda \hat{k} \hat{\lambda}^2 (\lambda\, 0, \lambda\, 0 | t\, 0) \left\{ \begin{array}{ccc} J & J & k \\ \lambda & \lambda & t \\ J_0 & J_0 & k_0 \end{array} \right\}
$$

$$
\times |M_\lambda^B(J_0 \to J)|^2 \, (t\,\tau, k\,q | k_0\, q_0)\, \rho_{k_0 q_0}(\alpha_0 J_0)\, Y_{t\tau}(\vartheta_Q, \varphi_Q) \qquad (2.68)
$$

The corresponding reduced statistical tensor is given by

$$
\mathcal{A}_{kq}(\alpha J; \mathbf{Q}) = \hat{f}\hat{k}(-1)^{J_0 + \lambda + J}
$$

$$
\times \frac{\displaystyle\sum_{k_0 q_0 t \tau}(\lambda\, 0, \lambda\, 0 | t\, 0) \left\{ \begin{array}{ccc} J & J & k \\ \lambda & \lambda & t \\ J_0 & J_0 & k_0 \end{array} \right\} (t\,\tau, k\,q | k_0\, q_0)\, Y_{t\tau}(\vartheta_Q, \varphi_Q)\, \mathcal{A}_{k_0 q_0}(\alpha_0 J_0)}{\displaystyle\sum_{k_0 q_0}(-1)^{k_0}\hat{k}_0^{-1}(\lambda\, 0, \lambda\, 0 | k_0\, 0) \left\{ \begin{array}{ccc} J_0 & J_0 & k_0 \\ \lambda & \lambda & J \end{array} \right\} Y_{k_0 q_0}(\vartheta_Q, \varphi_Q)\, \mathcal{A}_{k_0 q_0}(\alpha_0 J_0)}
$$

$$(2.69)$$

The reduced statistical tensors (2.69) obtained in the PWBA within rather general assumptions depend only on the polarization state of the initial atom, which is described by the reduced statistical tensors $\mathcal{A}_{k_0 q_0}(\alpha_0 J_0)$, and the direction of the momentum transfer $\mathbf{Q}$. They take an especially simple form in the coordinate system with $z \parallel \mathbf{Q}$:

$$
\mathcal{A}_{kq}(\alpha J; \mathbf{Q}) = \hat{f}\hat{k}(-1)^{J_0 + \lambda + J}
$$

$$
\times \frac{\displaystyle\sum_{k_0 t}\hat{t}\,(\lambda\, 0, \lambda\, 0 | t\, 0) \left\{ \begin{array}{ccc} J & J & k \\ \lambda & \lambda & t \\ J_0 & J_0 & k_0 \end{array} \right\} (t\, 0, k\, q | k_0\, q)\, \mathcal{A}_{k_0 q}(\alpha_0 J_0)}{\displaystyle\sum_{k_0}(-1)^{k_0}(\lambda\, 0, \lambda\, 0 | k_0\, 0) \left\{ \begin{array}{ccc} J_0 & J_0 & k_0 \\ \lambda & \lambda & J \end{array} \right\} \mathcal{A}_{k_0 0}(\alpha_0 J_0)}
$$

$$(2.70)$$

For an unpolarized initial state, $(k_0 = q_0 = 0)$, Eq. (2.69) yields

$$
\mathcal{A}_{kq}(\alpha J; \mathbf{Q}) = (-1)^{J_0 + J}\hat{f}\hat{\lambda}^2 (\lambda\, 0, \lambda\, 0 | k\, 0)
$$

$$
\times \left\{ \begin{array}{ccc} J & J & k \\ \lambda & \lambda & J_0 \end{array} \right\} \sqrt{\frac{4\pi}{2k+1}} Y_{kq}^*(\vartheta_Q, \varphi_Q) \qquad (2.71)
$$

and the polarization of the excited state is fully determined by the direction of the momentum transfer, the angular momenta of the initial and excited atomic states, and the value of λ. In the coordinate system with $z \parallel \mathbf{Q}$, Eq. (2.71) takes the form

$$\mathcal{A}_{kq}(\alpha J; \mathbf{Q}) = (-1)^{J_0+J} \hat{J} \hat{\lambda}^2 (\lambda 0, \lambda 0 | k 0) \begin{Bmatrix} J & J & k \\ \lambda & \lambda & J_0 \end{Bmatrix} \delta_{q0} \qquad (2.72)$$

explicitly demonstrating an axial symmetry of the excitation process with respect to the direction of the momentum transfer $\mathbf{Q}$ in the case of an unpolarized target. Although the right side of Eq. (2.72) is independent of the momentum transfer, we keep the argument $\mathbf{Q}$ on the left side to indicate the PWBA.

Integrating the statistical tensors (2.66) over the scattering angles with the help of the relation

$$d\Omega_{sc} = (p_0 p)^{-1} Q \, dQ \, d\varphi_Q \qquad (2.73)$$

and Eq. (A.27) gives the PWBA expression for the integral statistical tensor (2.48) that describes the polarization state of the excited atom when the scattered electron is not detected. For an unpolarized target it may be presented as

$$\rho_{kq}(\alpha J) = c\sqrt{\pi} (\hat{J}_0^2 p_0 p)^{-1} \sum_{\lambda \lambda'} (-1)^{J+J_0-(\lambda-\lambda')/2} \hat{\lambda} \hat{\lambda}' (\lambda 0, \lambda' 0 | k 0) \begin{Bmatrix} J & J & k \\ \lambda & \lambda' & J_0 \end{Bmatrix}$$

$$\times \int_{Q_{min}}^{Q_{max}} \frac{dQ}{Q^3} M_\lambda^B(J_0 \to J) M_{\lambda'}^{B*}(J_0 \to J) P_k(\cos \vartheta_Q) \qquad (2.74)$$

where the integration limits are $Q_{min} = p_0 - p$, $Q_{max} = p_0 + p$ and $\cos \vartheta_Q = (p_0^2 + Q^2 - p^2)/(2 p_0 Q)$. In the case when a single multipole contributes, Eq. (2.74) simplifies. In this case the reduced statistical tensors take the form

$$\mathcal{A}_{kq}(\alpha J) = (-1)^{J+J_0} \hat{J} \hat{\lambda}^2 (\lambda 0, \lambda 0 | k 0) \begin{Bmatrix} J & J & k \\ \lambda & \lambda & J_0 \end{Bmatrix} \frac{R_k(\lambda)}{R_0(\lambda)} \delta_{q0} \qquad (2.75)$$

where

$$R_k(\lambda) = \int_{Q_{min}}^{Q_{max}} \frac{dQ}{Q^3} \left| M_\lambda^B(J_0 \to J) \right|^2 P_k(\cos \vartheta_Q) \qquad (2.76)$$

At a high energy limit, the angles close to $\vartheta_Q = \pi/2$ dominate the integral in (2.76). Since $P_k(\cos 90°) = (-1)^{k/2} (k-1)!!/k!!$ for even k, the high-energy (non-relativistic) limit of the reduced statistical tensor in the PWBA is purely algebraic

$$\mathcal{A}_{kq}(\alpha J)|_{p_0 \to \infty} = (-1)^{J+J_0+k/2} \hat{J} \hat{\lambda}^2 (\lambda 0, \lambda 0 | k 0) \begin{Bmatrix} J & J & k \\ \lambda & \lambda & J_0 \end{Bmatrix} \frac{(k-1)!!}{k!!} \delta_{q0}$$

$$(2.77)$$

Note the remarkable relation between the integral reduced statistical tensors of an excited atomic state at high energies (2.77) and the statistical tensors (2.72) of the same atomic state when the scattered electron is detected:

$$\mathcal{A}_{k0}(\alpha J)\,|_{p_0 \to \infty} = (-1)^{k/2}\,\frac{(k-1)!!}{k!!}\,\mathcal{A}_{k0}(\alpha J; \mathbf{Q}) \qquad (2.78)$$

It is important that the tensors on the left are taken in the coordinate frame with the z-axis directed along the incident electron beam, while the tensors in the right are taken in the frame with the z-axis directed along the momentum transfer. Prerequisites for the relation (2.78) are axial symmetries for both processes (but with respect to different directions) and the independence of the reduced statistical tensors on the scattering amplitudes [see Eqs. (2.77) and (2.72)]. The relation (2.78) is independent of the quantum numbers of atomic states and is purely kinematic.

Often the LS-coupling approximation is used in analysis of polarization parameters of the excited atomic state. The above formalism allows us to obtain, for example, the statistical tensor of the orbital momentum L of the excited atomic state. A general approach that we will use is suitable for rigorous reduction to the LS-coupling approximation in other situations also. First we write down a straightforward generalization of Eq. (2.66) to the case of unfixed total angular momenta J_0 and J of initial and excited atomic states:

$$\rho_{kq}(LSJ, LSJ'; \mathbf{Q}) = \frac{1}{Q^4} \sum_{\substack{\lambda\lambda' k_0 q_0 t \tau \\ J_0 J_0'}} \hat{k}\hat{\lambda}\hat{\lambda}'\,(\lambda\,0, \lambda'\,0\,|t\,0) \begin{Bmatrix} J' & J & k \\ \lambda' & \lambda & t \\ J_0' & J_0 & k_0 \end{Bmatrix}$$

$$\times\, i^{-\lambda-\lambda'} M_\lambda^B(J_0 \to J) M_{\lambda'}^{B*}(J_0' \to J')$$

$$\times\, (t\,\tau, k\,q\,|\,k_0\,q_0)\,\rho_{k_0 q_0}(L_0 S_0 J_0, L_0 S_0 J_0')\,Y_{t\tau}(\vartheta_Q, \varphi_Q) \qquad (2.79)$$

Equation (2.79) is derived in full similarity with Eq. (2.66). Then consider the double statistical tensors of the orbital angular momentum L and spin S of the excited state $\rho_{k_L q_L k_S q_S}(LS)$, expressing them in terms of the statistical tensors (2.79) by Eq. (1.65). Similarly, in Eq. (2.79), uncouple orbital angular momentum L_0 and spin S_0 in the statistical tensor $\rho_{k_0 q_0}(L_0 S_0 J_0, L_0 S_0 J_0')$ of the initial state with the help of Eq. (1.64). The next step is a reduction of the PWBA amplitudes (2.67) by Eq. (A.74) to factorize the spin quantum numbers.

The final step is a summation over the quantum numbers of the total angular momenta, in the present case over J, J', J_0, J_0'. For this sometimes rather laborious step, equations from Section A.11 in combination with the symmetry properties of the nj-symbols (Sections A.9 and A.10) are used. As a result, the double statistical tensors $\rho_{k_L q_L k_S q_S}(LS)$ are found, which include complete information on the polarization parameters of the excited LS-multiplet. To find the statistical tensors of the orbital angular momentum L, the convolution procedure (1.69) should be used. As an example, we will write down a result for an initially unpolarized atomic multiplet and when only one multipole λ contributes to the excitation in the PWBA:

$$\rho_{k_L q_L}(L) = \frac{1}{Q^4}(-1)^{L+L_0}\hat{\lambda}^2\hat{k}_L^{-1}(\lambda\,0,\lambda\,0|k_L\,0)\begin{Bmatrix} L & L & k_L \\ \lambda & \lambda & L_0 \end{Bmatrix}$$

$$\times Y_{k_L q_L}^*(\vartheta_Q,\varphi_Q)\left|M_\lambda^B(L_0\to L)\right|^2 \tag{2.80}$$

Here the reduced multipole PWBA amplitude is defined similar to Eq. (2.67):

$$M_\lambda^B(L_0\to L) = \left\langle L\,\middle\|\,\sum_n j_\lambda(Qr_n)Y_\lambda(\vartheta_n,\varphi_n)\,\middle\|\,L_0\right\rangle \tag{2.81}$$

2.2.1.4. Polarization Parameters of an Atom at the Excitation Threshold

Near the excitation threshold, only a few partial waves of slow scattered electrons should be taken into account. The threshold limit for excitation of neutral atoms corresponds to the single s-wave of the scattered electron ($l = l' = 0$) in Eqs. (2.60)–(2.63). This simplifies general expressions for the polarization parameters of the excited atom and sometimes allows us even to write their exact threshold values within rather general assumptions. (Note that for ions, other partial waves of the scattered electron will also contribute to the threshold excitation, owing to the long-range attractive Coulomb potential of the ion.)

Consider the threshold excitation of neutral atoms, neglecting relativistic interactions, i.e., assuming that the total spin and total orbital angular momentum of the system $(e+A)$ are both conserved during scattering. This assumption is valid with high accuracy for atoms that are not heavy and allows us to take full advantage of the threshold approximation $l = l' = 0$. We are interested in the integral statistical tensors of the excited atom. This corresponds to the usual scheme of an experiment, since the angular distribution and/or spin state of a very slow scattered electron is difficult to analyze. The main features of the polarization parameters of the excited atom may be seen from a simple case of an unpolarized atomic target. Integrating Eq. (2.61) over the scattering angle and taking into account only the s-wave of the scattered electron, we obtain

$$\rho_{kq}(LSJ,LSJ') = \sum_{\substack{l_0 l_0' j_0 j_0' \, J_0 J_t J_t' \\ k_{l_0} k_{s_0}}} \sum (-1)^{J_0+j_0'+l_0'+J_t+J_t'+J+1/2}\hat{J}_t\hat{J}_t'\hat{j}_0\hat{j}_0'\hat{l}_0\hat{l}_0'\hat{k}_{l_0}\hat{k}_{s_0}$$

$$\times (k_{l_0}\,0,k_{s_0}\,q|k\,q)\,(l_0\,0,l_0'\,0|k_{l_0}\,0)\begin{Bmatrix} J_t' & j_0' & J_0 \\ j_0 & J_t & k \end{Bmatrix}\begin{Bmatrix} J_t' & J' & \frac{1}{2} \\ J & J_t & k \end{Bmatrix}\begin{Bmatrix} l_0 & \frac{1}{2} & j_0 \\ l_0' & \frac{1}{2} & j_0' \\ k_{l_0} & k_{s_0} & k \end{Bmatrix}$$

$$\times \rho_{k_{s_0}q}\left(\frac{1}{2}\right)\left\langle LSJ, l=0\,j=\frac{1}{2}:J_t\,\|\,T\,\|\,L_0S_0J_0, l_0j_0:J_t\right\rangle$$

$$\times \left\langle LSJ', l' = 0\, j' = \frac{1}{2} : J_t' \,\|\, T \,\|\, L_0 S_0 J_0, l_0' j_0' : J_t' \right\rangle^* \tag{2.82}$$

It is assumed that the atomic state is a multiplet state with a certain orbital angular momentum L and spin S. For further derivation, the amplitudes in Eq. (2.82) have to be transformed by Eq. (A.81) to the representation of the total orbital momentum $\mathbf{L}_t = \mathbf{L} + \mathbf{l}$ and the total spin $\mathbf{S}_t = \mathbf{S} + \mathbf{1/2}$ with subsequent application of the reduction formula (A.64) with $k_1 = k_2 = 0$. The latter is possible because the transition operator T acts as a scalar operator in the subspaces of the total spin and the total orbital angular momentum separately. Then summing Eq. (2.82) over quantum numbers J_0, J_t, J_t', j_0, and j_0' with the use of Eqs. (A.161) and (A.154) and transforming to the double statistical tensors (1.65), one obtains the desired result, which can be cast in the form

$$\rho_{k_L q_L k_S q_S}(LS) = (-1)^{L+L_0+k_L+S+S_0+1}\,\rho_{k_S q_S}\left(\frac{1}{2}\right)$$

$$\times \sum_{l_0 l_0' S_t S_t'} (-1)^{l_0'}\,\rho_{k_L q_L}(l_0, l_0') \left\{ \begin{matrix} L & L & k_L \\ l_0 & l_0' & L_0 \end{matrix} \right\} \left\{ \begin{matrix} S & S & k_S \\ S_t & S_t' & \frac{1}{2} \end{matrix} \right\} \left\{ \begin{matrix} \frac{1}{2} & \frac{1}{2} & k_S \\ S_t & S_t' & S_0 \end{matrix} \right\}$$

$$\times (-1)^{S_t+S_t'}\,\hat{S}_t \hat{S}_t'\, T_{l_0}^{LS_t}\, T_{l_0'}^{LS_t'\,*} \tag{2.83}$$

Here the statistical tensors $\rho_{k_L q_L}(l_0, l_0')$ of the orbital motion of the incident electron are given by Eq. (1.101) and a short notation for the transition matrix elements is introduced: $T_{l_0}^{LS_t} = \langle (Ll = 0)L_t = L, (S\frac{1}{2})S_t \,\|\, T \,\|\, (L_0 l_0)L_t = L, (S_0 \frac{1}{2})S_t\rangle$. Equation (2.83) shows that ranks k_L, k_S and projections q_L, q_S of the statistical tensors of the orbital angular momentum and spin of the excited atomic state coincide with the corresponding tensorial indices of the incoming electron. This qualitative result is clearly due to the conservation laws, even in a more general case, not only at the excitation threshold. Indeed, polarization of spin and orbital angular momentum of the system $(e + A)$ before scattering is completely defined by the corresponding parameters of the incoming electron beam because the spin and the orbital angular momentum of the initial atomic state are unpolarized.

The tensorial dimensions of the statistical tensors of the whole system $(e + A)$ are conserved during the scattering [see Eq. (1.58)]. Moreover, owing to the nonrelativistic approximation for the scattering amplitudes, the tensorial dimensions are conserved separately for spatial and spin subspaces. The unobserved scattered electron does not influence the symmetry of the process and the tensorial structure of polarization of the excited atomic state coincides with that for the whole $(e + A)$ system before scattering. As a result, polarization of the excited atomic state takes tensorial components of the incident electron beam corresponding to both orbital angular momentum and spin. Note that at the threshold, the above analysis is still valid if the scattered electron is detected at an arbitrary fixed angle, since the s-electron is primarily isotropic. This follows from Eq. (2.61), where at $l = l' = 0$ there is only an isotropic part ($k_l = 0$) in the angular distribution of scattered electrons, which is formally equivalent to integration over the scattering angles.

The reduced statistical tensors of the orbital angular momentum and spin of the excited multiplet as well as the statistical tensors of the total angular momentum can be found from the double statistical tensor (2.83) by applying the general rules (1.68) and (1.69). For example, for the reduced statistical tensors of the orbital angular momentum, we obtain

$$
\mathcal{A}_{k_L q_L}(L) = \delta_{q_L 0} \, (-1)^{k_L + L + L_0} \hat{L} \left(\sum_{l_0 S_t} \left| T_{l_0}^{LS_t} \right|^2 \right)^{-1}
$$

$$
\times \sum_{l_0 l_0'} \hat{l}_0 \hat{l}'_0 \, (l_0 0, l_0' 0 | k_L 0) \left\{ \begin{matrix} L & L & k_L \\ l_0 & l_0' & L_0 \end{matrix} \right\} \sum_{S_t} T_{l_0}^{LS_t} T_{l_0'}^{LS_t *} \tag{2.84}
$$

The tensors (2.84) are constants independent of the transition matrix elements (scattering amplitudes) if only one partial wave of the incident electron l_0 contributes to the excitation. This is realized, for example, in a frequent case of atoms initially in the S-state: noble gases, alkalis, alkaline earths, and other atoms. Putting $L_0 = 0$ into Eq. (2.84) gives

$$
\mathcal{A}_{k_L q_L}(L) = (-1)^L \hat{L} \, (L0, L0 | k_L 0) \, \delta_{q_L 0} \tag{2.85}
$$

The simple expression (2.85) for $L_0 = 0$ is essentially a result of the threshold approximation ($l = l' = 0$) and could be predicted from the conservation of projection of the orbital angular momentum. Indeed, before excitation, the projection of the total orbital angular momentum L_t on the direction of the incident electron beam vanishes. The atomic orbital angular momentum L is isotropic and the incident electron does not carry the projection of the orbital angular momentum on the direction of motion [see Eq. (1.100)]. After scattering, the slow s electron has no projection of orbital angular momentum and therefore the excited atom is forced to have $M_L = 0$. Then Eq. (2.85) immediately follows from the definition of the reduced statistical tensor (1.41).

So, the threshold excitation shows axial symmetry with respect to the direction of the incident electron beam, and the reduced statistical tensors (2.85) are independent of the scattering amplitudes. Then, similar to the discussion in Section 2.2.1.3, one can expect simple relations between the threshold limit of polarization parameters of the excited atomic state and their high-energy limit, provided the PWBA is valid at high energies. Integrating Eq. (2.80) over scattering angles, proceeding exactly as in the derivation of Eq. (2.77) and putting

$L_0 = 0$, we obtain the relation

$$\mathcal{A}_{k0}(L)\,|_{p_0 \to \infty} = (-1)^{k/2}\,\frac{(k-1)!!}{k!!}\,\mathcal{A}_{k0}(L)\,|_{p_0 \to p_{\mathrm{thr}}} \qquad (2.86)$$

where on the left is the high-energy limit of the integral reduced statistical tensor of the orbital angular momentum in the PWBA and on the right is the threshold value (2.85) of same tensor. In practice, the relationship (2.77) does not hold accurately because the polarization of the excited atomic state is influenced by a radiative cascade from higher-lying excited atomic states. This applies to the tensor on the left side of Eq. (2.77). Equation (2.86) shows that the reduced statistical tensor of the second rank $\mathcal{A}_{20}(L)$ should change sign at least once (odd number of nodes) when going from the excitation threshold to high energies, while tensor $\mathcal{A}_{40}(L)$ may not cross zero (or has an even number of nodes).

2.2.2. Angular Distribution and Polarization of Scattered Electrons

Now we consider another characteristic of the same process (2.36) of electron–atom scattering, namely, the angular distribution and the spin polarization of the scattered electrons. The starting point is again the density matrix of the final state of the total system $A^* + e_{sc}$, Eq. (2.39). However, this time we calculate the spin density matrix of the scattered electrons for the case where the final atom is not observed and therefore we should calculate the trace over the variables characterizing the excited atomic state:

$$\langle \mu \,|\, \rho_e^f(\mathbf{p}_0, \mathbf{p}) \,|\, \mu' \rangle = \sum_{M_0 M_0' \mu_0 \mu_0' M} T_{M_0 \mu_0 \to M\mu}(\mathbf{p}_0, \mathbf{p})$$
$$\times \langle \alpha_0 J_0 M_0 \,|\, \rho_A^i \,|\, \alpha_0 J_0 M_0' \rangle \langle \mu_0 \,|\, \rho_e^i \,|\, \mu_0' \rangle$$
$$\times T_{M_0' \mu_0' \to M\mu'}^*(\mathbf{p}_0, \mathbf{p}) \qquad (2.87)$$

Here for brevity we omitted the electron spin $s = \frac{1}{2}$ from notations of the electron spin density matrices.

2.2.2.1. Scattering of Unpolarized Electrons from Unpolarized Atoms

In this case, both density matrices on the right side of Eq. (2.87) describe unpolarized particles and are represented by Eq. (1.40). The density matrix of the scattered electrons (2.87) takes the form

$$\langle \mu \,|\, \rho_e^f(\mathbf{p}_0, \mathbf{p}) \,|\, \mu' \rangle = \frac{1}{2(2J_0 + 1)} \sum_{M_0 \mu_0 M} T_{M_0 \mu_0 \to M\mu}(\mathbf{p}_0, \mathbf{p})$$
$$\times T_{M_0 \mu_0 \to M\mu'}^*(\mathbf{p}_0, \mathbf{p}) \qquad (2.88)$$

The trace of this matrix gives the differential cross section, i.e., the angular distribution of the scattered electrons:

$$\frac{d\sigma}{d\Omega_{sc}} = \frac{1}{2(2J_0+1)} \frac{p_f}{p_i} \sum_{M_0\mu_0 M\mu} |T_{M_0\mu_0 \to M\mu}(\mathbf{p}_0,\mathbf{p})|^2 \tag{2.89}$$

To analyze the spin polarization of the scattered electrons, we normalize the density matrix (2.88) by the condition (1.7):

$$\langle \mu \,|\, \rho_e^f(\mathbf{p}_0,\mathbf{p}) \,|\, \mu' \rangle = \frac{\sum_{M_0\mu_0 M} T_{M_0\mu_0 \to M\mu}(\mathbf{p}_0,\mathbf{p}) \, T_{M_0\mu_0 \to M\mu'}^*(\mathbf{p}_0,\mathbf{p})}{\sum_{M_0\mu_0\mu M} |T_{M_0\mu_0 \to M\mu}(\mathbf{p}_0,\mathbf{p})|^2} \tag{2.90}$$

According to general arguments (see Section 1.3.2), the spin density matrix of an electron can always be presented in the form

$$\rho_e = \frac{1}{2}(I + \mathbf{P}_e\,\boldsymbol{\sigma}) \tag{2.91}$$

where $P_e = |\mathbf{P}_e|$ is the degree of polarization of the scattered electrons. In our case of scattering of an unpolarized electron beam from unpolarized atoms, the polarization vector of scattered electrons is always directed along the vector $\mathbf{n}$ normal to the scattering plane:

$$\mathbf{P}_e^f(\mathbf{p}_0,\mathbf{p}) = P_e^f(\mathbf{p}_0,\mathbf{p})\,\mathbf{n} \tag{2.92}$$

because $\mathbf{n} = \frac{\mathbf{p}_0 \times \mathbf{p}}{|\mathbf{p}_0 \times \mathbf{p}|}$ is the only axial vector that can be formed from the kinematic parameters of the scattering process in this case. [A similar situation has been discussed in Section 2.2.1.1 after Eq. (2.49).] So the electron spin density matrix can be presented as

$$\rho_e^f(\mathbf{p}_0,\mathbf{p}) = \frac{1}{2}\left[1 + P_e^f(\mathbf{p}_0,\mathbf{p})\left(\mathbf{n}\,\boldsymbol{\sigma}\right)\right] \tag{2.93}$$

Let us choose a coordinate frame with the z-axis along the incident electron beam and the x-axis lying in the scattering plane. This coordinate frame is shown in Figure 2.1 and is called the *collision frame*. In the collision frame, vector $\mathbf{P}_e^f(\mathbf{p}_0,\mathbf{p})$ has a y-component only, and the degree of polarization $P_e^f(\mathbf{p}_0,\mathbf{p}) \equiv P_y^f$ can be calculated according to Eq. (1.92) using the spin density matrix of the scattered electrons given by Eq. (2.93) and the Pauli matrix σ_y:

$$P_y^f(\mathbf{p}_0,\mathbf{p}) = \mathrm{Tr}\left[\rho_e^f(\mathbf{p}_0,\mathbf{p})\,\sigma_y\right] \tag{2.94}$$

Using Eq. (2.90) for the density matrix and the explicit form of the Pauli matrix (A.2), one obtains

$$P_y^f(\mathbf{p}_0,\mathbf{p}) = -2\frac{\sum_{M_0 M\mu_0} \mathrm{Im}\left(T_{M_0,\mu_0 \to M,1/2}(\mathbf{p}_0,\mathbf{p}) T_{M_0,\mu_0 \to M,-1/2}^*(\mathbf{p}_0,\mathbf{p})\right)}{\sum_{M_0 M\mu_0\mu} |T_{M_0\mu_0 \to M\mu}(\mathbf{p}_0,\mathbf{p})|^2} \tag{2.95}$$

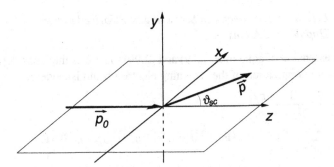

FIGURE 2.1. Collision coordinate frame.

Consider the particular case of the elastic scattering from an atom with $J_0 = 0$ and, also, the excitation transition $(J_0 = 0) \rightarrow (J = 0)$. The transition operator $T(\mathbf{p_0}, \mathbf{p})$, which should be a scalar with respect to the total angular momentum, has in this case a universal form dictated by the general symmetry arguments:

$$\sqrt{\frac{p_i}{p_f}} T(\mathbf{p_0}, \mathbf{p}) = a(E, \vartheta_{sc}) + b(E, \vartheta_{sc})(\mathbf{n}\boldsymbol{\sigma}) \qquad (2.96)$$

where $a(E, \vartheta_{sc})$ and $b(E, \vartheta_{sc})$ are the nonspin-flip and spin-flip components, respectively. This form is independent of the character of the interaction between the incoming electron and the target atom, including the interrelation of direct (nonexchange) and exchange scattering, independent of the wave functions of the atom, the role of relativistic effects, and so on. Writing the amplitudes $T_{M_0\mu\,0 \rightarrow M\mu}(\mathbf{p_0}, \mathbf{p})$ corresponding to the transition operator (2.96) and substituting them into Eqs. (2.89) and (2.95), we obtain the differential cross section and the polarization vector of scattered electrons averaged over polarization states of electrons in the incoming unpolarized beam:

$$\frac{d\sigma}{d\Omega_{sc}} = |a(E, \vartheta_{sc})|^2 + |b(E, \vartheta_{sc})|^2 \qquad (2.97)$$

$$\mathbf{P}_e^f = \frac{2\,\mathrm{Re}[a(E, \vartheta_{sc})b^*(E, \vartheta_{sc})]}{|a(E, \vartheta_{sc})|^2 + |b(E, \vartheta_{sc})|^2}\, \mathbf{n} \qquad (2.98)$$

This shows that the phenomenon of spin polarization of scattered electrons is due to the interference effect of *spin-flip* and *nonspin-flip scattering amplitudes*. Equations (2.93) and (2.98) give the spin density matrix of the scattered electrons:

$$\rho_e^f(\mathbf{p_0}, \mathbf{p}) = \frac{1}{2}\left[1 + \frac{2\,\mathrm{Re}[a(E, \vartheta_{sc})b^*(E, \vartheta_{sc})]}{|a(E, \vartheta_{sc})|^2 + |b(E, \vartheta_{sc})|^2}\,(\mathbf{n}\boldsymbol{\sigma})\right] \qquad (2.99)$$

2.2.2.2. Left–Right Asymmetry in Scattering of Polarized Electrons from Unpolarized Atoms

An extension of equations (2.89) and (2.90) that takes into consideration an arbitrary spin polarization of the incoming electron beam is evident:

$$\frac{d\sigma}{d\Omega_{sc}} = \frac{1}{2J_0+1}\frac{p_f}{p_i}$$

$$\times \sum_{M_0\mu_0\mu'_0 M\mu} T_{M_0\mu_0\to M\mu}(\mathbf{p}_0,\mathbf{p})\left\langle\mu_0\left|\rho^i_e\right|\mu'_0\right\rangle T^*_{M_0\mu'_0\to M\mu}(\mathbf{p}_0,\mathbf{p}) \qquad (2.100)$$

$$\left\langle\mu\left|\rho^f_e(\mathbf{p}_0,\mathbf{p})\right|\mu'\right\rangle$$

$$= \frac{\sum_{M_0\mu_0\mu'_0 M} T_{M_0\mu_0\to M\mu}(\mathbf{p}_0,\mathbf{p})\left\langle\mu_0\left|\rho^i_e\right|\mu'_0\right\rangle T^*_{M_0\mu'_0\to M\mu'}(\mathbf{p}_0,\mathbf{p})}{\sum_{M_0\mu_0\mu'_0\mu M} T_{M_0\mu_0\to M\mu}(\mathbf{p}_0,\mathbf{p})\left\langle\mu_0\left|\rho^i_e\right|\mu'_0\right\rangle T^*_{M_0\mu'_0\to M\mu}(\mathbf{p}_0,\mathbf{p})} \qquad (2.101)$$

To start, we consider scattering of an incoming electron beam polarized normally to the scattering plane:

$$\rho^i_e = \frac{1}{2}\left[1+P^i_e\sigma_y\right] \qquad (2.102)$$

To reveal the dependence of the angular distribution of scattered electrons on the degree of polarization of the incoming beam $P^i_e \equiv P^i_y$, we present the differential cross section (2.100) in the following form:

$$\frac{d\sigma}{d\Omega_{sc}} = \frac{1}{2(2J_0+1)}\frac{p_f}{p_i}\sum_{M_0M\mu_0\mu}\left|T_{M_0\mu_0\to M\mu}(\mathbf{p}_0,\mathbf{p})\right|^2$$

$$\times\left[1+P^i_y\frac{\sum_{M_0M\mu\mu_0\mu'_0\mu} T_{M_0\mu_0\to M\mu}(\mathbf{p}_0,\mathbf{p})\left\langle\mu_0\left|\sigma_y\right|\mu'_0\right\rangle T^*_{M_0\mu'_0\to M\mu}(\mathbf{p}_0,\mathbf{p})}{\sum_{M_0M\mu_0\mu}\left|T_{M_0\mu_0\to M\mu}(\mathbf{p}_0,\mathbf{p})\right|^2}\right]$$

$$(2.103)$$

Equation (2.103) shows that the cross section at a fixed scattering angle ϑ_{sc} depends on the sign of P^i_y and therefore shows the *spin up–spin down asymmetry* of scattering. Equivalently, one can speak about *left–right asymmetry* of scattering when the direction of the polarization of the incident electrons is fixed. Conventionally, the asymmetry parameter S_A is defined for a completely polarized incident beam as:

$$S_A \equiv S_A(P^i_y=1) = \frac{\frac{d\sigma}{d\Omega}(P_y=1)-\frac{d\sigma}{d\Omega}(P_y=-1)}{\frac{d\sigma}{d\Omega}(P_y=1)+\frac{d\sigma}{d\Omega}(P_y=-1)} \qquad (2.104)$$

Using Eqs. (2.104) and (2.103), one obtains:

$$S_A = \frac{2\sum_{M_0M\mu}\text{Im}\left[T_{M_0,1/2\to M\mu}(\mathbf{p}_0,\mathbf{p})\,T^*_{M_0,-1/2\to M\mu}(\mathbf{p}_0,\mathbf{p})\right]}{\sum_{M_0M\mu_0\mu}\left|T_{M_0\mu_0\to M\mu}(\mathbf{p}_0,\mathbf{p})\right|^2} \qquad (2.105)$$

For a partly polarized beam, when $|P_y^i| < 1$, the magnitude of the left–right asymmetry is proportional to P_y^i:

$$S_A(P_y^i) = S_A P_y^i \tag{2.106}$$

Equation (2.105) for asymmetry looks similar to Eq. (2.95) for the degree of polarization P_y^f of scattered electrons in the same process, but with an unpolarized incoming beam. However, they do not coincide with each other. As an exception, Eqs. (2.105) and (2.95) give identical results for elastic scattering and in a special case of inelastic scattering, $(J_0 = 0) \rightarrow (J = 0)$. Indeed, one has in these cases

$$S_A = P_e^f = \frac{2\,\mathrm{Re}[a(E,\vartheta_{sc})b^*(E,\vartheta_{sc})]}{|a(E,\vartheta_{sc})|^2 + |b(E,\vartheta_{sc})|^2} \tag{2.107}$$

2.2.2.3. Polarization of Scattered Electrons in Collisions of Polarized Electrons with Atoms

As a universal tool for calculating the polarization of scattered electrons for arbitrary polarizations of a target atom and an incoming electron beam, one can use Eq. (1.92):

$$\mathbf{P}_e^f(\mathbf{p}_0,\mathbf{p}) = \mathrm{Tr}\left[\rho_e^f(\mathbf{p}_0,\mathbf{p})\boldsymbol{\sigma}\right] \tag{2.108}$$

Here the the spin density matrix of the scattered electrons is presented in the general form by Eqs. (2.101) and (2.102) where $P_e^i \equiv P^i$ is the degree of polarization of the electron beam in the initial state. We illustrate such a calculation by the case of elastic electron scattering from atoms with $J_0 = 0$.

Let $\mathbf{P}_i$, $a \equiv a(E,\vartheta_{sc})$, and $b \equiv b(E,\vartheta_{sc})$ denote the polarization vector of the incoming beam and the nonspin-flip and spin-flip scattering amplitudes, respectively. The polarization vector of the scattered electrons $\mathbf{P}^f \equiv \mathbf{P}_e^f(\mathbf{p}_0,\mathbf{p})$ can be calculated as follows [see Eq. (1.17)]:

$$\mathbf{P}^f = \frac{\mathrm{Tr}\Big(\boldsymbol{\sigma}[a+b(\boldsymbol{\sigma}\mathbf{n})][1+\mathbf{P}^i\boldsymbol{\sigma}][a^*+b^*(\boldsymbol{\sigma}\mathbf{n})]\Big)}{\mathrm{Tr}\Big([a+b(\boldsymbol{\sigma}\mathbf{n})][1+\mathbf{P}^i\boldsymbol{\sigma}][a^*+b^*(\boldsymbol{\sigma}\mathbf{n})]\Big)} \tag{2.109}$$

This equation can be rearranged and presented in the form:

$$\mathbf{P}^f = \frac{1}{|a|^2+|b|^2+2\,\mathrm{Re}(ab^*)(\mathbf{P}^i\mathbf{n})} \tag{2.110}$$
$$\times \left[(|a|^2+|b|^2)\mathbf{P}^i+2\,\mathrm{Re}(ab^*)\mathbf{n}+2\,\mathrm{Im}(ab^*)[\mathbf{n}\times\mathbf{P}^i]-2|b|^2\left(\mathbf{P}^i-\mathbf{n}(\mathbf{P}^i\mathbf{n})\right)\right]$$

In Eq. (2.110) one can separate out the combination of amplitudes that gives the degree of polarization of scattered electrons when the incoming beam is not polarized [see Eq. (2.98)]. We denote it here as P_0^f, and the corresponding polarization vector as $\mathbf{P}_0^f$:

$$\mathbf{P}_0^f = \frac{2\,\mathrm{Re}(ab^*)}{|a|^2 + |b|^2}\,\mathbf{n} \tag{2.111}$$

By dividing both the numerator and denominator in Eq. (2.110) by $(|a|^2 + |b|^2)$, we can rewrite it in a more compact form:

$$\mathbf{P}^f = \frac{1}{1 + \mathbf{P}^i\,\mathbf{P}_0^f}$$
$$\times \left[\mathbf{P}^i + \mathbf{P}_0^f + \frac{2\,\mathrm{Im}(ab^*)}{|a|^2 + |b|^2}[\mathbf{n} \times \mathbf{P}^i] - \frac{2|b|^2}{|a|^2 + |b|^2}\left(\mathbf{P}^i - \mathbf{n}(\mathbf{P}^i\mathbf{n})\right)\right] \tag{2.112}$$

Consider two particular cases of the scattering process when the initial polarization vector lies in the scattering plane ($\mathbf{P}^i \perp \mathbf{n}$) and when it is directed along the normal to the plane ($\mathbf{P}^i \parallel \mathbf{n}$). In the first case

$$\mathbf{P}^f\big|_{\mathbf{P}^i\perp\mathbf{n}} = \mathbf{P}_0^f + \frac{2\,\mathrm{Im}(ab^*)}{|a|^2 + |b|^2}[\mathbf{n} \times \mathbf{P}^i] + \frac{|a|^2 - |b|^2}{|a|^2 + |b|^2}\mathbf{P}^i \tag{2.113}$$

The normal component of the polarization vector in the final state does not depend on the initial polarization and is exactly the same as when the incoming electron beam is not polarized. As for the component of $\mathbf{P}^f$ which is in the scattering plane, its deviation from the initial direction $\mathbf{P}^i$ depends on the ratio and relative phase of the scattering amplitudes $a(E, \vartheta_{sc})$ and $b(E, \vartheta_{sc})$. In the second case, when the initial polarization vector is directed normally to the scattering plane, the polarization vector in the final state is also oriented normally to this plane.

2.3. Direct Atomic Photoeffect

In the process of photoionization of atoms, there are two particles in the final state: a photoion and a photoelectron. Therefore, all correlation and polarization phenomena in photoionization can be described using an approach developed in the preceding section. However, owing to its importance in fundamental atomic physics and numerous applications, atomic photoionization deserves special consideration. Moreover, with the advent of high-brilliance synchrotron radiation sources, many correlation and polarization experiments have become feasible. Their analysis is sometimes rather complicated because of the complexity of the experimental arrangements. The formalism of statistical tensors and density matrix greatly simplifies the analysis and planning of such experiments. The theory of photoeffect is very well developed. There are numerous books and reviews in which various aspects of the theory, including the angular distribution and polarization of the photoionization products, are discussed. As a recent example we mention the collection of reviews in Ref. 16. Therefore, in this book we limit ourselves to the particular problems where the use of the described formalism provides a great advantage in obtaining a general parameterization of the cross sections in complicated kinematic conditions.

2.3.1. Angular Distribution of Photoelectrons in Photoionization of Polarized Atoms: General Expression

We consider the process of direct photoionization of an arbitrarily polarized atom:

$$\gamma + A(\alpha_0 J_0) \longrightarrow A^+(\alpha_f J_f) + e_{ph} \tag{2.114}$$

The initial atomic state is characterized by the total angular momentum J_0 and by other quantum numbers that we denote α_0. We assume that the photoion is produced in a state with a sharp angular momentum J_f and definite parity. An extension of the formalism to mixtures of states with different total angular momenta is straightforward and can be done as in Section 2.2.1.2.

To derive a general expression for the angular distribution of photoelectrons, we start from Eq. (1.151), which expresses the probability of some event in photoionization in terms of statistical tensors of the final state of the atomic system, $\rho^f_{kq}(\alpha J, \alpha' J')$, and efficiency tensors of the detecting system, $\varepsilon_{kq}(\alpha J, \alpha' J')$:

$$W = \sum_{\alpha \alpha' J J' kq} \rho^f_{kq}(\alpha J, \alpha' J')\, \varepsilon^*_{kq}(\alpha J, \alpha' J') \tag{2.115}$$

Here J and α are, respectively, the total angular momentum and other quantum numbers that characterize the atomic system after photoabsorption.

The statistical tensors of the final state may be expressed in terms of statistical tensors of the initial state and reduced matrix elements of the transition operator describing the photon–atom interaction [see Eqs. (1.58) and (2.2)]:

$$\rho^f_{kq}(\alpha J, \alpha' J') = \frac{1}{\hat{J}\hat{J}'} \sum_{\alpha_i \alpha'_i} \rho^i_{kq}(\alpha_i J, \alpha'_i J') \langle \alpha J \| T \| \alpha_i J \rangle$$
$$\times \langle \alpha' J' \| T \| \alpha'_i J' \rangle^* \tag{2.116}$$

where α_i denotes the quantum numbers of the initial photon + atom system. As in Section 2.1.1, we use the dipole approximation to calculate the transition matrix elements and express the initial state statistical tensor in terms of statistical tensors of the photon, $\rho^{\gamma}_{k_{\gamma}q_{\gamma}}$, and the initial atom, $\rho_{k_0 q_0}(\alpha_0 J_0)$, using Eqs. (1.64) and (1.66):

$$
\rho^{f}_{kq}(\alpha J, \alpha' J') = \sum_{k_0 q_0 k_{\gamma} q_{\gamma}} \hat{k}_0 \hat{k}_{\gamma} (k_0 q_0, k_{\gamma} q_{\gamma} | kq) \begin{Bmatrix} J_0 & 1 & J \\ J_0 & 1 & J' \\ k_0 & k_{\gamma} & k \end{Bmatrix}
$$
$$
\times \rho_{k_0 q_0}(\alpha_0 J_0) \rho^{\gamma}_{k_{\gamma} q_{\gamma}}
$$
$$
\times \langle \alpha_f J_f, lj : J \| D \| \alpha_0 J_0 \rangle \langle \alpha_f J_f, l'j' : J' \| D \| \alpha_0 J_0 \rangle^* \quad (2.117)
$$

The final state of the system is specified by the quantum numbers and total angular momentum of the final ion, $\alpha_f J_f$, and the orbital and total angular momenta of the photoelectron, l, j. We emphasize that the statistical tensor (2.117) is not normalized according to the standard conditions (1.49) or (1.48). Hereafter in Section 2.3, general factors in expressions derived with the use of Eq. (2.117) are in accordance with the normalization of the statistical tensors (2.117) until otherwise indicated.

The statistical tensors $\rho^{\gamma}_{k_{\gamma}q_{\gamma}}$ that describe the initial photon beam are determined by three Stokes parameters. We discussed their explicit form in Section 1.3.5. In the dipole approximation they are given, for example, in Table 1.1 on page 25. We note that here $k_{\gamma} \leq 2$. The polarization state of the initial atom can be, in principle, arbitrary; its statistical tensors are limited only by the general conditions (1.46) and (1.51) and by the condition

$$
0 \leq k_0 \leq 2J_0, \qquad -k_0 \leq q_0 \leq k_0 \quad (2.118)
$$

Several methods of preparation of the polarized atomic state exist, such as laser optical pumping and polarization of the atomic spin by a hexapole magnet. In all cases of practical interest, the prepared atomic state has axial symmetry. The symmetry axis, $\mathbf{A}$, is characterized by two angles (ϑ_a, φ_a) with respect to the laboratory fixed coordinate system. In a coordinate frame with the z-axis directed along this symmetry axis, which is called the *atomic frame*, only zero components of the statistical tensors exist (see Section 1.4.2), denoted by $\bar{\rho}_{k_0 0}(\alpha_0 J_0)$. Using the transformation properties of statistical tensors (1.44) and Eq. (A.51), one can express the tensors in the laboratory frame as follows:

$$
\rho_{k_0 q_0}(\alpha_0 J_0) = \sqrt{\frac{4\pi}{2k_0 + 1}} Y^*_{k_0 q_0}(\vartheta_a, \varphi_a) \bar{\rho}_{k_0 0}(\alpha_0 J_0) \quad (2.119)
$$

Now consider the second factor in Eq. (2.115), the efficiency tensor of the detectors. We are interested here in the angular distribution of photoelectrons irrespective of their polarization. If only photoelectrons are detected and the detector is not sensitive to their spin polarization, the efficiency tensor may be presented in the form of Eq. (1.171):

$$\varepsilon_{kq}\left(\alpha_f J_f, lj : J; \alpha_f J_f, l'j' : J'\right) = \frac{1}{4\pi}(-1)^{J+J_f+k-1/2}\hat{l}\hat{l}'\,\hat{j}\hat{j}'\hat{J}\hat{J}'\,(l0,l'0|k0)$$

$$\times \begin{Bmatrix} J & j & J_f \\ j' & J' & k \end{Bmatrix}\begin{Bmatrix} l & j & \frac{1}{2} \\ j' & l' & k \end{Bmatrix}\sqrt{\frac{4\pi}{2k+1}}Y_{kq}^*(\vartheta,\varphi) \quad (2.120)$$

Collecting now Eqs. (2.115), (2.117), (2.119), and (2.120), we obtain the general expression for the angular distribution of photoelectrons from arbitrarily polarized target atoms exposed to polarized radiation. With absolute normalization to the differential cross section it can be cast in the form

$$\frac{d\sigma}{d\Omega} = \pi\alpha\omega(3\hat{J}_0)^{-1}\sum_{k_0 k k_\gamma}\overline{p}_{k_0 0}(\alpha_0 J_0)B_{k_0 k k_\gamma}$$

$$\times F_{k_0 k k_\gamma}(\vartheta_a,\varphi_a,\vartheta_e,\varphi_e; P_1, P_2, P_3) \quad (2.121)$$

Here ω is the frequency of the ionizing photon, α is the fine structure constant, and

$$B_{k_0 k k_\gamma} = 3\,\hat{J}_0\sum_{\substack{jj'\\ll'jj'}}(-1)^{J+J_f+k_\gamma-1/2}\hat{J}\hat{J}'\,\hat{j}\hat{j}'\,\hat{l}\hat{l}'\,(l0,l'0|k0)$$

$$\times \begin{Bmatrix} j & l & \frac{1}{2} \\ l' & j' & k \end{Bmatrix}\begin{Bmatrix} j & J & J_f \\ J' & j' & k \end{Bmatrix}\begin{Bmatrix} J_0 & 1 & J \\ J_0 & 1 & J' \\ k_0 & k_\gamma & k \end{Bmatrix}M_{ljJ}M_{l'j'J'}^* \quad (2.122)$$

where we have introduced a short notation for the reduced dipole matrix element $M_{ljJ} \equiv \langle \alpha_f J_f, lj : J \Vert D \Vert \alpha_0 J_0 \rangle$.

The kinematic functions

$$F_{k_0 k k_\gamma}(\vartheta_a,\varphi_a,\vartheta_e,\varphi_e; P_1, P_2, P_3)$$
$$= \sum_{q_\gamma}4\pi\left\{Y_{k_0}(\vartheta_a,\varphi_a)\otimes Y_k(\vartheta_e,\varphi_e)\right\}_{k_\gamma q_\gamma}\rho_{k_\gamma q_\gamma}^\gamma(P_1, P_2, P_3) \quad (2.123)$$

depend on the direction of the photoelectron emission (ϑ_e, φ_e) and that of the target polarization (ϑ_a, φ_a), as well as on the Stokes parameters P_1, P_2, and P_3 characterizing the initial photon beam. The tensorial product of spherical harmonics is defined as usual, Eq. (A.36):

$$\left\{Y_{k_0}(\vartheta_a,\varphi_a)\otimes Y_k(\vartheta_e,\varphi_e)\right\}_{k_\gamma q_\gamma} = \sum_{q_0 q}(k_0 q_0, kq|k_\gamma q_\gamma)$$

$$\times Y_{k_0 q_0}(\vartheta_a,\varphi_a)Y_{kq}(\vartheta_e,\varphi_e) \quad (2.124)$$

The values $F_{k_0kk_\gamma}$ are scalar products independent of the coordinate frame because $\rho^\gamma_{k_\gamma q_\gamma}$ transforms as a covariant tensor of rank k_γ.

Expression (2.121) gives the general form of the angular distribution of photoelectrons produced by an arbitrarily polarized photon beam from a polarized target. Each term of the sum is factored into a statistical tensor $\overline{\rho}_{k_0 0}(\alpha_0 J_0)$ that describes the polarization of the target, a dynamical part $B_{k_0kk_\gamma}$ that contains the dipole amplitudes, and a kinematic part $F_{k_0kk_\gamma}$ which contains angle dependence and a description of the photon state and is independent of any particular dynamic properties or properties of the atom. The formula is valid for an arbitrary reference frame and for any particular geometry of the experiment.

We can rewrite this expression in a more traditional form by introducing the reduced statistical tensors of the initial state $\overline{A}_{k_0 q_0}(\alpha_0 J_0)$ [see Eq. (1.50)] and defining the *generalized anisotropy coefficients* as

$$\beta_{k_0kk_\gamma} = B_{k_0kk_\gamma} \bigg/ \sum_{ljJ}|M_{ljJ}|^2 \tag{2.125}$$

Then Eq. (2.121) can be presented as

$$\frac{d\sigma}{d\Omega} = \frac{\sigma^{(\text{iso})}}{4\pi}\left(1 + \sum_{k_0}\overline{A}_{k_0 0}(\alpha_0 J_0)\sum_{k_\gamma k}\beta_{k_0kk_\gamma}F_{k_0kk_\gamma}(\vartheta_a, \varphi_a, \vartheta_e, \varphi_e; P_1, P_2, P_3)\right) \tag{2.126}$$

Here $\sigma^{(\text{iso})}$ is the total photoionization cross section, which would be calculated if the initial state were randomly oriented,

$$\sigma^{(\text{iso})} = \frac{4\pi^2\,\alpha\,\omega}{3(2J_0 + 1)}\sum_{ljJ}|M_{ljJ}|^2 \tag{2.127}$$

Summation in Eq. (2.126) is over all possible values of k_0, k, and k_γ, except for the $k_0 = k = k_\gamma = 0$ term. Note, that the values of k_0, k, and k_γ are connected by a "triangle inequality":

$$|k_0 - k_\gamma| \le k \le k_0 + k_\gamma \tag{2.128}$$

In addition, they are limited by the conditions

$$k_0 \le 2J_0, \qquad k_\gamma \le 2, \qquad k = \text{even} \tag{2.129}$$

the latter being a consequence of the parity conservation in photoionization; it follows formally from the property (A.94) of the Clebsch–Gordan coefficient in Eq. (2.122). Therefore, the number of terms in sums in (2.126) is limited. The dynamic coefficients $\beta_{k_0 k k_\gamma}$ contain bilinear combinations of the dipole amplitudes and therefore contain complete information on the dynamics of the photoprocess.

The general expression (2.121) will be analyzed in the following section. We conclude the present section by discussing the partial photoionization cross section averaged over all angles (4π-geometry), which is equal to the partial photoion production cross section. Integration of Eq. (2.121) over all angles (ϑ_e, φ_e) gives $k = q = 0$ and $k_0 = k_\gamma$. Then the integrated cross section for polarized atoms has the general form

$$\sigma = \sigma^{(\text{iso})} \left(1 + \sum_{k_0 > 0} \overline{A}_{k_0 0}(\alpha_0 J_0) \beta_{k_0 0 k_0} \sum_{q_\gamma} \rho^\gamma_{k_0 q_\gamma} \sqrt{4\pi} Y_{k_0 q_\gamma}(\vartheta_a, \varphi_a)\right) \quad (2.130)$$

In particular, if the initial atom is aligned ($k_0 = 2$) along the direction perpendicular to the photon beam and if the photons are linearly polarized, we can choose the z-axis along the direction of photon polarization and study the cross section as a function of angle ϑ_a between the polarization and the alignment axis. Then the cross section (2.130) reduces to the simple form

$$\sigma = \sigma^{(\text{iso})} \left(1 + \beta' P_2(\cos \vartheta_a)\right) \quad (2.131)$$

where $\beta' = -\sqrt{\frac{10}{3}} \beta_{202} \overline{A}_{20}(\alpha_0 J_0)$. An explicit expression for the coefficient β' can be easily obtained from Eqs. (2.122) and (2.125):

$$\beta' = \overline{A}_{20}(\alpha_0 J_0) \sqrt{6}\, \hat{J}_0 \sum_{ljJ} (-1)^{J+J_0} \begin{Bmatrix} J_0 & 1 & J \\ 1 & J_0 & 2 \end{Bmatrix} |M_{ljJ}|^2 \bigg/ \left(\sum_{ljJ} |M_{ljJ}|^2\right) \quad (2.132)$$

Note that the coefficient β' contains only squares of the photoionization amplitudes, which is natural because it determines the total cross section.

2.3.2. Particular Cases: Photoionization of Atoms with Total Angular Momentum $J_0 = 0$, $1/2$, and 1

2.3.2.1. Unpolarized Target Atom

First we consider the simple case of an unpolarized target atom. If the initial state is unpolarized ($k_0 = 0$), and the ionizing photon is completely linearly polarized, circularly polarized, or unpolarized, then taking into account that $k_\gamma = k = $ even and $k_\gamma \leq 2$, we can easily obtain from Eq. (2.126) the conventional expressions for the angular distribution of photoelectrons. For example, in the case of completely linearly polarized photons ($\rho^\gamma_{20} = -2/\sqrt{6}$; according to Table 1.1, the z-axis is chosen along the photon polarization), we have

$$\frac{d\sigma}{d\Omega} = \frac{\sigma^{(\text{iso})}}{4\pi} \left(1 + \beta P_2(\cos \vartheta_e)\right) \quad (2.133)$$

with

$$\beta = -\sqrt{\frac{10}{3}}\beta_{022} \qquad (2.134)$$

In the case of completely circularly polarized or unpolarized photons ($\rho_{20}^{\gamma} = 1/\sqrt{6}$; according to Table 1.1 on p. 25, the z-axis is chosen along the photon beam), we get a similar expression when substituting $\beta \rightarrow -\beta/2$. The anisotropy parameter β can be expressed in terms of the reduced matrix elements M_{ljJ} using expressions (2.122), (2.125), and (2.134):

$$\beta = \sqrt{6}\left(\sum_{ljJ}|M_{ljJ}|^2\right)^{-1}\sum_{ll'jj'JJ'}(-1)^{J_f-J_0-1/2}\hat{J}\hat{J}'\hat{j}\hat{j}'\hat{l}\hat{l}'\,(l0,l'0|20)$$

$$\times\begin{Bmatrix} j & l & \frac{1}{2} \\ l' & j' & 2 \end{Bmatrix}\begin{Bmatrix} j & J & J_f \\ J' & j' & 2 \end{Bmatrix}\begin{Bmatrix} 1 & J & J_0 \\ J' & 1 & 2 \end{Bmatrix}M_{ljJ}\,M^*_{l'j'J'} \qquad (2.135)$$

2.3.2.2. Target Atom with $J_0 = 0$

In the simplest case of a target atom with $J_0 = 0$ (an atom with closed subshells), the angular distribution of photoelectrons naturally coincides with that given by Eq. (2.133) for unpolarized targets. However, the expression for the asymmetry parameter is simplified owing to the dipole selection rules that give $J = J' = 1$. In this case we have

$$\beta = \sqrt{6}\left(\sum_{lj}|M_{lj1}|^2\right)^{-1}\sum_{ll'jj'}(-1)^{J_f-J_0-1/2}\hat{j}\hat{j}'\hat{l}\hat{l}'\,(l0,l'0|20)$$

$$\times\begin{Bmatrix} j & l & \frac{1}{2} \\ l' & j' & 2 \end{Bmatrix}\begin{Bmatrix} j & 1 & J_f \\ 1 & j' & 2 \end{Bmatrix}M_{lj1}\,M^*_{l'j'1} \qquad (2.136)$$

Further simplifications of this expression are possible if one introduces some approximations for calculation of the dipole matrix elements, for example, a nonrelativistic approximation for the photoelectrons.

2.3.2.3. Oriented Target Atom with $J_0 = 1/2$

Now we consider the target atoms with $J_0 = \frac{1}{2}$. These atoms can be oriented and their orientation may be characterized by the reduced statistical tensor of the first rank, $\overline{\mathcal{A}}_{10}$ ($k_0 = 1$). The orientation of the target affects the angular distribution of photoelectrons. According to conditions (2.128) and (2.129), besides β_{022}, which gives the anisotropy parameter β, three other anisotropy coefficients can be nonzero: β_{122}, β_{101}, and β_{121}. To illustrate the effect of orientation, we discuss some simple particular cases.

If the photon beam is linearly polarized (the z-axis is chosen along the electric vector $\mathbf{E}$) and the electric vector is perpendicular to the target orientation ($\mathbf{E} \perp \mathbf{J}_0$), then the photoionization cross section can be written as

$$\frac{d\sigma}{d\Omega} = \frac{\sigma^{(iso)}}{4\pi} \left(1 + \beta P_2(\cos \vartheta_e) + \overline{\mathcal{A}}_{10}\beta_2 \sin 2\vartheta_e \sin \varphi_e \right) \tag{2.137}$$

where β is the anisotropy coefficient for the photoelectron angular distribution from an unpolarized target, whereas $\beta_2 = -i\sqrt{\frac{15}{4}}\beta_{122}$ characterizes the anisotropy associated with the target polarization. The additional term in the angular distribution is proportional to the degree of orientation $\overline{\mathcal{A}}_{10}$. It disappears if the photoelectron linear momentum $\mathbf{p}_e$ is collinear with any of the two vectors ($\mathbf{E}$ and $\mathbf{J}_0$) or if the three vectors lie in one plane.

If the target atom is oriented along the photon polarization ($\mathbf{J}_0 \parallel \mathbf{E}$), then the photoelectron angular distribution is exactly the same as for the unpolarized target:

$$\frac{d\sigma}{d\Omega} = \frac{\sigma^{(iso)}}{4\pi}(1 + \beta P_2(\cos \vartheta_e)) \tag{2.138}$$

Now consider a circularly polarized photon beam. Choosing the z-axis along the beam direction $\mathbf{p}_\gamma$ and considering the case of target atom orientation along the beam, $\mathbf{J}_0 \parallel \mathbf{p}_\gamma$, we have

$$\frac{d\sigma^{(\pm)}}{d\Omega} = \frac{\sigma^{(iso)}}{4\pi} \left(1 - \frac{\beta}{2}P_2(\cos \vartheta_e) \pm \overline{\mathcal{A}}_{10}[\Delta + \beta_3 P_2(\cos \vartheta_e)] \right) \tag{2.139}$$

where $\Delta = \sqrt{\frac{3}{2}}\beta_{101}$ and $\beta_3 = -\sqrt{3}\beta_{121}$. Here superscripts $(+)$ and $(-)$ stand for right and left circularly polarized light, respectively. In this case, atomic orientation changes not only the differential cross section but also the total cross section. On the other side, the additional term in the differential cross section has exactly the same angular dependence as the conventional term. Therefore, this term changes the value of the asymmetry parameter only.

Finally, if the target atom is oriented perpendicular to the beam, $\mathbf{J}_0 \perp \mathbf{p}_\gamma$, the angular distribution of photoelectrons has the form

$$\frac{d\sigma^{(\pm)}}{d\Omega} = \frac{\sigma^{(iso)}}{4\pi}$$
$$\times \left(1 - \frac{\beta}{2}P_2(\cos \vartheta_e) + \overline{\mathcal{A}}_{10}\left[\pm\frac{3}{4}\beta_3 \sin 2\vartheta_e \cos \varphi_e - \frac{1}{2}\beta_2 \sin 2\vartheta_e \sin \varphi_e \right] \right) \tag{2.140}$$

Again the additional term is proportional to the target orientation. It contains two contributions: one is independent of the circular polarization of the photon, the other changes sign in transition from left to right circularly polarized photons.

The coefficient β_2 also governs the angular distribution of photoelectrons produced by unpolarized light when the target atom is oriented perpendicular to the beam:

$$\frac{d\sigma}{d\Omega} = \frac{\sigma^{(iso)}}{4\pi} \left(1 - \frac{\beta}{2} P_2(\cos\vartheta_e) - \overline{\mathcal{A}}_{10} \frac{1}{2}\beta_2 \sin 2\vartheta_e \sin\varphi_e \right) \tag{2.141}$$

Here again the additional term in the angular distribution disappears if any two of the three vectors $\mathbf{A}$, $\mathbf{p}_\gamma$, and $\mathbf{p}_e$ are collinear or all three lie in one plane.

Comparing the angular distributions of photoelectrons from oriented targets given by Eqs. (2.137) and (2.139)–(2.141) with that for an unpolarized target, Eq. (2.133), we see that experiments with oriented targets $\left(J_0 = \frac{1}{2}\right)$ can give three additional observable parameters that are determined by the photoionization amplitudes. Therefore, such experiments yield more information about the process of photoionization.

2.3.2.4. Aligned Target Atom with $J_0 = 1$

If the target atoms have $J_0 = 1$, they can be not only oriented but also aligned; their alignment is characterized by the second-rank statistical tensor $\overline{\mathcal{A}}_{20}$. More terms can thus contribute to the angular distribution of photoelectrons. To illustrate the influence of the alignment, we consider again some representative cases.

If the target is aligned ($k_0 = 2$) and the photon beam is linearly polarized ($k_\gamma = 0, 2$), then according to conditions (2.128) and (2.129), the following anisotropy coefficients can be nonzero: β_{022}, β_{220}, β_{202}, β_{222}, and β_{242}. The first coefficient is proportional to the usual anisotropy coefficient β, Eq. (2.135). Other coefficients enter the terms that modify the angular distribution owing to the presence of the initial alignment. Suppose that the atom is aligned along the photon polarization vector ($\mathbf{A} \parallel \mathbf{E}$). Then the angular distribution of photoelectrons is axially symmetric with respect to the photon polarization that we choose to be the z-axis. The angular distribution can be presented in the form

$$\left(\frac{d\sigma}{d\Omega}\right)^{\parallel} = \frac{\sigma^{(iso)}}{4\pi} \left(1 + \beta P_2(\cos\vartheta_e) + \overline{\mathcal{A}}_{20}[a_0^{\parallel} + a_2^{\parallel}\cos^2\vartheta_e + a_4^{\parallel}\cos^4\vartheta_e] \right) \tag{2.142}$$

Coefficients $a_i^{\parallel}$ are linear combinations of the above-mentioned $\beta_{k_0 k k_\gamma}$ generalized anisotropy parameters. The term in square brackets appears as a result of the alignment of the target. It is symmetric with respect to $\vartheta_e = 90°$, as it should be from symmetry considerations.

Now consider the case where the alignment is perpendicular to the photon polarization, $A \perp E$. We assume for simplicity that the photoelectron detector is moving in the plane containing both vectors A and E. Then the angular distribution has a form similar to (2.142):

$$\left(\frac{d\sigma}{d\Omega}\right)^{\perp} = \frac{\sigma^{(\text{iso})}}{4\pi}\left(1 + \beta P_2(\cos\vartheta_e) + \overline{\mathcal{A}}_{20}[a_0^{\perp} + a_2^{\perp}\cos^2\vartheta_e + a_4^{\perp}\cos^4\vartheta_e]\right)$$

(2.143)

but the coefficients in the square brackets are different from those in Eq. (2.142). Naturally, the angular distribution in this case when considered out of the plane is not axially symmetrical.

2.3.3. Circular and Linear Dichroism in Photoelectron Emission

Studying the angular distributions of photoelectrons gives the generalized anisotropy coefficients, $\beta_{k_0 k k_\gamma}$, which contain dynamic information on the photoionization amplitudes. In principle, a study of the angular distribution of photoelectrons from polarized atoms provides sufficient information to obtain experimental values of the amplitudes and their relative phase shifts, i.e., it constitutes a *complete experiment* (or *perfect experiment*) [23, 24]. However, in practice it is difficult to extract all $\beta_{k_0 k k_\gamma}$ coefficients from the experiment. One way to overcome the difficulty is to use a particular geometry in which the contributions of some terms disappear (some examples were discussed in the preceding section). The other way is to consider the difference between two angular distributions corresponding to two light polarizations or two directions of the atomic polarizations. This type of experiment is usually referred to as studying *dichroisms* of different kinds in the angular distribution of photoelectrons. Expressions for all kind of dichroisms can be obtained from the general equations of Section 2.3.1.

Consider, for example, the difference between the intensities of photoelectrons ejected at a definite angle by right and left circularly polarized light; this is called *circular dichroism* in the angular distribution (CDAD). For states with angular momentum $J_0 = \frac{3}{2}$ only five of thirteen parameters $\beta_{k_0 k k_\gamma}$ contribute to CDAD. In the simplest case of $J_0 = 1/2$, according to Eqs. (2.139) and (2.140), the CDAD is determined only by the parameter β_3 if the target atom is oriented perpendicular to the photon beam:

$$\frac{d\sigma^{(+)}}{d\Omega} - \frac{d\sigma^{(-)}}{d\Omega} = \frac{\sigma^{(\text{iso})}}{4\pi}\overline{\mathcal{A}}_{10}\frac{3}{2}\beta_3\sin 2\vartheta_e \cos\varphi_e$$

(2.144)

and by two parameters Δ and β_3 if it is oriented along the beam:

$$\frac{d\sigma^{(+)}}{d\Omega} - \frac{d\sigma^{(-)}}{d\Omega} = \frac{\sigma^{(iso)}}{4\pi} 2\overline{\mathcal{A}}_{10} [\Delta + \beta_3 P_2(\cos\vartheta_e)] \qquad (2.145)$$

Another possibility is to measure the difference in photoelectron currents produced by linearly polarized light with two mutually perpendicular polarizations, which is called *linear dichroism* in the angular distribution (LDAD). As an example we can consider a linear dichroism for the case of an aligned $J_0 = 1$ target atom. Defining the linear dichroism as the difference in the intensities of photoelectrons for parallel and perpendicular mutual orientations of photon polarization vector and alignment direction, we have from Eqs. (2.142) and (2.143):

$$\left(\frac{d\sigma}{d\Omega}\right)^{\parallel} - \left(\frac{d\sigma}{d\Omega}\right)^{\perp} = \frac{\sigma^{(iso)}}{4\pi} \overline{\mathcal{A}}_{20} [a_0 + a_2 \cos^2\vartheta_e + a_4 \cos^4\vartheta_e] \qquad (2.146)$$

We note that the detector is rotating in the plane containing the photon polarization and target alignment directions. The coefficients $a_i = a_i^{\parallel} - a_i^{\perp}$ are introduced in Eq. (2.146), where $a_i^{\parallel}$ and $a_i^{\perp}$ are from Eqs. (2.142) and (2.143). LDAD is proportional to the alignment of the initial state and is symmetric with respect to $\vartheta_e = 90°$.

Sometimes it is more convenient to invert not the light polarization, but the direction of the target polarization. The corresponding difference of the cross sections is called *magnetic dichroism* (MDAD). For example, for linearly polarized light and an oriented $J_0 = 1/2$ target with $\mathbf{J}_0 \perp \mathbf{E}$ [see Eq. (2.137)], MDAD is expressed as

$$\frac{d\sigma^{\uparrow}}{d\Omega} - \frac{d\sigma^{\downarrow}}{d\Omega} = \frac{\sigma^{(iso)}}{4\pi} 2\overline{\mathcal{A}}_{10} \beta_2 \sin 2\vartheta_e \sin\varphi_e \qquad (2.147)$$

Here MDAD is proportional to the atomic orientation and its value is determined by one generalized anisotropy coefficient, β_2, only.

2.3.4.   Spin Polarization of Photoelectrons

One can suggest several approaches for deriving the spin polarization of the reaction products. In Section 2.2.1 we used the straightforward method of calculating the density matrix of the excited atom by taking a trace of the total density matrix of the system after scattering over the unobserved variables. Another method was used in Section 2.2.2, where we obtained the spin polarization of scattered electrons by calculating the average value of the Pauli operators (spin-tensors). In this section we use the third method, which is useful mainly for photons or spin-1/2 particles and is based on the physical characteristics of the components of the polarization vector for these simple cases. Thus, the electron spin polarization component P_a along some direction **a** can be considered as a ratio of the difference of the number of electrons having spin projections $+1/2$ (N_1) and $-1/2$ (N_2) and of the total number of electrons. For an ideal detector, N_1 and N_2 are proportional to the probabilities of registration of the corresponding electrons by a spin-sensitive detector:

$$P_a = \frac{N_1 - N_2}{N_1 + N_2} = \frac{W(+1/2) - W(-1/2)}{W(+1/2) + W(-1/2)} \tag{2.148}$$

In order to calculate the probabilities, we can use Eq. (2.115), where the efficiency tensor should be taken for the spin-sensitive detector, Eq. (1.164). In this way one can obtain, in principle, the general expression for the spin polarization of photoelectrons emitted in a certain direction from a polarized target atom by an arbitrarily polarized photon. This expression is, however, too complicated for analysis. In this book we analyze the simple case of an unpolarized target only. For an unpolarized target, the statistical tensor of the initial state is given by Eq. (1.49):

$$\rho_{k_0 q_0}(\alpha_0 J_0) = (\hat{J}_0)^{-1} \delta_{k_0 0} \delta_{q_0 0} \tag{2.149}$$

Inserting it into Eq. (2.117), we have the following statistical tensor of the final state:

$$\rho_{kq}^f(\alpha J, \alpha' J') = (-1)^{1+J+k+J_0} (\hat{J}_0)^{-2} \begin{Bmatrix} 1 & J & J_0 \\ J' & 1 & k \end{Bmatrix} \rho_{kq}^\gamma M_{ljJ} M_{l'j'J'}^* \tag{2.150}$$

[see the note after Eq. (2.117) on the normalization of the statistical tensor]. The efficiency tensor for detecting an electron in the final state while the photoion is not observed was obtained in Section 1.5.3, Eq. (1.169). Now we choose the laboratory frame with the z-axis directed along the photon beam ($z \parallel \mathbf{p}_\gamma$) while the xz-plane contains the photoelectron emission direction $\mathbf{p}_e$ (reaction frame). If we consider the spin polarization of the photoelectron in the same reference frame, then the efficiency tensor (1.169) with $D_{q_s q_s'}^{k_s}(0,0,0) = \delta_{q_s q_s'}$ can be written as

$$\varepsilon_{kq}\left(\alpha_f J_f, lj : J; \alpha_f J_f, l'j' : J'\right) = \frac{1}{\sqrt{4\pi}} (-1)^{l'} \hat{l} \hat{l'} \hat{j} \hat{j'} \hat{J} \hat{J'}$$

$$\times \sum_{k_l k_s} (-1)^{j'+J+J_f+k} \hat{k}_s \, (l0, l'0 | k_l 0) \begin{Bmatrix} J & j & J_f \\ j' & J' & k \end{Bmatrix} \begin{Bmatrix} l & \frac{1}{2} & j \\ l' & \frac{1}{2} & j' \\ k_l & k_s & k \end{Bmatrix}$$

$$\times \sum_{q_l q_s} (k_l q_l, k_s q_s | kq) \varepsilon_{k_s q_s}\left(\frac{1}{2}\right) Y_{k_l q_l}^*(\vartheta_e, \varphi_e = 0) \tag{2.151}$$

Substituting expression (2.150) for $\rho_{kq}^f(\alpha J, \alpha' J')$ in Eq. (2.115) and using Eq. (2.151) for the efficiency tensor of a spin-sensitive detector of electrons, one obtains:

$$
W = \frac{1}{\sqrt{4\pi}} \sum_{ll'jj'JJ'kq} (-1)^{1-J_0+J_f+j'+l'} \hat{l}\hat{l}' \hat{j}\hat{j}' \hat{J}\hat{J}' (\hat{J}_0)^{-2} \begin{Bmatrix} 1 & J & J_0 \\ J' & 1 & k \end{Bmatrix} M_{ljJ} M^*_{l'j'J'}
$$

$$
\times \rho_{kq}^\gamma \sum_{k_l k_s} \hat{k}_s \, (l\,0, l'\,0 \,|\, k_l\, 0) \begin{Bmatrix} J & j & J_f \\ j' & J' & k \end{Bmatrix} \begin{Bmatrix} l & \frac{1}{2} & j \\ l' & \frac{1}{2} & j' \\ k_l & k_s & k \end{Bmatrix}
$$

$$
\times \sum_{q_l q_s} (k_l\, q_l, k_s\, q_s \,|\, k\, q) \, \varepsilon_{k_s q_s}\left(\tfrac{1}{2}\right) Y^*_{k_l q_l}(\vartheta_e, \varphi_e = 0) \tag{2.152}
$$

We assume further that the photon beam is also unpolarized; then the nonzero components of the photon statistical tensors are $\rho_{00}^\gamma = 1/\sqrt{3}$ and $\rho_{20}^\gamma = 1/\sqrt{6}$ only. Now it is easy to show that the only component of the photoelectron spin polarization is P_y, i.e., the electron spin is perpendicular to the reaction plane containing $\mathbf{p}_\gamma$ and $\mathbf{p}_e$. For example, if a detector selects electrons with spins along the z-axis, then in Eq. (1.164) for $\varepsilon_{k_s q_s}\left(\tfrac{1}{2}\right)$, one substitutes $P_z = 1, P_x = P_y = 0$, and gets $\varepsilon_{k_s q_s}\left(\tfrac{1}{2}\right) = \sqrt{2}\delta_{k_s 1}\delta_{q_s 0}$. Owing to parity conservation in photoabsorption, l and l' should be of the same parity and k_l is even. Taking into account that k is also even, we see that the corresponding Clebsch–Gordan coefficient $(k_l\, 0, k_s\, 0 \,|\, k\, 0)$ in Eq. (2.152) is zero. Therefore, the probability of finding electrons polarized along the z-axis is zero. The same is valid for the x-component of the polarization vector.

The formally obtained result that the photoelectron spin is directed perpendicular to the reaction plane is consistent with a simple symmetry consideration. In fact, the considered experiment is characterized by two vectors — the initial photon and the final electron momenta ($\mathbf{p}_\gamma$ and $\mathbf{p}_e$). Spin is an axial vector. The only axial vector that can be constructed from the two vectors is their vectorial product $\mathbf{p}_\gamma \times \mathbf{p}_e$. Therefore, the spin polarization of photoelectrons may be directed only along this vector, i.e., perpendicular to the reaction plane.

The value of the P_y component can be easily obtained from Eq. (2.152), taking into account expressions (1.164) for the efficiency tensors of electron spin $\varepsilon_{k_s q_s}\left(\tfrac{1}{2}\right)$ and the definition

$$
P_y = \frac{W(P_y = 1) - W(P_y = -1)}{W(P_y = 1) + W(P_y = -1)} \tag{2.153}
$$

where $W(P_y = \pm 1)$ is the probability (2.152) calculated with $P_x = P_z = 0, P_y = \pm 1$ in the spin part of the efficiency tensors $\varepsilon_{k_s q_s}\left(\tfrac{1}{2}\right)$. The resulting expression can be presented in the following form:

$$P_y = \frac{\xi \sin 2\vartheta}{1 + \beta P_2(\cos \vartheta)} \tag{2.154}$$

where β is the anisotropy parameter (2.135) and ξ is expressed in terms of photoionization amplitudes as follows:

$$\xi = 3\sqrt{\frac{15}{2}} \sum_{ljJ \leq l'j'J'} (-1)^{j'+l-J_f+J_0+1} \hat{l}\hat{l}' \hat{j}\hat{j}' \hat{J}\hat{J}' \, (l0,l'0|20) \begin{Bmatrix} j & j' & 2 \\ J' & J & J_f \end{Bmatrix}$$

$$\times \begin{Bmatrix} J' & 1 & J_0 \\ 1 & J & 2 \end{Bmatrix} \begin{Bmatrix} l & \frac{1}{2} & j \\ l' & \frac{1}{2} & j' \\ 2 & 1 & 2 \end{Bmatrix} \mathrm{Im}(M_{ljJ} M^*_{l'j'J'}) \Big/ \left(\sum_{ljJ} |M_{ljJ}|^2 \right) \tag{2.155}$$

From this expression it follows that spin polarization of photoelectrons arises from the interference of partial photoionization amplitudes. It vanishes if according to selection rules, only one partial wave contributes to the cross section. For example, in a nonrelativistic consideration in photoionization of the ns subshell, only p-wave photoelectrons are emitted and their spin polarization is zero. Finally, the spin polarization is determined by the sine of the phase differences between the photoionization amplitudes. Thus it gives some additional information in comparison with the photoelectron angular distribution, which depends on the cosine of the phase differences.

2.3.5. Polarization Parameters of Photoions

Similarly to photoelectrons, the photoions produced in atomic photoionization can be polarized also. Their polarization state influences the angular distribution and polarization of the subsequent Auger electron emission (if inner shells are photoionized) or fluorescence. Both the anisotropy and polarization of the photoinduced Auger electrons and fluorescence were widely investigated experimentally. The interpretation of these experiments is usually based on the *two-step approximation* in which excitation and decay of the atomic system are considered as independent events. As will be shown in Chapter 3, the angular distribution and polarization of the decay products are determined by the polarization state of the decaying atom or ion. Thus knowledge of the statistical tensors of the photoions is essential for the theoretical description of these processes.

In order to obtain the statistical tensors of the photoion, we can start again from expression (2.115). As earlier, the statistical tensors of the system in the final state are determined by Eq. (2.117). However, now the efficiency tensor is different since in the case considered the photoelectron is not detected. The general expression for the efficiency tensor of the final system photoion + photoelectron is given by Eq. (1.162):

$$\varepsilon_{kq}\left(\alpha_f J_f, lj : J; \alpha_f J_f, l'j' : J'\right) = \sum_{k_f q_f k_e q_e} \hat{j}\hat{j}'\hat{k}_f\hat{k}_e \left(k_f q_f, k_e q_e \,|\, kq\right) \begin{Bmatrix} J_f & j & J \\ J_f & j' & J' \\ k_f & k_e & k \end{Bmatrix}$$

$$\times \varepsilon_{k_f q_f}\left(\alpha_f J_f\right) \varepsilon_{k_e q_e}\left(lj, l'j'\right) \tag{2.156}$$

When the photoelectron is not detected, the efficiency tensor for the electron can be written as

$$\varepsilon_{k_e q_e}\left(j, j'\right) = \hat{j}\delta_{k_e 0}\delta_{q_e 0}\delta_{jj'} \tag{2.157}$$

Inserting Eq. (2.157) into Eq. (2.156) we obtain

$$\varepsilon_{kq}\left(\alpha J, \alpha' J'\right) = (-1)^{J'+J_f+k+j}\hat{j}\hat{j}' \begin{Bmatrix} J & J_f & j \\ J_f & J' & k \end{Bmatrix} \varepsilon_{kq}\left(\alpha_f J_f\right) \tag{2.158}$$

Here we assumed that the photoion is characterized by a sharp angular momentum J_f. Now we can combine Eqs. (2.115), (2.117), and (2.158) and obtain the probability of registering any state of the photoion as follows:

$$W = \sum_{\alpha\alpha'JJ'kq} \sum_{k_0 q_0 k_\gamma q_\gamma} (-1)^{J'+J_f+k+j}\hat{k}_0\hat{k}_\gamma\hat{j}\hat{j}' \left(k_0 q_0, k_\gamma q_\gamma \,|\, kq\right)$$

$$\times \begin{Bmatrix} J & J_f & j \\ J_f & J' & k \end{Bmatrix} \begin{Bmatrix} J_0 & 1 & J \\ J_0 & 1 & J' \\ k_0 & k_\gamma & k \end{Bmatrix} \overline{\rho}_{k_0 q_0}\left(\alpha_0 J_0\right)\rho^\gamma_{k_\gamma q_\gamma}\varepsilon^*_{kq}\left(\alpha_f J_f\right)M_{ljJ}M^*_{l'jJ'} \tag{2.159}$$

Comparing this expression with the standard relation:

$$W = \sum_{kq}\rho_{kq}\left(\alpha_f J_f\right)\varepsilon^*_{kq}\left(\alpha_f J_f\right) \tag{2.160}$$

we obtain the following expression for the statistical tensor of the photoion:

$$\rho_{kq}\left(\alpha_f J_f\right) = \sqrt{3}\hat{j}_0\left(\sum_{ljJ}|M_{ljJ}|^2\right)^{-1}\sum_{k_0 k_\gamma} A_{k_0 k k_\gamma}$$

$$\times \overline{P}_{k_0 0}(\alpha_0 J_0)\sum_{q_0 q_\gamma}\left(k_0 q_0, k_\gamma q_\gamma \,|\, kq\right)\rho^\gamma_{k_\gamma q_\gamma}\frac{\sqrt{4\pi}}{\hat{k}_0}Y^*_{k_0 q_0}\left(\vartheta_a, \varphi_a\right) \tag{2.161}$$

where

$$A_{k_0 k k_\gamma} = \sqrt{3} \hat{J}_0 \hat{k}_0 \hat{k}_\gamma \sum_{ljJJ'} (-1)^{J'+J_f+k+j} \hat{f} \hat{f}'$$

$$\times \left\{ \begin{matrix} J & J_f & j \\ J_f & J' & k \end{matrix} \right\} \left\{ \begin{matrix} J_0 & 1 & J \\ J_0 & 1 & J' \\ k_0 & k_\gamma & k \end{matrix} \right\} M_{ljJ} M_{ljJ'}^* \qquad (2.162)$$

Here we have used Eq. (2.119) for the statistical tensor of the initial polarized atomic state. The statistical tensor (2.161) is normalized in the standard way (1.49).

Equation (2.161) can be obtained by another standard approach by applying the convolution formula (1.69) to the statistical tensor of the final state in photoionization given by Eq. (2.117).

Equations (2.161) and (2.162) give the general expression for the statistical tensors of a photoion produced by an arbitrarily polarized photon from a polarized initial atomic state. In the next section we consider some particular cases, but one comment can be made in the general case. One can see that in expression (2.162), partial waves of the electron enter incoherently (i.e., $l = l'$, $j = j'$); thus there is no interference between different partial waves. This is a clear consequence of the integration over all of the directions of electron emission (the photoelectron is not detected) and corresponds to a taking of the trace over quantum numbers of an unobserved system.

2.3.5.1. Unpolarized Target Atom

In the case of an unpolarized target atom, Eq. (2.161) can be substantially simplified. In fact, for an unpolarized target, $k_0 = q_0 = 0$ and $\bar{\rho}_{k_0 0}(\alpha_0 J_0) = 1/\hat{J}_0$. Then $k = k_\gamma, q = q_\gamma$ and

$$\rho_{kq}(\alpha_f J_f) = \sqrt{3} \left(\sum_{ljJ} |M_{ljJ}|^2 \right)^{-1} A_{0kk} \rho_{kq}^\gamma \qquad (2.163)$$

where

$$A_{0kk} = \sqrt{3} \sum_{ljJJ'} (-1)^{J+J'+j+J_f+J_0+1} \hat{f} \hat{f}' \left\{ \begin{matrix} J & J_f & j \\ J_f & J' & k \end{matrix} \right\} \left\{ \begin{matrix} 1 & J & J_0 \\ J' & 1 & k \end{matrix} \right\} M_{ljJ} M_{ljJ'}^* \qquad (2.164)$$

We see that the polarization state of the photoion is completely determined by the polarization state of the photon beam and their statistical tensors are proportional. This can be easily understood in the context of conservation of the tensorial structure for the set of the statistical tensors of the whole system. A similar situation was discussed for atomic excitation by electrons in Section 2.2.1.4. As a result, in photoionization of an atom by an unpolarized or linearly polarized beam, the photoion can only be aligned, not oriented, and the rank of the tensor cannot be larger than 2 (in dipole approximation). Circularly polarized light can produce orientation of the photoion.

An especially simple expression is obtained for an initial atom with $J_0 = 0$. In this case $J = J' = 1$ and only squares of the dipole amplitudes determine the statistical tensors:

$$A_{0kk} = \sqrt{3} \sum_{lj} (-1)^{j+J_f+k+1} \begin{Bmatrix} 1 & J_f & j \\ J_f & 1 & k \end{Bmatrix} |M_{lj1}|^2 \qquad (2.165)$$

2.3.5.2. Polarized Target Atom

If the target atom is polarized (oriented and/or aligned), then in general the photoion is both aligned and oriented for any type of photon polarization and even for unpolarized light. However, the alignment and orientation of the photoion are determined by different parts of the photoionization amplitudes. Moreover, in some kinematic conditions, alignment or orientation can vanish. To see it clearer, consider Eq. (2.162) for $A_{k_0kk_\gamma}$ coefficients. If we interchange the angular momenta J and J', each term in the sum over JJ' gets an additional phase $(-1)^{k_0+k+k_\gamma}$. Therefore Eq. (2.162) can be rewritten in the following form:

$$A_{k_0kk_\gamma} = \sqrt{3}\,\hat{J}_0\hat{k}_0\hat{k}_\gamma \sum_{ljJJ'} (-1)^{J'+J_f+k+j} \hat{J}\hat{J}' \begin{Bmatrix} J & J_f & j \\ J_f & J' & k \end{Bmatrix} \begin{Bmatrix} J_0 & 1 & J \\ J_0 & 1 & J' \\ k_0 & k_\gamma & k \end{Bmatrix}$$

$$\times \frac{1}{2} \Bigg([1+(-1)^{k_0+k+k_\gamma}] \operatorname{Re}(M_{ljJ}M^*_{ljJ'})$$

$$+ [1-(-1)^{k_0+k+k_\gamma}]\, i\, \operatorname{Im}(M_{ljJ}M^*_{ljJ'}) \Bigg) \qquad (2.166)$$

If $k_0 + k + k_\gamma$ is even, only the real part of the product $M_{ljJ}M^*_{ljJ'}$ determines the statistical tensor of the photoion, while for odd values of $k_0 + k + k_\gamma$, the imaginary part of $M_{ljJ}M^*_{ljJ'}$ is essential. Consider, for example, an aligned target ($k_0 = $ even) ionized by linearly polarized light ($k_\gamma = $ even). The alignment of the photoion is determined by $\operatorname{Re}(M_{ljJ}M^*_{ljJ'})$, but orientation is produced only if there is a phase difference between amplitudes M_{ljJ} and $M_{ljJ'}$. If the target is oriented ($k_0 = $ odd) and the photon beam is linearly polarized, then the orientation of the photoion is always proportional to the orientation of the target. The alignment of the photoion is influenced by the target orientation only if $\operatorname{Im}(M_{ljJ}M^*_{ljJ'}) \neq 0$. Other cases can be analyzed in a similar way.

Equation (2.161) also provides the possibility of predicting the dependence of the photoion statistical tensors on the kinematic conditions. For illustration we consider alignment of the photoion produced by the circularly polarized light from an aligned target ($k_0 = 2$). According to the above consideration, there is a circular dichroism in this case, i.e., the photoion alignment is different for right and left circularly polarized light. The term responsible for the circular dichroism contains the coefficient A_{221}. However, the circular dichroism vanishes if the alignment axis coincides with the photon beam direction. In fact, in this case $q_0 = q_\gamma = 0$ (the z-axis is along the beam), and the corresponding Clebsch–Gordan coefficient $(20, 10 | 20)$ in Eq. (2.161) is zero.

2.3.5.3. Differential Statistical Tensors of Photoions

Expression (2.161) gives the statistical tensors of the photoion when the photoelectron is not detected. However, in modern experiments, the photoelectron is often registered at a certain direction in coincidence with the ion decay products. For interpretation of such experiments, it is desirable to know the statistical tensors of the photoion differential with respect to the direction of the photoelectron ejection. To derive the necessary expression, one can use the same procedure as described above [Eqs. (2.156)–(2.161)]. The only difference is that instead of Eq. (2.157) for the efficiency tensor of a photoelectron, one has to use the general expression (1.104) for an electron moving in the direction (ϑ_e, φ_e) (we assume that the detector is not sensitive to the electron spin polarization). Equivalently, one can use Eq. (1.155) to convolute the statistical tensor of the final state (2.117) with the efficiency tensor of an unpolarized photoelectron [see Eq. (1.104)]. Leaving this derivation to the reader, we consider a simpler problem that is encountered in practice more often, namely, the particular case of an unpolarized target atom.

Starting again from expression (2.115) for probability, we substitute Eq. (2.117) for the statistical tensor of the final state. However, here we take into account that the initial atom is unpolarized, thus $\rho_{k_0 q_0} (\alpha_0 J_0) = \hat{J}_0^{-1} \delta_{k_0 0} \delta_{q_0 0}$ and therefore

$$\rho_{kq}^f (\alpha J, \alpha' J') = (-1)^{1+J+J_0+k} \hat{J}_0^{-2} \begin{Bmatrix} 1 & J & J_0 \\ J' & 1 & k \end{Bmatrix} \rho_{kq}^\gamma M_{ljJ} M_{l'j'J'}^* \qquad (2.167)$$

On the other hand, the efficiency tensor includes both the photoion and the electron, [Eq. (1.162)] where the electron efficiency tensor corresponds to an angle-resolved measurement by a spin-insensitive detector [Eqs. (1.104) and (1.105)]:

$$\varepsilon_{k_e q_e} (lj, l'j') = \frac{(-1)^{j'+1/2}}{\sqrt{4\pi}} \hat{k}_e^{-1} \hat{l} \hat{l}' \hat{j} \hat{j}' \, (l0, l'0 | k_e 0)$$

$$\times \begin{Bmatrix} j & l & \frac{1}{2} \\ l' & j' & k_e \end{Bmatrix} Y_{k_e q_e}^* (\vartheta_e, \varphi_e) \qquad (2.168)$$

Substituting Eq. (2.167) and Eq. (1.162) with Eq. (2.168) into Eq. (2.115) and comparing the result with Eq. (2.160), we get the following expression for the differential statistical tensor of a photoion:

$$\rho_{k_f q_f}(\alpha_f J_f; \mathbf{n}_e) = \frac{1}{3(2J_0+1)} \sum_{k_e q_e k q} \sqrt{\frac{4\pi}{2k_e+1}} \, (k_f q_f, k_e q_e | k q)$$

$$\times Y_{k_e q_e}(\vartheta_e, \varphi_e) \rho_{kq}^{\gamma} B(k_e, k_f, k) \tag{2.169}$$

where coefficients $B(k_e, k_f, k)$ are determined as

$$B(k_e, k_f, k) = \frac{3}{4\pi} \hat{k}_f \hat{k}_e \sum_{ll'jj'JJ'} \hat{l}\hat{l}'\,\hat{j}\hat{j}'\,\hat{J}\hat{J}'(-1)^{J+J_0+k+j'-1/2}$$

$$\times (l0, l'0 | k_e 0) \begin{Bmatrix} j & l & \frac{1}{2} \\ l' & j' & k_e \end{Bmatrix} \begin{Bmatrix} 1 & J & J_0 \\ J' & 1 & k \end{Bmatrix} \begin{Bmatrix} J_f & j & J \\ J_f & j' & J' \\ k_f & k_e & k \end{Bmatrix} M_{ljJ} M_{l'j'J'}^* \tag{2.170}$$

For any particular transition, the number of coefficients $B(k_e, k_f, k)$ is limited by the conditions $k_f \leq 2J_f$, $k \leq 2$, $|k_f - k| \leq k_e \leq (k_f + k)$.

2.3.6. Direct Double Photoionization

Consider the direct double photoionization of an atom:

$$\gamma + A(\alpha_0 J_0) \longrightarrow A^{2+}(\alpha_f J_f) + e_1 + e_2$$

We assume that the initial atom is unpolarized and the final doubly ionized atom has a sharp angular momentum J_f. Two ejected electrons are assumed to be detected in coincidence, and the angular correlation between them is studied. The formalism of statistical tensors permits one to derive the double ionization cross section in exactly the same way as that used for a single ionization. The only difference is that here we use the efficiency tensor, which describes the detection of two electrons. We assume also that the detectors are not sensitive to spin polarization of the electrons and no polarization characteristics of the ion $A^{2+}(\alpha_f, J_f)$ are measured.

The efficiency tensor of the total system $A^{2+}(\alpha_f, J_f) + e_1 + e_2$ is calculated according to the general rule (1.149) using the efficiency tensor of the ion $\varepsilon_{k_f q_f}(\alpha_f J_f, \alpha_f J_f) = \hat{J}_f \delta_{k_f 0} \delta_{q_f 0}$ and that of the electron detecting system

$\varepsilon_{kq}(l_1 j_1, l_2 j_2 : j; l_1' j_1', l_2' j_2' : j)$ given by Eq. (1.179):

$$\varepsilon_{kq}\left(\alpha_f J_f, (l_1 j_1, l_2 j_2 : j) : J; \alpha_f J_f, (l_1' j_1', l_2' j_2' : j') : J'\right)$$

$$= \frac{1}{4\pi}(-1)^{J+j'+J_f+k+j_1'+j_2'+1} \hat{l}_1 \hat{l}_1' \hat{l}_2 \hat{l}_2' \hat{j}_1 \hat{j}_1' \hat{j}_2 \hat{j}_2' \hat{j} \hat{j}' \hat{J} \hat{J}'$$

$$\times \begin{Bmatrix} J & j & J_f \\ j' & J' & k \end{Bmatrix} \sum_{k_1 k_2} (l_1 0, l_1' 0 | k_1 0)\, (l_2 0, l_2' 0 | k_2 0) \begin{Bmatrix} j_1 & j_2 & j \\ j_1' & j_2' & j' \\ k_1 & k_2 & k \end{Bmatrix}$$

$$\times \begin{Bmatrix} l_1 & j_1 & \frac{1}{2} \\ j_1' & l_1' & k_1 \end{Bmatrix} \begin{Bmatrix} l_2 & j_2 & \frac{1}{2} \\ j_2' & l_2' & k_2 \end{Bmatrix} \{Y_{k_1}(\vartheta_1, \varphi_1) \otimes Y_{k_2}(\vartheta_2, \varphi_2)\}_{kq}^* \qquad (2.171)$$

Starting from the general equation (2.115) and taking into account Eq. (2.150) for the statistical tensors of a photoexcited atom in the final state which was initially unpolarized, one can obtain the following expression for a triply differential cross section of double photoionization:

$$\frac{d^3\sigma}{d\Omega_1 d\Omega_2 d\varepsilon_1} = \sum_{k_1 k_2 k q} A_{k_1 k_2 k} 4\pi \{Y_{k_1}(\vartheta_1, \varphi_1) \otimes Y_{k_2}(\vartheta_2, \varphi_2)\}_{kq}\, \rho_{kq}^{\gamma}(P_1, P_2, P_3)$$

$$= \sum_{k_1 k_2 k} A_{k_1 k_2 k} F_{k_1 k_2 k}(\vartheta_1, \varphi_1, \vartheta_2, \varphi_2; P_1, P_2, P_3) \qquad (2.172)$$

where ϑ_1, φ_1, and ϑ_2, φ_2 are the angles of ejection of the two electrons; $\rho_{kq}^{\gamma}(P_1, P_2, P_3)$ is the statistical tensor describing the photon beam; the kinematic factor $F_{k_1 k_2 k}(\vartheta_1, \varphi_1, \vartheta_2, \varphi_2; P_1, P_2, P_3)$ is determined by Eq. (2.123); and the dynamic coefficients $A_{k_1 k_2 k}$ are given by the following equation:

$$A_{k_1 k_2 k} = \frac{1}{(4\pi)^2} \sum_{\substack{JJ' l_1 l_1' l_2 l_2' \\ j_1 j_1' j_2 j_2' jj'}} (-1)^{j'+j_1'+j_2'+J_f-J_0}(\hat{J}_0)^{-2} \hat{l}_1 \hat{l}_1' \hat{l}_2 \hat{l}_2' \hat{j}_1 \hat{j}_1' \hat{j}_2 \hat{j}_2' \hat{j} \hat{j}' \hat{J} \hat{J}'$$

$$\times (l_1 0, l_1' 0 | k_1 0)\, (l_2 0, l_2' 0 | k_2 0) \begin{Bmatrix} 1 & J & J_0 \\ J' & 1 & k \end{Bmatrix} \begin{Bmatrix} J & j & J_f \\ j' & J' & k \end{Bmatrix}$$

$$\times \begin{Bmatrix} l_1 & j_1 & \frac{1}{2} \\ j_1' & l_1' & k_1 \end{Bmatrix} \begin{Bmatrix} l_2 & j_2 & \frac{1}{2} \\ j_2' & l_2' & k_2 \end{Bmatrix} \begin{Bmatrix} j_1 & j_2 & j \\ j_1' & j_2' & j' \\ k_1 & k_2 & k \end{Bmatrix}$$

$$\times \langle J_f(j_1 j_2)j : J \| D \| J_0 \rangle \langle J_f(j_1' j_2')j' : J' \| D \| J_0 \rangle^* \qquad (2.173)$$

The coefficients $A_{k_1 k_2 k}$ contain the double photoionization amplitudes

$$\langle J_f(j_1 j_2)j : J \| D \| J_0 \rangle$$

and we use the angular momentum coupling scheme $\mathbf{J}_f + (\mathbf{j}_1 + \mathbf{j}_2)\mathbf{j} = \mathbf{J}$.

Let us discuss some properties of the angular correlation function. We note that the kinematic part contains bipolar spherical harmonics and formally coincides with that obtained for photoionization of polarized atoms, Eqs. (2.121) and (2.123). However, the physics of the two processes are obviously different and this is reflected in the dynamic coefficients. For example, in single photoionization, owing to parity conservation, the value of k is always even. The photoelectron partial waves are of the same parity. In double ionization, the parity conservation determines the parity of the wave function of the electron pair, i.e., $(l_1 + l_2)$ should be of a certain parity. By inspecting the Clebsch–Gordan coefficients in Eq. (2.173), we conclude that the summation over k_1 and k_2 in Eq. (2.172) is limited by the condition $k_1 + k_2 = $ even.

One interesting property of double photoionization can be easily obtained from the general expression (2.172). If the ionization is produced by circularly polarized light, the angular correlation function is different for the right and left circular polarizations, i.e., there is a circular dichroism in the angular distribution of photoelectrons, even for unpolarized targets. In fact, consider photoionization by circularly polarized light. If the z-axis is chosen along the direction of the photon beam, then only three components of the photon statistical tensor are nonzero: $\rho_{00}^\gamma = 1/\sqrt{3}$, $\rho_{10}^\gamma = P_3/\sqrt{2}$, and $\rho_{20}^\gamma = 1/\sqrt{6}$ (see Table 1.1 on p. 25), where P_3 is the Stokes parameter. The circular dichroism, defined as the difference between cross sections for right ($P_3 = +1$) and left ($P_3 = -1$) circularly polarized light, can now be obtained from Eq. (2.172) as

$$\Delta = \sqrt{2} \sum_{k_1 k_2} 4\pi A_{k_1 k_2 1} \{Y_{k_1}(\vartheta_1, \varphi_1) \otimes Y_{k_2}(\vartheta_2, \varphi_2)\}_{10} \qquad (2.174)$$

By analyzing the Clebsch–Gordan coefficients in $A_{k_1 k_2 1}$, we conclude that $k_1 = k_2$ and the double sum in Eq. (2.174) reduces to

$$\Delta = \sqrt{2} \sum_k 4\pi A_{kk1} \{Y_k(\vartheta_1, \varphi_1) \otimes Y_k(\vartheta_2, \varphi_2)\}_{10} \qquad (2.175)$$

Some general properties of circular dichroism can be obtained from this formula:

1. Circular dichroism disappears if the direction $\mathbf{n}_1 \equiv \{\vartheta_1, \varphi_1\}$ is parallel or antiparallel to $\mathbf{n}_2 \equiv \{\vartheta_2, \varphi_2\}$, or if two vectors $\mathbf{n}_1$, $\mathbf{n}_2$ and the direction of the photon beam are coplanar.

2. The coefficients A_{kk1} are purely imaginary.

3. Integration over one of the directions, $\mathbf{n}_1$ or $\mathbf{n}_2$, leads to $\Delta = 0$.

2.4. Direct Ionization by Particle Impact

In this section we consider the process of direct ionization of an atom by particle impact. This case is more complicated from the point of view of kinematics than the excitation of the atom considered in Section 2.2 since there are three particles in the final state: the ion, the ejected electron, and the scattered particle. A variety of correlation and polarization experiments are possible even when only one particle is detected, especially if the projectile beam and/or the target is polarized. The angular distributions of all three particles are measured as well as their polarization states. (The polarization state of the residual ion, which is often produced in an excited state, can be determined by measuring, for example, the angular distribution or polarization of fluorescence.) Even more sophisticated correlation experiments are performed with the use of coincidence technique, when two or all three of the final state particles are detected. The formalism of density matrix and statistical tensors is especially convenient for the analysis of such complicated experiments.

2.4.1. Density Matrix for Direct Ionization in Fast Electron–Atom Collisions

We begin with a general case of electron-impact ionization. Consider the process in which an initially polarized electron beam ionizes a polarized atomic target:

$$e_i(\mathbf{p}_i) + A(\alpha_0 J_0) \rightarrow e_f(\mathbf{p}_f) + e_1(\mathbf{p}_1) + A^+(\alpha_1 J_1) \qquad (2.176)$$

Before the collision, the target atom is in a state with the total angular momentum J_0 while J_1 denotes the total angular momentum of the residual ion A^+. Sets of other quantum numbers characterizing the atom and the ion are denoted by α_0 and α_1, respectively. The linear momenta of the electrons involved in the process (2.176) are denoted by $\mathbf{p}_i$, $\mathbf{p}_f$, and $\mathbf{p}_1$. We choose the z-axis of the laboratory frame along the direction of the incident beam $\mathbf{p}_i$. Two electrons in continuum are indistinguishable except for their energies. Therefore, one cannot specify which of the electrons, e_f or e_1, is "scattered" or "ejected." Conventionally, that which is of higher energy is treated as scattered. Usually, except for some special cases, it cannot lead to confusion. With this reservation we denote the "scattered" electron by e_f and the "ejected" one by e_1. The energy conservation law

$$E_i = E_f + E_1 + \varepsilon_B \qquad (2.177)$$

relates the kinetic energies of the incoming and outgoing electrons to the binding energy ε_B of the ejected electron in the target atom.

Following the same procedure as in Section 2.2.1.1, we can write the density matrix of the final state in ionization using the spin projection representation:

$$
\begin{aligned}
&\left\langle \alpha_1 J_1 M_1, \mu_f, \mu_1 \,\middle|\, \rho^f(\mathbf{p}_i, \mathbf{p}_f, \mathbf{p}_1) \,\middle|\, \alpha_1 J_1 M_1', \mu_f', \mu_1' \right\rangle \\
&= \sum_{M_0 M_0', \mu_i \mu_i'} T_{M_0 \mu_i \to M_1 \mu_f \mu_1}(\mathbf{p}_i, \mathbf{p}_f, \mathbf{p}_1) \, T^*_{M_0' \mu_i' \to M_1' \mu_f' \mu_1'}(\mathbf{p}_i, \mathbf{p}_f, \mathbf{p}_1) \\
&\quad \times \left\langle \alpha_0 J_0 M_0 \,\middle|\, \rho_A^i \,\middle|\, \alpha_0 J_0 M_0' \right\rangle \left\langle \mu_i \,\middle|\, \rho_e^i \,\middle|\, \mu_i' \right\rangle
\end{aligned}
\tag{2.178}
$$

where ρ_A^i and ρ_e^i are the density operators describing the target atom and the projectile electron, respectively; μ_i, μ_f, and μ_1 are the spin projections of the incoming, scattered, and ejected electrons, respectively; and the ionization amplitude is denoted by

$$
T_{M_0 \mu_i \to M_1 \mu_f \mu_1}(\mathbf{p}_i, \mathbf{p}_f, \mathbf{p}_1) \equiv \left\langle \alpha_1 J_1 M_1, \mathbf{p}_f \mu_f, \mathbf{p}_1 \mu_1 \,\middle|\, T \,\middle|\, \alpha_0 J_0 M_0, \mathbf{p}_i \mu_i \right\rangle
\tag{2.179}
$$

We have suppressed the spin $\frac{1}{2}$ of all electrons from the notations of matrix elements to shorten the expressions. Formally, the density matrix completely describes the ionization process. Any observable in any possible experiment can be expressed in terms of the matrix elements (2.178).

Consider, for example, the double differential cross section $d^2\sigma/dE_f d\Omega_f$ for the energy and angular distribution of scattered electrons in conditions when the ejected electrons are not detected. The scattered electron is described by the density matrix that is the trace of the density matrix (2.178) over all variables characterizing the residual ion and the ejected electron, including integration over all directions of its emission:

$$
\begin{aligned}
&\left\langle \mu_f \,\middle|\, \rho_e^f(\mathbf{p}_i, \mathbf{p}_f) \,\middle|\, \mu_f' \right\rangle = \sum_{M_0 M_0' \mu_i \mu_i' M_1 \mu_1} \int d\Omega_1 \, T_{M_0 \mu_i \to M_1 \mu_f \mu_1}(\mathbf{p}_i, \mathbf{p}_f, \mathbf{p}_1) \\
&\quad \times \left\langle \alpha_0 J_0 M_0 \,\middle|\, \rho_A^i \,\middle|\, \alpha_0 J_0 M_0' \right\rangle \left\langle \mu_i \,\middle|\, \rho_e^i \,\middle|\, \mu_i' \right\rangle T^*_{M_0' \mu_i' \to M_1 \mu_f' \mu_1}(\mathbf{p}_i, \mathbf{p}_f, \mathbf{p}_1)
\end{aligned}
\tag{2.180}
$$

If the detector is not sensitive to spin polarization of the scattered electrons, then the differential cross section can be written as a trace of the density matrix (2.180) over $\mu_f = \mu_f'$:

$$
\begin{aligned}
&\frac{d^2\sigma}{dE_f d\Omega_f}(\vartheta_f, \varphi_f) = \frac{p_f}{p_i} \sum_{M_0 M_0' \mu_i \mu_i' M_1 \mu_1 \mu_f} \int d\Omega_1 \, T_{M_0 \mu_i \to M_1 \mu_f \mu_1}(\mathbf{p}_i, \mathbf{p}_f, \mathbf{p}_1) \\
&\quad \times \left\langle \alpha_0 J_0 M_0 \,\middle|\, \rho_A^i \,\middle|\, \alpha_0 J_0 M_0' \right\rangle \left\langle \mu_i \,\middle|\, \rho_e^i \,\middle|\, \mu_i' \right\rangle T^*_{M_0' \mu_i' \to M_1 \mu_f \mu_1}(\mathbf{p}_i, \mathbf{p}_f, \mathbf{p}_1)
\end{aligned}
\tag{2.181}
$$

The latter expression is valid for arbitrary polarization states of the target atom and the projectile. If the target atoms and the projectiles are unpolarized, Eq. (2.181) simplifies to the following expression:

$$\frac{d^2\sigma}{dE_f d\Omega_f}(\vartheta_f, \varphi_f) = \frac{p_f}{p_i} \frac{1}{2(2J_0+1)}$$

$$\times \sum_{M_0\mu_i M_1\mu_1\mu_f} \int d\Omega_1 |T_{M_0\mu_i \to M_1\mu_f\mu_1}(\mathbf{p}_i, \mathbf{p}_f, \mathbf{p}_1)|^2 \quad (2.182)$$

The polarization of the scattered electrons is given by Eq. (1.92) with account for the normalization:

$$\mathbf{P}_e^f = \text{Tr}(\rho_e^f \boldsymbol{\sigma}) / \text{Tr}(\rho_e^f) \quad (2.183)$$

In complete analogy with the case of excitation considered in Section 2.2.2, one can analyze some general properties of the angular distribution and spin polarization of scattered electrons for the case of ionization.

2.4.2. Angular Distribution of Ejected Electrons

As we noted in the preceding section, the terms "scattered" and "ejected" electrons are rather conventional. Two electrons in the final state are distinguished only by their energies. Therefore, the formal description of the angular distribution and spin polarization given above for the scattered electron is valid also for the ejected electron. However, in practical calculations, different methods are used to describe fast and slow electrons. The latter are often described in the partial wave formalism using to advantage the fact that only a very limited number of partial waves contribute to the cross sections, which makes the calculations feasible. On the other hand, the partial wave expansion gives full advantage in the use of the statistical tensor formalism. In the following paragraphs we apply the angular momentum representation of the density matrix and the statistical tensor formalism in order to describe the angular distribution of ejected electrons in ionization of polarized atoms.

In accordance with the cases of practical interest (see Section 2.3), we assume that the polarized target atom state is axially symmetric with respect to the axis $\mathbf{A}$, which is characterized by two angles (ϑ_a, φ_a) in the laboratory frame. The latter indicates that the angular momentum density matrix of the initial atomic state is diagonal in the atomic frame with the z-axis directed along the axis $\mathbf{A}$. Then the statistical tensor $\rho_{k_0 q_0}(\alpha_0 J_0)$ of the target atom may be determined in the laboratory frame by Eq. (2.119).

Discussing ionization, we can use the results obtained earlier for atomic excitation by electron impact if we formally consider the subsystem $A^+(\alpha_1 J_1) + e_1$ as an excited atomic state with one electron in continuum. The only principal difference between this case and excitation of the discrete atomic state is that the angular momentum of the subsystem $A^+ + e_1$ in the final state is not fixed. Nevertheless, we can use the expression for the statistical tensor of the final atomic state $\rho_{kq}(\alpha J, \alpha' J')$, obtained in Section 2.2.1 [Eq. (2.60)]. In the following discussion we assume that the detector of electrons, positioned at the angles (ϑ_e, φ_e) in the laboratory frame, is not sensitive to spin polarization, and that the projectile electrons are not polarized. This means that $k_s = q_s = 0$ and $k_{s_0} = q_{s_0} = 0$ in Eq. (2.60). In addition, the scattered electron is not detected and therefore $k_l = q_l = 0$. Taking all this into account, one can reduce Eq. (2.60) to the following form with the obvious change of notations:

$$\rho_{kq}(\alpha J, \alpha' J') = \sum_{l_i l_i' j_i j_i' J_t J_t' k_0 k_e l_f j_f} \sum \hat{J}_t \hat{J}_t' \hat{j}_i \hat{j}_i' \hat{k}_0 \hat{k}_e \frac{1}{2} (-1)^{J_t' + J + j_i' + j_f + l_i + 1/2 + k + k_e}$$

$$\times (k_0 q, k_e 0 | k q) \begin{Bmatrix} J_t & J & j_f \\ J' & J_t' & k \end{Bmatrix} \begin{Bmatrix} l_i' & j_i' & \frac{1}{2} \\ j_i & l_i & k_e \end{Bmatrix} \begin{Bmatrix} J_0 & j_i & J_t \\ J_0 & j_i' & J_t' \\ k_0 & k_e & k \end{Bmatrix}$$

$$\times \rho_{k_0 q_0}(\alpha_0 J_0) \rho_{k_e 0}(l_i l_i') \langle \alpha J, l_f j_f : J_t \| T \| \alpha_0 J_0, l_i j_i : J_t \rangle$$

$$\times \langle \alpha' J', l_f j_f : J_t' \| T \| \alpha_0 J_0, l_i' j_i' : J_t' \rangle^* \tag{2.184}$$

To obtain the cross section we have to use the standard Eq. (1.151) with the statistical tensors (2.184) and efficiency tensors (2.120).

Substituting Eq. (2.119) for the statistical tensor of the initial state, and Eq. (1.101) for the statistical tensor of the orbital motion of the initial electron moving along the z-axis, into (2.184) and convoluting it with the relevant efficiency tensor (1.171) yields the final expression, which can be cast in the form

$$\frac{d^2\sigma}{dE d\Omega}(\vartheta_e, \varphi_e) = \sum_{k_0 k k_e} \bar{\rho}_{k_0 0}(\alpha_0 J_0) A_{k_0 k k_e} \left\{ Y_{k_0}(\vartheta_a, \varphi_a) \otimes Y_k(\vartheta_e, \varphi_e) \right\}_{k_e 0} \tag{2.185}$$

where $\bar{\rho}_{k_0 0}(\alpha_0 J_0)$ are the statistical tensors of the initial polarized atom in the coordinate frame with the z-axis directed along the A-axis, and

$$A_{k_0 k k_e} = (-1)^{k_e + k} \hat{J}_0 \sum_{\substack{l_f j_f l_i l_i' j_i j_i' \\ \alpha \alpha' J J' J_t J_t'}}' (-1)^{J_1 + j_i' - j_f - J_t'} \hat{l}_i \hat{l}_i' \hat{j}_i \hat{j}_i' \hat{l}_1 \hat{l}_1' \hat{j}_1 \hat{j}_1' \hat{J} \hat{J}' \hat{J}_t \hat{J}_t'$$

$$\times (l_i 0, l_i' 0 | k_e 0)(l_1 0, l_1' 0 | k 0) \begin{Bmatrix} j_i & l_i & \frac{1}{2} \\ l_i' & j_i' & k_e \end{Bmatrix} \begin{Bmatrix} J_t & J & j_f \\ J' & J_t' & k \end{Bmatrix}$$

$$\times \begin{Bmatrix} j_1 & l_1 & \frac{1}{2} \\ l_1' & j_1' & k \end{Bmatrix} \begin{Bmatrix} j_1 & J & J_1 \\ J' & j_1' & k \end{Bmatrix} \begin{Bmatrix} J_0 & j_i & J_t \\ J_0 & j_i' & J_t' \\ k_0 & k_e & k \end{Bmatrix}$$

$$\times \langle \alpha J, l_f j_f : J_t \| T \| \alpha_0 J_0, l_i j_i : J_t \rangle$$

$$\times \left\langle \alpha' J', l_f j_f : J'_t \left\| T \right\| \alpha_0 J_0, l'_i j'_i : J'_t \right\rangle^* \tag{2.186}$$

A prime at the summation sign indicates that only terms with $\alpha_1 J_1 = \alpha'_1 J'_1$ are included in the sum over α, α'. Equation (2.185) contains the tensorial products of the spherical harmonics [see Eq. (2.124)]. The reduced matrix elements in Eq. (2.186) and hence the parameters $A_{k_0 k k_e}$ include information on the dynamics of the process and depend on the values of p_i, p_f, and p_1, where p_i and p_1 are fixed in the experiment and p_f depends on the state of the residual ion $\alpha_1 J_1$ in accordance with the conservation of energy.

Equation (2.185) is very similar to Eq. (2.121) in which the photoionization of polarized atoms has been considered. The difference is that here we must use the partial wave expansion for the incident electron instead of the fixed multipolarity of the incident dipole photon as in Section 2.3 and perform a corresponding summation over l_i, l'_i. Therefore, k_e can be of high value ($k_e \leq l_i + l'_i$), not restricted by 2 as in the case of photoionization.

Equation (2.185) is the main result of this section. It factorizes the geometrical factor and reduces the problem to calculation of the dynamic coefficients $A_{k_0 k k_e}$. Note that the above consideration is also valid in the more general case of multiple ionization:

$$e_i + A(\alpha_0 J_0) \longrightarrow e_f + e_1 + \left[A^{(n+1)+} + e_2 + \cdots + e_{n+1} \right] \tag{2.187}$$

when only one of the electrons, e_1, is detected in the final state. The only difference is that $\alpha_1 J_1$ should be the quantum numbers of the system in square brackets in Eq. (2.187). Therefore, all the above equations are valid for the reaction (2.187) provided the set of quantum numbers α_1, α is appropriately chosen together with the corresponding angular momenta couplings.

In the triple sum (2.185) over k_0, k, and k_e, the range of k_0 is limited by the condition $0 \leq k_0 \leq 2J_0$, whereas the sum over k and k_e is in principle infinite. However, in real calculations, the sum converges due to the diminishing contributions of higher multipoles. From the parity conservation in the amplitudes of Eq. (2.186), it follows that indices k and k_e must have the same parity. This statement is valid both for process (2.176) and a more general process (2.187). It is essential only that a single electron be detected in the final state and therefore the identical quantum numbers of unobserved particles are contained in the bra states in Eq. (2.186).

When changing primed and unprimed momenta, the expression under summation in (2.186) changes to the complex conjugate with an additional phase factor $(-1)^{k_0}$. Therefore, $A_{k_0 k k_e}$ obeys the relation $A_{k_0 k k_e} = (-1)^{k_0} A^*_{k_0 k k_e}$ and the coefficients with even k_0 are real whereas those with odd k_0 are imaginary.

Defining the generalized anisotropy coefficients

$$\beta_{k_0 k k_e} = \frac{A_{k_0 k k_e}}{A_{000}} = A_{k_0 k k_e} \left/ \sum_{\substack{l_f j_f l_i j_i \\ \alpha J J_t}}' |\langle \alpha J; l_f j_f : J_t \| T \| \alpha_0 J_0, l_i j_i : J_t \rangle|^2 \right. \quad (2.188)$$

we can rewrite the general expression (2.185) in a form that reveals the dual origin of the angular anisotropy:

$$\frac{d^2\sigma}{dEd\Omega}(\vartheta_e, \varphi_e) = \frac{\sigma^{iso}(E)}{4\pi} \left(1 + \sum_{k>0} \beta_k P_k(\cos\vartheta_e) \right.$$

$$\left. + 4\pi \sum_{\substack{k_0>0 \\ k k_e}} \overline{\mathcal{A}}_{k_0 0}(\alpha_0 J_0) \beta_{k_0 k k_e} \left\{ Y_{k_0}(\vartheta_a, \varphi_a) \otimes Y_k(\vartheta_e, \varphi_e) \right\}_{k_e 0} \right) \quad (2.189)$$

Here $\sigma^{iso}(E) \equiv d\sigma^{iso}/dE$ is the "isotropic" cross section, i.e., the cross section for the unpolarized target integrated over the angle of ejection. The anisotropy coefficient β_k is a particular case of the generalized anisotropy coefficients (2.188):

$$\beta_k = \hat{k} \beta_{0kk} \quad (2.190)$$

The second term in the parentheses in Eq. (2.189) represents the anisotropy of the angular distribution due to a pure collisional mechanism, i.e., the anisotropy from an unpolarized target. The last term includes a combined effect of the two factors: polarization of the initial state and the anisotropy induced by the electron impact. The former factor is represented by the reduced statistical tensors $\overline{\mathcal{A}}_{k_0 0}(\alpha_0 J_0)$ and the latter by the generalized anisotropy coefficients $\beta_{k_0 k k_e}$. Among other things, representation (2.189) is useful because of the fundamental possibility of experimentally distinguishing the contribution from the three terms in Eq. (2.189). For example, only the last term in Eq. (2.189) depends on the polarization of the initial state and in the case of polarization by laser optical pumping, its contribution changes as a function of the target density because of depolarization due to resonance trapping.

For isotropic targets ($k_0 = 0$), the last term in Eq. (2.189) vanishes and Eq. (2.189) reduces to the conventional form

$$\left(\frac{d^2\sigma}{dEd\Omega} \right)^{iso} = \frac{\sigma^{iso}(E)}{4\pi} \left(1 + \sum_{k>0} \beta_k P_k(\cos\vartheta_e) \right) \quad (2.191)$$

The angular distribution (2.191) is axially symmetric with respect to the initial electron beam direction and in general it shows the forward–backward asymmetry due to the contribution of partial waves of different parity. For a polarized target [see Eq. (2.189)], the axial symmetry is, naturally, broken. However, in a particular case of target polarization along an electron beam ($\vartheta_a = \varphi_a = 0$), the angular distribution (2.189) reduces to the axially symmetric form, similar to Eq. (2.191), but with coefficients depending on the degree of target polarization.

The energy differential cross section of ionization from a polarized target, integrated over the ejection angle, $\frac{d\sigma}{dE}(\vartheta_a)$, does not coincide with σ^{iso} because of the third term in Eq. (2.189), which contains terms with $k = 0$. The general expression for $\frac{d\sigma}{dE}(\vartheta_a)$ follows from Eq. (2.189) and is of the form

$$\frac{d\sigma}{dE}(\vartheta_a) = \sigma^{iso}(E)\left(1 + \sum_{k=2,4,\ldots} \overline{\mathcal{A}}_{k0}(\alpha_0 J_0)\beta_k' P_k(\cos\vartheta_a)\right) \qquad (2.192)$$

where

$$\beta_k' = \hat{k}\beta_{k0k} \qquad (2.193)$$

In Eq. (2.192) k obeys the restriction $k \leq 2J_0$; therefore the integrated cross section (2.192) is independent of ϑ_a and coincides with $\sigma^{iso}(E)$ for $J_0 < 1$. When the atomic axis **A** makes the magic angle with the incident beam, the second term vanishes for $J_0 < 2$.

2.4.3. Density Matrix and Polarization Parameters of Ions Produced in Direct Ionization Processes

The density matrix and statistical tensors of the residual ion characterize its polarization state, which determines the angular distribution and polarization of subsequent fluorescence or autoionization electrons (see Chapter 3). We consider two types of experiment: coincidence and noncoincidence ones. In the former type, the scattered or ejected electron is detected in a certain direction (ϑ_e, φ_e) with respect to the initial beam (z-axis) in coincidence with the ion decay products. Then the angular distribution or polarization of the decay products is determined by the ion statistical tensors that are differential with respect to this direction. In another, noncoincidence experiment only a particle (photon or electron) emitted by the residual ion is detected. Here its angular distribution is determined by the integral statistical tensors of the ion.

The density matrix of the ion can be easily obtained from the total density matrix of the final state (2.178) according to the general approach [see Section 1.1.2, Eq. (1.38)] as a trace over unobserved variables. Suppose that both the target atom and the projectile beam are unpolarized. In addition, we assume that one of the electrons, for example, the scattered one, is detected in a certain direction by a spin-insensitive detector. Then the density matrix of the ion can be written as

$$\left\langle \alpha_1 J_1 M_1 \left| \rho_A^f(\mathbf{p}_i, \mathbf{p}_f) \right| \alpha_1 J_1 M_1' \right\rangle = \frac{1}{2(2J_0 + 1)}$$

$$\times \sum_{M_0\mu \, i\mu \, f\mu \, 1} \int d\Omega_1 T_{M_0\mu \, i \to M_1\mu \, f\mu \, 1}(\mathbf{p}_i, \mathbf{p}_f, \mathbf{p}_1)$$

$$\times T^*_{M_0\mu \, i \to M_1'\mu \, f\mu \, 1}(\mathbf{p}_i, \mathbf{p}_f, \mathbf{p}_1) \qquad (2.194)$$

The corresponding statistical tensors can be calculated according to definition (1.41):

$$\rho_{kq}(\alpha_1 J_1; \mathbf{p}_i, \mathbf{p}_f) = \sum_{M_1 M_1'} (-1)^{J_1 - M_1'} (J_1 M_1, J_1 - M_1' | kq)$$

$$\times \left\langle \alpha_1 J_1 M_1 \left| \rho_A^f(\mathbf{p}_i, \mathbf{p}_f) \right| \alpha_1 J_1 M_1' \right\rangle \qquad (2.195)$$

This is the statistical tensor differential with respect to the direction of scattering $\mathbf{p}_f$. After integrating over all directions, we obtain the integral statistical tensor:

$$\rho_{kq}(\alpha_1 J_1) = \int d\Omega_f \rho_{kq}(\alpha_1 J_1; \mathbf{p}_i, \mathbf{p}_f) \qquad (2.196)$$

Sometimes it is convenient to expand the wave functions of electrons in partial waves; then the statistical tensors can be expressed in terms of ionization amplitudes in the total angular momentum representation. We can start again from expression (2.60) for the statistical tensor of the $(e_1 + A^+)$ subsystem. If the initial atom and the projectile are unpolarized, this expression reduces to Eq. (2.62) (we suppose also that the detector of electrons is not sensitive to their spin directions), which we can write in the considered case as follows (for a sharp initial state $\alpha_0 J_0$):

$$\rho_{kq}(\alpha J, \alpha' J') = \sum_{\substack{l_i l_i' j_i j_i' \\ l_f l_f' j_f j_f'}} \sum_{\substack{J_t J_t' k k_l}} \hat{J}_t \hat{J}_t' \hat{j}_i \hat{j}_i' \hat{\tilde{k}} \hat{\tilde{k}} \hat{j}_f \hat{j}_f' \hat{l}_i \hat{l}_i' \hat{l}_f \hat{l}_f'$$

$$\times (-1)^{J_0 + J_t + j_f' + \tilde{k} + k} (kq, \tilde{k}0 | k_l q) (l_i 0, l_i' 0 | \tilde{k} 0) (l_f 0, l_f' 0 | k_l 0)$$

$$\times \begin{Bmatrix} J_t' & j_i' & J_0 \\ j_i & J_t & \tilde{k} \end{Bmatrix} \begin{Bmatrix} l_f' & j_f' & \frac{1}{2} \\ j_f & l_f & k_l \end{Bmatrix} \begin{Bmatrix} l_i' & j_i' & \frac{1}{2} \\ j_i & l_i & \tilde{k} \end{Bmatrix} \begin{Bmatrix} J & J' & k \\ j_f & j_f' & k_l \\ J_t & J_t' & \tilde{k} \end{Bmatrix} Y_{k_l q}(\vartheta_{sc}, \varphi_{sc})$$

$$\times \left\langle \alpha J, l_f j_f : J_t \| T \| \alpha_0 J_0, l_i j_i : J_t \right\rangle$$

$$\times \left\langle \alpha' J', l_f' j_f' : J_t' \| T \| \alpha_0 J_0, l_i' j_i' : J_t' \right\rangle^* \qquad (2.197)$$

As was shown in Section 1.2.4, the statistical tensor of the subsystem (in our case, ion A^+) can be expressed in terms of the statistical tensor of the entire $(e_1 + A^+)$ system according to Eq. (1.69):

$$\rho_{kq}\left(J_1, J_1'\right) = \sum_{JJ'j_1} (-1)^{J'+J_1+j_1+k} \hat{f}\hat{f}' \begin{Bmatrix} J_1 & J_1' & k \\ J' & J & j_1 \end{Bmatrix} \rho_{kq}\left(J_1 j_1 J; J_1' j_1 J'\right) \quad (2.198)$$

Combining this expression and Eq. (2.197) one obtains

$$\rho_{kq}\left(\alpha_1 J_1\right) = \sum_{\substack{JJ'j_1\,l_il_i'j_ij_i' \\ l_fl_f'j_fj_f'}} \sum_{J_tJ_t'\tilde{k}k_l} \hat{f}\hat{f}'\hat{J}_t\hat{J}_t'\hat{j}_i\hat{j}_i'\hat{\tilde{k}}\hat{\tilde{k}}\hat{j}_f\hat{j}_f'\hat{l}_i\hat{l}_i'\hat{l}_f\hat{l}_f'$$

$$\times (-1)^{J'+J_1+J_0+J_t+j_1+j_f'+\tilde{k}} \left(kq, \tilde{k}0 \,|\, k_l q\right) \left(l_i 0, l_i' 0 \,|\, \tilde{k} 0\right) \left(l_f 0, l_f' 0 \,|\, k_l 0\right)$$

$$\times \begin{Bmatrix} J_1 & J_1' & k \\ J' & J & j_1 \end{Bmatrix} \begin{Bmatrix} J_t' & j_i' & J_0 \\ j_i & J_t & \tilde{k} \end{Bmatrix} \begin{Bmatrix} l_f' & j_f' & \frac{1}{2} \\ j_f & l_f & k_l \end{Bmatrix} \begin{Bmatrix} l_i' & j_i' & \frac{1}{2} \\ j_i & l_i & \tilde{k} \end{Bmatrix} \begin{Bmatrix} J & J' & k \\ j_f & j_f' & k_l \\ J_t & J_t' & \tilde{k} \end{Bmatrix}$$

$$\times Y_{k_l q}\left(\vartheta_{sc}, \varphi_{sc}\right) \left\langle \alpha J, l_f j_f : J_t \,\|\, T \,\|\, \alpha_0 J_0, l_i j_i : J_t \right\rangle$$

$$\times \left\langle \alpha' J', l_f' j_f' : J_t' \,\|\, T \,\|\, \alpha_0 J_0, l_i' j_i' : J_t' \right\rangle^* \quad (2.199)$$

The integral statistical tensors can be obtained by integrating Eq. (2.199) over scattering angles. This gives $k_l = q = 0$ and therefore, $k = \tilde{k}, j_f = j_f', l_f = l_f'$. Taking this into account, we finally get the following expression:

$$\rho_{kq}\left(\alpha_1 J_1\right) = \delta_{q0}\sqrt{4\pi} \sum_{\substack{JJ'j_1\,l_il_i'j_ij_i'\,J_tJ_t'l_fj_f}} \sum \hat{f}\hat{f}'\hat{J}_t\hat{J}_t'\hat{j}_i\hat{j}_i'\hat{l}_i\hat{l}_i'$$

$$\times (-1)^{J+J'+J_1+J_0+J_t+J_t'+j_1-j_f+k+1/2} \left(l_i 0, l_i' 0 \,|\, k 0\right)$$

$$\times \begin{Bmatrix} J_1 & J_1' & k \\ J' & J & j_1 \end{Bmatrix} \begin{Bmatrix} J_t' & j_i' & J_0 \\ j_i & J_t & k \end{Bmatrix} \begin{Bmatrix} l_i' & j_i' & \frac{1}{2} \\ j_i & l_i & k \end{Bmatrix} \begin{Bmatrix} J_t & J_t' & k \\ J' & J & j_f \end{Bmatrix}$$

$$\times \left\langle \alpha J, l_f j_f : J_t \,\|\, T \,\|\, \alpha_0 J_0, l_i j_i : J_t \right\rangle$$

$$\times \left\langle \alpha' J', l_f j_f : J_t' \,\|\, T \,\|\, \alpha_0 J_0, l_i' j_i' : J_t' \right\rangle^* \quad (2.200)$$

As would be expected when both electrons in the final state are not detected, the ion state is axially symmetric with respect to the initial beam direction (z-axis). This is reflected in the fact that only $q = 0$ components of the statistical tensors are nonzero. In addition, one can show that owing to parity conservation, the rank of the integral tensor can only be even. The contribution of different partial waves of the scattered and ejected electrons is incoherent.

Decay of Polarized States

In this chapter we consider a broad class of processes of decay of a prepared quantum state, such as spontaneous emission of electromagnetic radiation (radiative decay) and Auger decay (nonradiative decay). The angular distributions and the polarization of the decay products, which are often studied in modern experiments, depend on the dynamics of the decay as well as the polarization properties of the initial decaying state. Here we consider some properties of the decay products and their relation to the initial state of polarization.

3.1. Nonradiative Decay

3.1.1. Density Matrix and Statistical Tensors after Two-Particle Decay

Consider the process

$$A(\alpha_A J_A) \longrightarrow B(\alpha_B J_B) + b(\alpha_b J_b) \qquad (3.1)$$

when an initial system A with the total angular momentum J_A decays into two parts: a residual system B with the total angular momentum J_B and a "radiation" b with the total angular momentum J_b. As usual α_A, α_B, and α_b are sets of all other quantum numbers specifying the states. Consider a heavy particle, atom or ion, as the residual system B, and the light particle, electron or photon, as the "radiation" b. Moreover, in most cases, one can consider B as a particle with an infinite mass. In general, J_A, J_B, and J_b may be not fixed; however, we start from a simple case of the decay of an atomic state with a sharp value of J_A.

Let $\langle \alpha_A J_A M_A \,|\, \rho \,|\, \alpha_A J_A M_A' \rangle$ be the spin density matrix of the decaying system A, and $\rho_{k_A q_A}(\alpha_A J_A)$ the corresponding statistical tensors. We introduce the *decay amplitudes*, i.e., the matrix elements $\langle f \,|\, T \,|\, i \rangle$ of the operator describing the transition from the initial state, $|i\rangle = |\alpha_A J_A M_A\rangle$, to the final state

$|f\rangle = |\alpha_B J_B, \alpha_b J_b : JM\rangle$. Now we can apply the general Eq. (1.57) to relate the density matrix of the system $B + b$ to the density matrix of the initial system A:

$$\langle \alpha_B J_B, \alpha_b J_b : JM \,|\, \rho \,|\, \alpha_B' J_B', \alpha_b' J_b' : J'M' \rangle = \sum_{M_A M_A'} \langle \alpha_A J_A M_A \,|\, \rho \,|\, \alpha_A J_A M_A' \rangle$$
$$\times \langle \alpha_B J_B, \alpha_b J_b : JM \,|\, T \,|\, \alpha_A J_A M_A \rangle \langle \alpha_A J_A M_A' \,|\, T^+ \,|\, \alpha_B' J_B', \alpha_b' J_b' : J'M' \rangle \quad (3.2)$$

The operator T is a scalar in the space of the total angular momentum of the system. Applying the Wigner–Eckart theorem (A.62) for the scalar operator:

$$\langle \alpha_B J_B, \alpha_b J_b : JM \,|\, T \,|\, \alpha_A J_A M_A \rangle = \frac{1}{\hat{J}} \delta_{J J_A} \delta_{M M_A} \langle \alpha_B J_B, \alpha_b J_b : J \,\|\, T \,\|\, \alpha_A J_A \rangle \quad (3.3)$$

one obtains from Eq. (3.2) the relation

$$\langle \alpha_B J_B, \alpha_b J_b : JM \,|\, \rho \,|\, \alpha_B' J_B', \alpha_b' J_b' : J'M' \rangle = \frac{1}{\hat{J}^2} \delta_{J J_A} \delta_{M M_A} \delta_{J' J_A} \delta_{M' M_A'}$$
$$\times N \langle \alpha_B J_B, \alpha_b J_b : J \,\|\, T \,\|\, \alpha_A J_A \rangle \langle \alpha_B' J_B', \alpha_b' J_b' : J' \,\|\, T \,\|\, \alpha_A J_A \rangle^*$$
$$\times \langle \alpha_A J_A M_A \,|\, \rho \,|\, \alpha_A J_A M_A' \rangle \quad (3.4)$$

We have introduced the *reduced decay amplitudes* $\langle \alpha_B J_B, \alpha_b J_b : J \,\|\, T \,\|\, \alpha_A J_A \rangle$. Using the definition (1.41) and Eq. (3.4) one can write

$$\rho_{kq} \left(\alpha_B J_B, \alpha_b J_b : J; \alpha_B' J_B', \alpha_b' J_b' : J' \right) = \delta_{J J_A} \delta_{J' J_A} \frac{1}{\hat{J}_A^2} \rho_{kq} \left(\alpha_A J_A \right)$$
$$\times \langle \alpha_B J_B, \alpha_b J_b : J \,\|\, T \,\|\, \alpha_A J_A \rangle \langle \alpha_B' J_B', \alpha_b' J_b' : J' \,\|\, T \,\|\, \alpha_A J_A \rangle^* \quad (3.5)$$

which is a special case of Eq. (1.58). As it was emphasized in Section 1.2.3 the final system contains the statistical tensors of the total angular momentum with only those ranks k and projections q which have been present in the initial decaying state. In other words, the type of symmetry of the whole system is conserved in the decay.

3.1.2. Angular Distribution of Ejected Electrons

Consider the angular distribution of ejected electrons in the process of Auger decay or autoionization:

$$A^*(\alpha J) \longrightarrow A^+(\alpha_f J_f) + e \quad (3.6)$$

when the detector is not sensitive to spin polarization of the electron and the residual ion in (3.6) is not detected. To write the angular distribution, we again use the standard expression (1.151):

$$W_{\alpha_f J_f}(\vartheta, \varphi) = \sum_{J J' k q} \rho_{kq}(J, J') \, \varepsilon_{kq}^*(J, J') \quad (3.7)$$

and substitute Eq. (3.5) for $\rho_{kq}(J,J')$ and Eq. (1.171) for $\varepsilon_{kq}(J,J')$. We obtain

$$
W_{\alpha_f J_f}(\vartheta,\varphi) = \sum_{\substack{ll'jj' \\ kq}} (-1)^{J+J_f+k-1/2} \hat{l}\hat{l'}\,\hat{j}\hat{j'} \,(l0,l'0|k0)
$$

$$
\times \begin{Bmatrix} J & j & J_f \\ j' & J & k \end{Bmatrix} \begin{Bmatrix} l & j & \frac{1}{2} \\ j' & l' & k \end{Bmatrix}
$$

$$
\times \langle \alpha_f J_f, lj : J \| V \| \alpha J \rangle \langle \alpha_f J_f, l'j' : J \| V \| \alpha J \rangle^*
$$

$$
\times \rho_{kq}(\alpha J) \sqrt{\frac{4\pi}{2k+1}} Y_{kq}(\vartheta,\varphi) \tag{3.8}
$$

Here the matrix elements $\langle \alpha_f J_f, lj : J \| V \| \alpha J \rangle$ are used instead of $\langle \alpha_f J_f, lj : J \| T \| \alpha J \rangle$ to point out that usually, in the decay problem, one deals with the electron–electron interaction V. In applications, the operator V is most often the Coulomb interaction.

If several final levels are not resolved by the detector, the angular distribution is an incoherent sum of angular distributions from the decays (3.6) into unresolved levels with different angular momenta J_f, for example, a sum over the unresolved fine-structure components:

$$
W(\vartheta,\varphi) = \sum_{\alpha_f J_f} W_{\alpha_f J_f}(\vartheta,\varphi) \tag{3.9}
$$

Equation (3.8) can be written in an alternative convenient form

$$
W_{\alpha_f J_f}(\vartheta,\varphi) = \frac{W_0}{4\pi} \left[1 + \sum_{k=2,4,\ldots}^{k_{max}} \sqrt{\frac{4\pi}{2k+1}} \,\alpha_k \sum_{q=-k}^{k} \mathcal{A}_{kq}(\alpha J) Y_{kq}(\vartheta,\varphi) \right] \tag{3.10}
$$

where $\mathcal{A}_{kq}(\alpha J)$ are the reduced statistical tensors of the decaying state and α_k are the *intrinsic anisotropy parameters* for the electron emission:

$$
\alpha_k = N_\alpha^{-1}(-1)^{J+J_f+k-1/2} f \sum_{ll'jj'} \hat{l}\hat{l'}\,\hat{j}\hat{j'} \,(l0,l'0|k0) \begin{Bmatrix} J & j & J_f \\ j' & J & k \end{Bmatrix} \begin{Bmatrix} l & j & \frac{1}{2} \\ j' & l' & k \end{Bmatrix}
$$

$$
\times \langle \alpha_f J_f, lj : J \| V \| \alpha J \rangle \langle \alpha_f J_f, l'j' : J \| V \| \alpha J \rangle^* \tag{3.11}
$$

where

$$
N_\alpha = \sum_{lj} |\langle \alpha_f J_f, lj : J \| V \| \alpha J \rangle|^2 \tag{3.12}
$$

The factor W_0 is the total probability of the decay $\alpha J \to \alpha_f J_f$, integrated over the ejection angles. The anisotropy parameters α_k (3.11) contain information about the dynamics of the decay, while the tensors $\mathcal{A}_{kq}(\alpha J)$ describe polarization properties of the decaying state. For an unpolarized decaying state $(k = q = 0)$, the angular distribution is isotropic.

In general, for a transition $\alpha J \to \alpha_f J_f$ between states with a certain spin and parity, the ejected electron can have various total angular momenta j and orbital angular momenta l. Owing to parity conservation in the decay, l and l' are of the same parity. As a result, k is even in Eqs. (3.8) and (3.10) because of the property of the Clebsch–Gordan coefficient with zero projections (A.94). The angular distribution (3.10) is sensitive to the phase difference between the matrix elements of the decay, in contrast to the integrated probability W_0, which depends only on the square module of the decay matrix elements. In the sum (3.11), the coefficients are symmetric with respect to the interchange $lj \to l'j'$; therefore the anisotropy parameters α_k may be presented as

$$
\begin{aligned}
\alpha_k &= N_\alpha^{-1} \sum_{lj \leq l'j'} a_k(lj, l'j') \, \mathrm{Re}\left(\langle \alpha_f J_f, lj : J \| V \| \alpha J \rangle \langle \alpha_f J_f, l'j' : J \| V \| \alpha J \rangle^*\right) \\
&= N_\alpha^{-1} \sum_{lj \leq l'j'} a_k(lj, l'j') \, \cos\left(\delta_{lj} - \delta_{l'j'}\right) \\
&\quad \times |\langle \alpha_f J_f, lj : J \| V \| \alpha J \rangle| |\langle \alpha_f J_f, l'j' : J \| V \| \alpha J \rangle|
\end{aligned}
\tag{3.13}
$$

Note that α_k reduces to pure algebraic values if only one decay channel with fixed l and j contributes to the decay:

$$
\alpha_k = (-1)^{J+J_f+k-1/2} \hat{f}l^2 \hat{j}^2 (10,10|k0) \begin{Bmatrix} J & j & J_f \\ j & J & k \end{Bmatrix} \begin{Bmatrix} l & j & \frac{1}{2} \\ j & l & k \end{Bmatrix}
\tag{3.14}
$$

This is realized in a case of practical importance when the final angular momentum is zero, $J_f = 0$. Then $j = J$ and Eq. (3.14) for the anisotropy parameters simplifies to

$$
\alpha_k = (-1)^{j+1/2} \hat{j}l^2 (10,10|k0) \begin{Bmatrix} l & j & \frac{1}{2} \\ j & l & k \end{Bmatrix}
\tag{3.15}
$$

Let us discuss further some general properties of the angular distribution (3.10). It is invariant with respect to the inversion: $\vartheta \to \pi - \vartheta$, $\varphi \to \varphi + \pi$ [see (A.18)] because it contains the spherical harmonics of even rank. The angular distribution (3.10) is not affected by the odd tensors of the decaying system; thus only the alignment of the initial state αJ is important for the angular distribution, but not the orientation. The maximal complexity of the angular distribution is $k_{\max} \leq 2J$. Therefore the decay of the states with $J < 1$ is always isotropic, provided parity is conserved.

When the decaying state has an axis of symmetry, the angular distribution depends only on the angle ϑ between the symmetry axis and the direction of the linear momentum of the ejected electron:

$$W_{\alpha_f J_f}(\vartheta) = \frac{W_0}{4\pi}\left[1 + \sum_{k=2,4,\dots}^{k_{max}} \alpha_k \mathcal{A}_{k0}(\alpha J)P_k(\cos\vartheta)\right] \qquad (3.16)$$

In this case the angular distribution includes only the Legendre polynomials $P_k(\cos\vartheta)$ with even k, and is symmetric with respect to the reflection in the plane perpendicular to the symmetry axis: $\vartheta \rightarrow \pi - \vartheta$. (In nuclear physics this plane symmetry is broken for the angular distributions of electrons from β-decay of polarized nuclei due to parity nonconservation in weak interactions; the effect of parity nonconservation has been observed also in atomic processes, but it manifests itself in very special conditions and is extremely small.)

Consider the decay of a state with plane symmetry and definite parity. Such a case may be realized, for example, in measurements of the angular distribution of Auger electrons produced by particle impact in coincidence experiments where the scattered particle and the Auger electrons are measured in coincidence. Then the scattering plane is a symmetry plane, and according to Section 1.4, the following statistical tensors are nonzero (see Table 1.2 on p. 30): $\mathcal{A}_{11} = \mathcal{A}_{1-1}$ (imaginary), $\mathcal{A}_{20}, \mathcal{A}_{21} = -\mathcal{A}_{2-1}, \mathcal{A}_{22} = \mathcal{A}_{2-2}$ (real) and so on. (Here we choose the z-axis along the beam and the x-axis is in the scattering plane.) As only even-rank statistical tensors influence the angular distribution (3.10), we immediately find that it can be written in the form

$$W_{\alpha_f J_f}(\vartheta, \varphi) = \frac{W_0}{4\pi}\left\{1 + \sum_{k=2,4,\dots}^{k_{max}} \alpha_k\left[\mathcal{A}_{k0}(\alpha J)P_k(\cos\vartheta)\right.\right.$$
$$\left.\left. +2\sum_{q>0}^{k} \mathcal{A}_{kq}(\alpha J)(-1)^q \sqrt{\frac{(k-q)!}{(k+q)!}}P_k^q(\cos\vartheta)\cos q\varphi\right]\right\} \qquad (3.17)$$

In a particular case, if the detector is rotated in the plane perpendicular to the beam (as is the case in many real experiments), the angular distribution has the simple form

$$W = \sum_{\kappa=0,2,\dots}^{k_{max}} A_\kappa \cos\kappa\varphi \qquad (3.18)$$

If the detector is rotated in the scattering plane, the angular distribution has the form

$$W = \sum_{\kappa=0}^{\kappa_{max}} A_\kappa \cos^{2\kappa}\theta + \sum_{\kappa=1}^{\kappa_{max}} B_\kappa \sin^{2\kappa}\theta \qquad (3.19)$$

where the maximal power κ is limited by the angular momentum of the decaying state ($\kappa_{max} = J$).

3.1.3. *Spin Polarization of Ejected Electrons*

As was pointed out in Chapter 2, there are several approaches to the problem of spin polarization of reaction products. The same approaches are valid for the decay products as well. The most straightforward and general method is to find the spin density matrix or the corresponding statistical tensors of the particle as a function of the direction of its ejection. This can be done by taking a trace of the total density matrix of the system after the decay over the unobserved variables. We use this approach here to determine the spin polarization of ejected electrons.

The problem is to calculate the statistical tensors of the spin of an electron ejected in a well-defined direction $\mathbf{n}$:

$$\rho_{k_s q_s}\left(\frac{1}{2};\mathbf{n}\right) = \sum_{\mu\mu'} (-1)^{1/2-\mu'} \left(\frac{1}{2}\mu,\frac{1}{2}-\mu'\bigg|k_s q_s\right) \left\langle \mathbf{n},\frac{1}{2}\mu \,|\rho|\, \mathbf{n},\frac{1}{2}\mu'\right\rangle$$

(3.20)

Expanding the electron wave function in partial waves [see Eqs. (1.97), (1.98)] and coupling the orbital and spin angular momenta, one obtains

$$\rho_{k_s q_s}\left(\frac{1}{2};\mathbf{n}\right) = \sum_{\substack{lml'm' \\ \mu\mu'}} (-1)^{1/2-\mu'} \left(\frac{1}{2}\mu,\frac{1}{2}-\mu'\bigg|k_s q_s\right)$$

$$\times \langle \mathbf{n}|lm\rangle \left\langle lm\frac{1}{2}\mu\,|\rho|\,l'm'\frac{1}{2}\mu'\right\rangle \langle l'm'|\mathbf{n}\rangle$$

$$= \sum_{\substack{lml'm' \\ \mu\mu'}} \sum_{\substack{jmj'm'_j \\ kq}} (-1)^{1/2-\mu'} \left(\frac{1}{2}\mu,\frac{1}{2}-\mu'\bigg|k_s q_s\right) \left(lm,\frac{1}{2}\mu\bigg|jm_j\right)$$

$$\times \left(l'm',\frac{1}{2}\mu'\bigg|j'm'_j\right)(-1)^{j'-m'_j}\left(jm_j,j'-m'_j|kq\right)$$

$$\times \rho_{kq}\left(lj,l'j'\right)Y_{lm}(\vartheta,\varphi)Y^*_{l'm'}(\vartheta,\varphi)$$

(3.21)

In the latter equation we have used the relation (1.42) between the density matrix and statistical tensors as well as the explicit form of the wave function $\langle \mathbf{n}|lm\rangle$ [Eq. (A.25)]. After summation over μ,μ',m_j,m'_j [Eq. (A.91)], using the expansion (A.29), and subsequent summation over mm', the statistical tensor (3.21) takes the form

$$\rho_{k_s q_s}\left(\frac{1}{2};\mathbf{n}\right) = \frac{1}{\sqrt{4\pi}} \sum_{\substack{ll'jj' \\ k_l k}} (-1)^l \hat{k}\hat{j}\hat{j}'\hat{l}\hat{l}' \,(l0,l'0|k_l 0) \begin{Bmatrix} l & \frac{1}{2} & j \\ l' & \frac{1}{2} & j' \\ k_l & k_s & k \end{Bmatrix}$$

$$\times \sum_{qq_l} (kq,k_l q_l|k_s q_s)\rho_{kq}\left(lj,l'j'\right)Y^*_{k_l q_l}(\vartheta,\varphi)$$

(3.22)

The right side of Eq. (3.22) has a necessary tensor dimension: It contains the tensor product of two conjugate tensors and therefore transforms as a conjugate tensor of the rank k_s. Now we find $\rho_{kq}(lj,l'j)$ in terms of the statistical tensors of the initial decaying state $\rho_{kq}(\alpha J)$. This may be achieved using Eqs. (3.5) and (1.68), which in our case have the forms

$$\rho_{kq}\left(\alpha_f J_f, lj : J; \alpha_f J_f, l'j' : J\right) = \hat{J}^{-2}\rho_{kq}(\alpha J)$$
$$\times \left\langle \alpha_f J_f, lj : J \| V \| \alpha J \right\rangle \left\langle \alpha_f J_f, l'j' : J \| V \| \alpha J \right\rangle^* \qquad (3.23)$$

and

$$\rho_{kq}\left(lj, l'j'\right) = \sum_{JJ'}(-1)^{J+J_f+j'+k}\hat{J}\hat{J}' \left\{ \begin{array}{ccc} j & j' & k \\ J' & J & J_f \end{array} \right\}$$
$$\times \rho_{kq}\left(\alpha_f J_f, lj : J; \alpha_f J_f, l'j' : J'\right) \qquad (3.24)$$

The resulting expression for the statistical tensor of the electron spin is obtained by substitution of Eq. (3.24) with Eq. (3.23) into Eq. (3.22):

$$\rho_{k_s q_s}\left(\frac{1}{2}; \mathbf{n}\right) = \frac{1}{\sqrt{4\pi}} \sum_{\substack{ll'jj' \\ k_l k}} (-1)^{l+J+J_f+j'+k}\hat{k}\hat{j}\hat{j}'\hat{l}\hat{l}'\,(l0,l'0|k_l 0)$$

$$\times \left\{ \begin{array}{ccc} j & j' & k \\ J & J & J_f \end{array} \right\} \left\{ \begin{array}{ccc} l & \frac{1}{2} & j \\ l' & \frac{1}{2} & j' \\ k_l & k_s & k \end{array} \right\} \left\langle \alpha_f J_f, lj : J \| V \| \alpha J \right\rangle \qquad (3.25)$$

$$\times \left\langle \alpha_f J_f, l'j' : J \| V \| \alpha J \right\rangle^* \sum_{qq_l}(kq,k_l q_l | k_s q_s)\rho_{kq}(\alpha J)Y^*_{k_l q_l}(\vartheta, \varphi)$$

The angles ϑ, φ characterize the direction of the electron linear momenta in the laboratory frame, in which the statistical tensors of the decaying system, $\rho_{kq}(\alpha J)$, are given.

The spin polarization of the ejected electrons can be nonzero only if the initial state is polarized (oriented and/or aligned). Remember that we consider the case where only the ejected electron is detected. Indeed, for the unpolarized initial state ($k = q = 0$), Eq. (3.25) simplifies considerably:

$$\rho_{k_s q_s}\left(\frac{1}{2}; \mathbf{n}\right) = \frac{1}{\sqrt{4\pi}}\hat{J}^{-1}\rho_{00}(\alpha J)Y^*_{k_s q_s}(\vartheta, \varphi)\sum_{ll'j}(-1)^{j+1/2}\hat{k}_s^{-1}\hat{l}\hat{l}'\,(l0,l'0|k_s 0)$$

$$\times \left\{ \begin{array}{ccc} l & l' & k_s \\ \frac{1}{2} & \frac{1}{2} & j \end{array} \right\} \left\langle \alpha_f J_f, lj : J \| V \| \alpha J \right\rangle \left\langle \alpha_f J_f, l'j : J \| V \| \alpha J \right\rangle^* \qquad (3.26)$$

Polarization of the ejected electron is described by the components of the tensor (3.26) with the rank $k_s = 1$. However, the value of $k_s = 1$ is possible only if

$l' = l \pm 1$, as follows from the properties of the Clebsch–Gordan coefficient with zero projections in (3.26), i.e., l and l' are of different parity. As discussed earlier, this is impossible for the level αJ with definite parity and assuming parity conservation in the decay; therefore the electron ejected from an isotropic target cannot be polarized under the usual conditions. The result is obvious from symmetry considerations. If the initial system is unpolarized, then only one vector (the linear momentum of the ejected electron) characterizes the system, from which it is impossible to build an axial vector (polarization).

If the initial state of the decaying system is aligned along some direction **A**, we can choose this direction as the z-axis. Then the general formula (3.25) can be applied with $q = 0$. Consider again the spin polarization, i.e., the tensor with $k_s = 1$. From parity conservation it follows that l and l' are of the same parity and k_l is even. Now it is easy to show that the z-component of the polarization $\rho_{10}\left(\frac{1}{2}\right) = 0$. Indeed, for the aligned initial state, $k =$ even and the Clebsch–Gordan coefficient, $(k0, k_l 0 \,|\, 10)$ is zero. Therefore, only components $\rho_{1\pm1}\left(\frac{1}{2}\right) \neq 0$. Since $k_l + k_s + k =$ odd, the $9j$-symbol in Eq. (3.25) can be nonzero only if $j \neq j'$ [see Eq. (A.121)]. Therefore the spin polarization in the case of an aligned decaying system also is a pure interference effect. It disappears in a single-channel decay.

Finally, choosing the reference frame in such a way that the xz-plane contains both the alignment direction and the emission direction and using Eq. (1.95), connecting the statistical tensors $\rho_{1\pm1}\left(\frac{1}{2}\right)$ with the components of the polarization vector, we obtain $P_x = 0$ and the only nonzero component of the polarization vector is P_y. Thus the spin polarization of emitted electrons is perpendicular to the plane containing the alignment axis and the direction of electron emission. Spin polarization of electrons emitted from the aligned decaying state is due to the dynamics of decay and is connected with the phase difference between the decay amplitudes (see also the discussion in Section 2.1.2). This type of spin polarization of emitted electrons is called *dynamic polarization*.

In the case of an oriented target, the spin polarization of the emitted electrons can have x- and z-components in the chosen reference frame. In this case ($k = 1$) a nonzero spin polarization can also exist in a single-channel decay independently of particular decay matrix elements. The spin polarization of the electrons emitted from an oriented decaying state basically exists because of the conservation of the total angular momentum and its projection (or, equivalently, the conservation of tensorial components for statistical tensors of the whole system). Therefore, the corresponding mechanism of production of the spin polarization is called *polarization transfer*.

After integration of (3.25) over the angles of ejection, one obtains an integral spin polarization of the total electron flux from the decay $J \to J_f$:

$$\rho_{kq}\left(\frac{1}{2}\right) = \sum_{ljj'} (-1)^{l+J+J_f+j'+j+1/2} \, \hat{j}\hat{j}' \begin{Bmatrix} j & j' & k \\ J & J & J_f \end{Bmatrix} \begin{Bmatrix} \frac{1}{2} & \frac{1}{2} & k \\ j & j' & l \end{Bmatrix}$$

$$\times \left\langle \alpha_f J_f, lj : J \left\| V \right\| \alpha J \right\rangle \left\langle \alpha_f J_f, lj' : J \left\| V \right\| \alpha J \right\rangle^* \rho_{kq}(\alpha J) \qquad (3.27)$$

Exactly the same formula for the integral polarization (3.27) can be obtained directly from the expression (1.68), which relates the statistical tensor of one of the two angular momenta (in our case spin) with that of their sum (total angular momentum):

$$\rho_{kq}\left(\frac{1}{2}\right) = \sum_{ljj'}(-1)^{j+l+k+1/2}\,\hat{j}\hat{j}' \left\{ \begin{array}{ccc} \frac{1}{2} & \frac{1}{2} & k \\ j & j' & l \end{array} \right\} \rho_{kq}(lj, lj') \qquad (3.28)$$

Substituting Eq. (3.24) with (3.23) into (3.28), one gets Eq. (3.27). This more elegant approach is possible when we are not interested in the dependence of the spin polarization on the direction of ejection. We note that the net (integral) spin polarization of decay electrons (3.27) can exist only if the decaying state is oriented ($k = 1$). The aligned ($k = 2$) state gives zero integral spin polarization. Formally it follows from the property (A.99) for the second $6j$-symbol in (3.27).

In this section we considered the spin polarization of the ejected electrons as a function of the ejection angle. On the other hand, it is sometimes necessary to consider the angular distribution measured by a detector sensitive to spin polarization of the ejected electrons. For this purpose one can follow the procedure outlined in Section 3.1.2 with the use of Eq. (1.169) instead of Eq. (1.171) to calculate the efficiency tensor of the detector.

3.2.   Radiative Decay

3.2.1.   Angular Distribution of Emitted Photons

Consider the angular distribution of photons radiated from an excited atomic state:

$$A^*(\alpha J) \longrightarrow A(\alpha_f J_f) + \gamma \qquad (3.29)$$

By analogy to the previous case of electron emission, we start from Eq. (3.7) and substitute Eq. (3.5) for the statistical tensor of the final system, taking into account that this time the decay is due to the electromagnetic interaction. Since the final system contains a photon, we use Eq. (1.174) for the efficiency tensor $\varepsilon_{kq}(J, J')$. As a result we obtain

$$W_{\alpha_f J_f} = \sum_{\substack{pp'LL' \\ kq}} (-1)^{L'+J+J_f+k} \left\{ \begin{array}{ccc} J & L' & J_f \\ L & J & k \end{array} \right\} \varepsilon_{kq}^*(pL, p'L') \rho_{kq}(\alpha J)$$

$$\times \left\langle \alpha_f J_f, pL : J \left\| H_{\text{int}} \right\| \alpha J \right\rangle \left\langle \alpha_f J_f, p'L' : J \left\| H_{\text{int}} \right\| \alpha J \right\rangle^* \qquad (3.30)$$

where H_{int} is the interaction of an atom with the electromagnetic field, $\rho_{kq}(\alpha J)$ are the statistical tensors of the decaying state, and $\varepsilon_{kq}(pL, p'L')$ are the efficiency tensors of the photon detector, pL characterizing the type and multipolarity of the photon. In Eq. (3.30) we assume that $\rho_{kq}(\alpha J)$ and $\varepsilon_{kq}(pL, p'L')$ are given in a common coordinate frame. If the detector cannot resolve several lines corresponding to different final atomic states, the angular distribution is an incoherent sum of several contributions:

$$W = \sum_{\alpha_f J_f} W_{\alpha_f J_f} \tag{3.31}$$

In the following paragraphs we consider only the allowed (electric dipole) transitions. In the dipole approximation ($p = p' = 0$, $L = L' = 1$), the reduced matrix element of electromagnetic interaction in the photon frame is proportional to the matrix element of the dipole operator [compare Eq. (2.6)]. To obtain the angular distribution of the dipole photons, we also transform $\rho_{kq}(\alpha J)$ into the photon frame by using Eq. (1.44). The resulting expression for the dipole radiation can be cast in the form

$$W_{\alpha_f J_f}(\varphi, \vartheta, \psi) = \frac{W_0}{4\pi}\left[1 + \sqrt{\frac{3}{2}}\sum_{k=1,2}\alpha_k^\gamma\sum_q(\varepsilon_{kq}^{\text{det}})^*\sum_{q'}\mathcal{A}_{kq'}(\alpha J)D_{q'q}^{k*}(\varphi, \vartheta, \psi)\right] \tag{3.32}$$

where the efficiency tensors of the photon detector, $\varepsilon_{kq}^{\text{det}}$, are given by Eq. (1.176) and the reduced statistical tensors $\mathcal{A}_{kq}(\alpha J)$ are given in a laboratory frame. The Euler angles φ, ϑ, ψ characterize the rotation from the laboratory to the photon frame; they are shown in Figure 1.2 on p. 38. The intrinsic anisotropy parameters for photoemission are given by

$$\alpha_k^\gamma = \sqrt{\frac{3}{2}}\hat{J}(-1)^{J+J_f+k+1}\begin{Bmatrix} J & J & k \\ 1 & 1 & J_f \end{Bmatrix} \tag{3.33}$$

Table 3.1 shows the constants α_k^γ for different pairs J and J_f connected by the dipole transitions. In contrast to the electron emission [compare Eq. (3.11)], the coefficients α_k^γ are independent of the decay matrix elements because in the dipole approximation the photon emission always proceeds through a single channel, $p = 0, L = 1$. The coefficient W_0 in Eq. (3.32) represents the total radiation probability for the transition $\alpha J \rightarrow \alpha_f J_f$.

If the detector is not sensitive to the photon polarization, the angular distribution of photons can be obtained from Eq. (3.32) using the efficiency tensors

$$\varepsilon_{00}^{\text{det}} = 2/\sqrt{3}; \qquad \varepsilon_{20}^{\text{det}} = \sqrt{2/3} \tag{3.34}$$

Table 3.1. Anisotropy Parameters α_k^γ ($k = 1, 2$) Characterizing the Angular Distribution and Polarization of Emitted Photons [see Eq. (3.33)]

J	J_f	α_1^γ	α_2^γ	J	J_f	α_1^γ	α_2^γ
0	1	0	0	1/2	1/2	$1/\sqrt{3}$	0
1	0	$1/\sqrt{2}$	$1/\sqrt{2}$	1/2	3/2	$-1/2\sqrt{3}$	0
1	1	$1/2\sqrt{2}$	$-1/2\sqrt{2}$	3/2	1/2	$\sqrt{5}/2\sqrt{3}$	1/2
1	2	$-1/2\sqrt{2}$	$1/10\sqrt{2}$	3/2	3/2	$1/\sqrt{15}$	$-2/5$
2	1	$\sqrt{3}/2\sqrt{2}$	$\sqrt{7}/2\sqrt{10}$	3/2	5/2	$-\sqrt{3}/2\sqrt{5}$	1/10
2	2	$1/2\sqrt{6}$	$-\sqrt{7}/2\sqrt{10}$	5/2	3/2	$\sqrt{7}/2\sqrt{5}$	$\sqrt{7}/5\sqrt{2}$
2	3	$-1/\sqrt{6}$	$1/\sqrt{70}$	5/2	5/2	$1/\sqrt{35}$	$-4\sqrt{2}/5\sqrt{7}$
3	2	$1/\sqrt{3}$	$\sqrt{3}/5$	5/2	7/2	$-\sqrt{5}/2\sqrt{7}$	$1/2\sqrt{14}$
3	3	$1/4\sqrt{3}$	$-\sqrt{3}/4$	7/2	5/2	$3/2\sqrt{7}$	$\sqrt{3}/2\sqrt{7}$
3	4	$-\sqrt{3}/4$	$1/4\sqrt{3}$	7/2	7/2	$1/3\sqrt{7}$	$-2/\sqrt{21}$
4	3	$\sqrt{5}/4$	$\sqrt{11}/4\sqrt{7}$	7/2	9/2	$-\sqrt{7}/6$	$\sqrt{7}/10\sqrt{3}$
4	4	$1/4\sqrt{5}$	$-\sqrt{77}/20$	9/2	7/2	$\sqrt{11}/6$	$\sqrt{11}/2\sqrt{30}$
4	5	$-1/\sqrt{5}$	$\sqrt{7}/5\sqrt{11}$	9/2	9/2	$1/3\sqrt{11}$	$-4\sqrt{2}/\sqrt{165}$

in accordance with Eq. (1.176):

$$W_{\alpha_f J_f}(\vartheta, \varphi) = \frac{W_0}{4\pi}\left[1 + \alpha_2^\gamma \sqrt{\frac{4\pi}{5}} \sum_{q=-2}^{2} \mathcal{A}_{2q}(\alpha J) Y_{2q}(\vartheta, \varphi)\right] \tag{3.35}$$

In the dipole approximation, the statistical tensors of the decaying atom of rank $k > 2$ do not affect the angular distribution. This is true both for polarization-sensitive and polarization-insensitive detectors, and it is the main qualitative difference between the angular distributions in decay with ejection of a dipole photon and in decay with ejection of an electron. Other properties of the angular distribution (3.35) are similar to those of the angular distribution of the ejected electrons discussed in the preceding section. In particular cases, expressions (3.16)–(3.19) are valid for the photon emission as well, with the only restriction $k \le 2$. Naturally, coefficients α_k in those equations should be replaced by the coefficients α_k^γ determined by Eq. (3.33). For example, if the decaying state is aligned along some direction (chosen as the z-axis of the reference frame), the angular distribution of the dipole photons has the simple form

$$W_{\alpha_f J_f}(\vartheta) = \frac{W_0}{4\pi}\left[1 + \alpha_2^\gamma \mathcal{A}_{20}(\alpha J) P_2(\cos\vartheta)\right] \tag{3.36}$$

Anisotropy of the angular distribution for the polarization-insensitive detector exists only if the decaying state is aligned, $k = 2$. Therefore, the angular distribution (3.35) is isotropic for $J < 1$. Orientation of the decaying state, i.e., the statistical tensors with odd ranks, does not affect the angular distribution. For a decaying state with axial symmetry, Eq. (3.36) gives the angular distribution of the type

$W \sim a + b\cos^2 \vartheta$, which is axially symmetric with respect to the symmetry axis of the initial state and also symmetric with respect to reflection in the plane perpendicular to this axis.

3.2.2. Polarization of Emitted Photons

To find the polarization of the photons radiated in a definite direction (ϑ, φ), we start from the photon density matrix in the helicity representation $\langle \{\vartheta\varphi\}, \lambda \, | \, \hat{\rho} \, | \, \{\vartheta, \varphi\}, \lambda' \rangle$. First we express this density matrix in terms of the statistical tensors of the photon. To do so we make a transformation to the representation of the total angular momentum:

$$\langle \{\vartheta, \varphi\}, \lambda \, | \, \rho \, | \, \{\vartheta, \varphi\}, \lambda' \rangle = \sum_{\substack{pLM \\ p'L'M'}} \langle \{\vartheta\varphi\}, \lambda \, | \, pLM \rangle$$

$$\times \langle pLM \, | \, \rho \, | \, p'L'M' \rangle \langle p'L'M' \, | \, \{\vartheta\varphi\}, \lambda' \rangle \quad (3.37)$$

and use Eqs. (1.122) and (1.123) together with the summation formula for the product of two D-functions (A.49). We obtain

$$\langle \{\vartheta, \varphi\}, \lambda \, | \, \rho \, | \, \{\vartheta, \varphi\}, \lambda' \rangle = \frac{1}{8\pi} \sum_{\substack{pLp'L' \\ k_\gamma q_\gamma}} \lambda^p \lambda'^{p'} \hat{L}\hat{L}'(-1)^{L'+1}$$

$$\times (L\lambda, L' - \lambda' \, | \, k_\gamma \lambda - \lambda') \rho_{k_\gamma q_\gamma} (pL, p'L') D_{\lambda - \lambda' q_\gamma}^{k_\gamma *} (0, \vartheta, \varphi) \quad (3.38)$$

We have taken into account in Eq. (3.38) that q_γ is even (see Section 1.3.5.2). The statistical tensors $\rho_{k_\gamma q_\gamma} (pL, p'L')$ for the emitted photon can be found by using Eq. (1.68), which relates them to the statistical tensors of the entire system "atom + photon," and further by using Eq. (3.5) connecting the latter with the initial state statistical tensors, as was done in the case of electron decay [see Eqs. (3.23) and (3.24)] with obvious changes in notations. The resulting expression for the density matrix is an incoherent sum of the contributions from different unresolved fine-structure levels, $\alpha_f J_f$:

$$\langle \{\vartheta, \varphi\}, \lambda \, | \, \rho \, | \, \{\vartheta, \varphi\}, \lambda' \rangle = \sum_{\alpha_f J_f} \langle \{\vartheta, \varphi\}, \lambda \, | \, \rho \, | \, \{\vartheta, \varphi\}, \lambda' \rangle_{\alpha_f J_f} \quad (3.39)$$

where

$$\langle \{\vartheta, \varphi\}, \lambda \, | \, \rho \, | \, \{\vartheta, \varphi\}, \lambda' \rangle_{\alpha_f J_f}$$

$$= \frac{1}{8\pi} \sum_{\substack{pLp'L' \\ kq}} \lambda^p \lambda'^{p'} \hat{L}\hat{L}'(-1)^{J+J_f+k+q+1} (L\lambda, L' - \lambda' \, | \, kq)$$

$$\times \left\{ \begin{matrix} L & L' & k \\ J & J & J_f \end{matrix} \right\} \rho_{kq}(\alpha J) D^{k*}_{\lambda - \lambda' q}(0, \vartheta, \varphi)$$

$$\times \langle \alpha_f J_f, pL : J \| H_{\text{int}} \| \alpha J \rangle \langle \alpha_f J_f, p'L' : J \| H_{\text{int}} \| \alpha J \rangle^* \qquad (3.40)$$

Here $\rho_{kq}(\alpha J)$ are given in the laboratory frame and (ϑ, φ) characterize the direction of the photon in this frame. The problem is solved: the density matrix (3.40) defines all polarization properties of the ejected photon. For example, the Stokes parameters of the radiation can be found from (3.40) using the definition (1.106):

$$P_1 = -\frac{\langle \lambda = +1 | \rho | \lambda' = -1 \rangle + \langle \lambda = -1 | \rho | \lambda' = +1 \rangle}{\langle \lambda = +1 | \rho | \lambda' = +1 \rangle + \langle \lambda = -1 | \rho | \lambda' = -1 \rangle} \qquad (3.41)$$

$$P_2 = i \frac{\langle \lambda = -1 | \rho | \lambda' = +1 \rangle - \langle \lambda = +1 | \rho | \lambda' = -1 \rangle}{\langle \lambda = +1 | \rho | \lambda' = +1 \rangle + \langle \lambda = -1 | \rho | \lambda' = -1 \rangle} \qquad (3.42)$$

$$P_3 = \frac{\langle \lambda = +1 | \rho | \lambda' = +1 \rangle - \langle \lambda = -1 | \rho | \lambda' = -1 \rangle}{\langle \lambda = +1 | \rho | \lambda' = +1 \rangle + \langle \lambda = -1 | \rho | \lambda' = -1 \rangle} \qquad (3.43)$$

The denominator in Eqs. (3.41)–(3.43), which is the trace of the density matrix (3.40), normalizes the photon density matrix. We perform further analysis in the dipole approximation. We choose the photon frame $z \| \mathbf{k}$ [i.e., $\vartheta = \varphi = 0$ in Eq. (3.40)], and write the tensor $\rho_{kq}(\alpha J)$ in the photon frame using Eq. (1.44). Then Eq. (3.40) for the dipole photon takes the form

$$\langle \{\vartheta, \varphi\}, \lambda | \rho | \{\vartheta, \varphi\}, \lambda' \rangle_{\alpha_f J_f} = \frac{3}{8\pi} |\langle \alpha_f J_f \| \hat{D} \| \alpha J \rangle|^2$$

$$\times \sum_{kq} (-1)^{J + J_f + k + 1} \left\{ \begin{matrix} 1 & 1 & k \\ J & J & J_f \end{matrix} \right\} (1\lambda, 1 - \lambda' | kq)$$

$$\times \sum_{q'} \rho_{kq'}(\alpha J) D^{k*}_{q'q}(\varphi, \vartheta, \psi)$$

$$= \frac{1}{8\pi} |\langle \alpha_f J_f \| \hat{D} \| \alpha J \rangle|^2 \hat{J}^{-1} \rho_{00}(\alpha J)$$

$$\times \left[\delta_{\lambda\lambda'} + \sqrt{6} \sum_{\substack{k=1,2 \\ q}} \alpha_k^\gamma (1\lambda, 1 - \lambda' | kq) \sum_{q'} \mathcal{A}_{kq'}(\alpha J) D^{k*}_{q'q}(\varphi, \vartheta, \psi) \right]$$

$$\qquad (3.44)$$

where the coefficients α_k are given by Eq. (3.33). The Stokes parameters of the emitted photon for the transition $\alpha J \to \alpha_f J_f$ in the dipole approximation follow from Eqs. (3.41)–(3.44):

$$P_1 = -\frac{\sqrt{\frac{3}{2}} \alpha_2^\gamma \sum_q \mathcal{A}_{2q}(\alpha J) \left[D^{2*}_{q2}(\varphi, \vartheta, \psi) + D^{2*}_{q-2}(\varphi, \vartheta, \psi) \right]}{1 + \alpha_2^\gamma \sqrt{\frac{4\pi}{5}} \sum_q \mathcal{A}_{2q}(\alpha J) Y_{2q}(\vartheta, \varphi)} \qquad (3.45)$$

$$P_2 = i\,\frac{\sqrt{\frac{3}{2}}\alpha_2^\gamma\sum_q \mathcal{A}_{2q}(\alpha J)\left[D_{q-2}^{2*}(\varphi,\vartheta,\psi)-D_{q2}^{2*}(\varphi,\vartheta,\psi)\right]}{1+\alpha_2^\gamma\sqrt{\frac{4\pi}{5}}\sum_q \mathcal{A}_{2q}(\alpha J)Y_{2q}(\vartheta,\varphi)} \tag{3.46}$$

$$P_3 = \frac{\alpha_1^\gamma\sqrt{4\pi}\sum_q \mathcal{A}_{1q}(\alpha J)Y_{1q}(\vartheta,\varphi)}{1+\alpha_2^\gamma\sqrt{\frac{4\pi}{5}}\sum_q \mathcal{A}_{2q}(\alpha J)Y_{2q}(\vartheta,\varphi)} \tag{3.47}$$

The denominator in Eqs. (3.45)–(3.47) is a normalized angular distribution for the photon emission [see Eq. (3.35)].

The Stokes parameter P_3 depends only on the direction of the radiation, whereas P_1 and P_2 are also functions of an angle ψ. This reflects the dependence of the latter two Stokes parameters on the orientation of the x and y coordinate axes in the photon frame, as was discussed in Section 1.3.5.1.

Now we illustrate how to derive the polarization of the emitted photon by another approach, starting from the angular distribution (3.32). To obtain the Stokes parameter P_3, one can use Eq. (1.109). First we obtain the intensities $W_{+1}(\vartheta,\varphi)$ and $W_{-1}(\vartheta,\varphi)$ by substituting in (3.32) the corresponding values of $\varepsilon_{kq}^{\text{det}}$ for the right and left circularly polarized light, respectively [$P_1 = P_2 = 0$, $P_3 = \pm 1$, see Eq. (1.176)]. Then we construct the ratio (1.109):

$$P_3(\vartheta,\varphi) = P_c(\vartheta,\varphi) = \frac{W_{+1}(\vartheta,\varphi) - W_{-1}(\vartheta,\varphi)}{W_{+1}(\vartheta,\varphi) + W_{-1}(\vartheta,\varphi)} \tag{3.48}$$

which now depends on the emission angles (ϑ,φ). The dependence of the intensities on the angle ψ disappears because the efficiency tensor of the detector has only zero projection. The ratio (3.48) takes exactly the form (3.47). To consider P_1 and P_2, we first define somehow an x-axis of the detector (x^{det}), which is perpendicular to the wave vector of the photon $\mathbf{k}$ ($z^{\text{det}} \parallel \mathbf{k}$), by the Euler angle ψ (Figure 1.2 on p. 38). Then we calculate the angular distribution (3.32), with the efficiency tensors (1.176) corresponding to the linear polarization along the axis x^{det}, i.e., $P_1 = +1$, $P_2 = P_3 = 0$. Let us call this angular distribution $W_\parallel(\vartheta,\varphi,\psi)$. Next we calculate (3.32) for the direction of photon polarization along the axis y^{det} ($P_1 = -1$, $P_2 = P_3 = 0$) and call the corresponding angular distribution $W_\perp(\vartheta,\varphi,\psi)$.

Now we construct the ratio

$$\begin{aligned}
P_L(\vartheta,\varphi,\psi) &= \frac{W_\parallel(\vartheta,\varphi,\psi) - W_\perp(\vartheta,\varphi,\psi)}{W_\parallel(\vartheta,\varphi,\psi) + W_\perp(\vartheta,\varphi,\psi)} \\
&= -\frac{\sqrt{\frac{3}{2}}\alpha_2^\gamma\sum_q \mathcal{A}_{2q}(\alpha J)\left[D_{q2}^{2*}(\varphi,\vartheta,\psi)+D_{q-2}^{2*}(\varphi,\vartheta,\psi)\right]}{1+\alpha_2^\gamma\sqrt{\frac{4\pi}{5}}\sum_q \mathcal{A}_{2q}(\alpha J)Y_{2q}(\vartheta,\varphi)}
\end{aligned} \tag{3.49}$$

This coincides exactly with Eq. (3.45) for the Stokes parameter P_1, as it should because of definition (1.114). Expression (3.46) for P_2 can be derived in a way similar to that for P_1, but the easiest way is to replace in Eq. (3.49) or (3.45) $\psi \rightarrow \psi + 45°$ in accordance with Eq. (1.115). Then Eq. (3.46) follows immediately from the explicit form of the D-functions (A.44).

General expression (3.49) is simplified considerably in one important particular case when the decaying state is aligned along some direction which we choose as the z-axis. In this case, only $q = 0$ components of the alignment tensor $\mathcal{A}_{2q}(\alpha J)$ are nonzero. Usually the degree of linear polarization is measured in the direction perpendicular to the axis of the alignment (i.e., $\vartheta = 90°$, $\varphi = 0$), and the intensities for two directions of a polarimeter axis: parallel (antiparallel) and perpendicular to the axis of the alignment (i.e., $\psi = 0$). Then it is easy to obtain the following relation from Eq. (3.49):

$$P \equiv \frac{W_\| - W_\perp}{W_\| + W_\perp} = \frac{3\alpha_2^\gamma \mathcal{A}_{20}(\alpha J)}{\alpha_2^\gamma \mathcal{A}_{20}(\alpha J) - 2} \tag{3.50}$$

Comparing this equation with Eq. (3.36), we see that both the anisotropy and linear polarization of the emitted photon are determined by the same product, $\alpha_2^\gamma \mathcal{A}_{20}(\alpha J) = \beta$. Therefore the measurements of the angular distribution and polarization are absolutely equivalent in this case. Combining Eqs. (3.36) and (3.50), one can express the polarization in terms of the anisotropy and vice versa:

$$P = \frac{3\beta}{\beta - 2} \tag{3.51}$$

$$W(\vartheta) = \frac{W_0}{4\pi}\left[1 + \frac{2P}{P - 3}P_2(\cos\vartheta)\right] \tag{3.52}$$

These, or equivalent, equations are used, for example, for correction of the cross sections in experiments on atomic excitation by a particle impact [25].

3.3. Polarization State of the Residual Atomic System after Decay

Let us consider the residual atomic system, $A^+(\alpha_f J_f)$ or $A(\alpha_f J_f)$ in Eq. (3.6) or (3.29), after emission of an electron or a photon, and find the statistical tensors of this residual system provided the particle was ejected in the direction **n**. We concentrate first on the electron emission (3.6). To calculate the statistical tensors of the residual atomic system, one should take the trace over the spin variables of the ejected particle in the final-state statistical tensor of the whole system. The derivation is straightforward by applying a standard technique very similar to that used in the derivations of Eqs. (3.25) and (3.26). According to definition (1.41), the statistical tensors of the residual atom can be expressed in terms of the density matrix of the final state:

$$\rho_{k_f q_f}(\alpha_f J_f; \mathbf{n}) = \sum_{M_f M'_f \mu} (-1)^{J_f - M'_f} (J_f M_f, J_f - M'_f | k_f q_f)$$

$$\times \left\langle \alpha_f J_f M_f, \mathbf{n} \frac{1}{2}\mu \,|\, \rho \,|\, \alpha_f J_f M'_f, \mathbf{n} \frac{1}{2}\mu \right\rangle \qquad (3.53)$$

Using the partial wave expansion of the electron wave function, we obtain

$$\rho_{k_f q_f}(\alpha_f J_f; \mathbf{n}) = \sum_{M_f M'_f \mu} \sum_{ll'mm'} (-1)^{J_f - M'_f} (J_f M_f, J_f - M'_f | k_f q_f)$$

$$\times \left\langle \alpha_f J_f M_f, lm, \frac{1}{2}\mu \,|\, \rho \,|\, \alpha_f J_f M'_f, l'm' \frac{1}{2}\mu \right\rangle$$

$$\times Y_{lm}(\vartheta, \varphi) Y^*_{l'm'}(\vartheta, \varphi) \qquad (3.54)$$

Here ϑ and φ are angles that determine the direction of the electron emission. The general scheme of further transformations can be as follows: coupling of the orbital angular momentum and spin of the electron to the total angular momentum j with the projection m_j; use of the expansion (A.29) for the product of two spherical harmonics; summation over m, m', μ [Eq. (A.90)]; coupling of the angular momenta of the ion core J_f and the ejected electron j into the total angular momentum J_t and projection M_t; summation over m_j, m'_j, M_f, M'_f [Eq. (A.91)]; finally, transformation of the density matrix of the total angular momentum to the corresponding statistical tensors. After all of the transformations, the statistical tensor (3.54) takes the form

$$\rho_{k_f q_f}(\alpha_f J_f; \mathbf{n}) = \frac{1}{\sqrt{4\pi}} \sum_{\substack{k_l k J_t J'_t \\ ll'jj'}} (-1)^{k+k_f+j'+1/2} \hat{f}_t \hat{f}'_t \hat{k} \hat{j} \hat{j}' \hat{l} \hat{l}' (l0, l'0 | k_l 0) \left\{ \begin{matrix} j & j' & k_l \\ l' & l & \frac{1}{2} \end{matrix} \right\}$$

$$\times \left\{ \begin{matrix} J_f & j & J_t \\ J_f & j' & J'_t \\ k_f & k_l & k \end{matrix} \right\} \sum_{qq_l} (kq, k_l q_l | k_f q_f)$$

$$\times \rho_{kq}(\alpha_f J_f, lj : J_t; \alpha_f J_f, l'j' : J'_t) Y^*_{k_l q_l}(\vartheta, \varphi) \qquad (3.55)$$

The statistical tensor of the final state in representation of the total angular momentum is connected to the statistical tensors of the decaying state $\rho_{kq}(\alpha J)$ by Eq. (3.5), which takes the form

$$\rho_{kq}(\alpha_f J_f, lj : J_t; \alpha_f J_f, l'j' : J'_t) = \delta_{JJ_t} \delta_{JJ'_t} \rho_{kq}(\alpha J) \frac{1}{\hat{J}^2}$$

$$\times \langle \alpha_f J_f, lj : J \| V \| \alpha J \rangle \langle \alpha_f J_f, l'j' : J \| V \| \alpha J \rangle^* \qquad (3.56)$$

Equation (3.55) together with Eq. (3.56) gives the desired result, which expresses the statistical tensor of the residual system in terms of the statistical tensors of the initial decaying atom, decay matrix elements, and the direction of the electron ejection:

$$
\rho_{k_f q_f}(\alpha_f J_f; \mathbf{n}) = \frac{1}{\sqrt{4\pi}} \sum_{\substack{k_l k \\ l l' j j'}} (-1)^{k+k_f+j'+1/2} \hat{k} \hat{j} \hat{j}' \hat{l} \hat{l}' \, (l0, l'0 | k_l 0) \left\{ \begin{array}{ccc} j & j' & k_l \\ l' & l & \frac{1}{2} \end{array} \right\}
$$

$$
\times \left\{ \begin{array}{ccc} J_f & j & J \\ J_f & j' & J \\ k_f & k_l & k \end{array} \right\} \sum_{q q_l} (kq, k_l q_l | k_f q_f) \rho_{kq}(\alpha J) Y^*_{k_l q_l}(\vartheta, \varphi)
$$

$$
\times \langle \alpha_f J_f, lj : J \| V \| \alpha J \rangle \langle \alpha_f J_f, l'j' : J \| V \| \alpha J \rangle^* \tag{3.57}
$$

Two particular cases of the general expression (3.57) are of interest. For an unpolarized initial atom ($k = q = 0$), Eq. (3.57) takes the form

$$
\rho_{k_f q_f}(\alpha_f J_f; \mathbf{n}) = \frac{1}{\sqrt{4\pi}} \hat{J}^{-1} \rho_{00}(\alpha J) Y^*_{k_f q_f}(\vartheta, \varphi)
$$

$$
\times \sum_{l l' j j'} (-1)^{J+J_f+j+j'+1/2} \hat{j} \hat{j}' \hat{l} \hat{l}' \hat{k}_f^{-1} \, (l0, l'0 | k_f 0) \left\{ \begin{array}{ccc} j & j' & k_f \\ l' & l & \frac{1}{2} \end{array} \right\}
$$

$$
\times \left\{ \begin{array}{ccc} j & j' & k_f \\ J_f & J_f & J \end{array} \right\} \langle \alpha_f J_f, lj : J \| V \| \alpha J \rangle \langle \alpha_f J_f, l'j' : J \| V \| \alpha J \rangle^* \tag{3.58}
$$

The direction of ejection is the symmetry axis of the residual ion: After putting $\vartheta = 0$ into the spherical harmonic in Eq. (3.58), only tensors with $q_f = 0$ survive [see Eq. (A.20)]. Moreover, the odd ranks k_f are not possible for the residual ion [as follows from the properties of the Clebsch–Gordan coefficient in Eq. (3.58)] provided parity is conserved in the decay. Therefore, the residual ion turns out to be aligned along the direction of electron emission. For $J_f < 1$, the final ion is isotropic.

Another important case is an ensemble of the residual ions corresponding to all possible directions of the ejected electrons. The integration of Eq. (3.57) over the angles of the electron emission gives

$$
\rho_{kq}(\alpha_f J_f) = \rho_{kq}(\alpha J) \sum_{lj} (-1)^{J+J_f+j+k}
$$

$$
\times \left\{ \begin{array}{ccc} J & J & k \\ J_f & J_f & j \end{array} \right\} |\langle \alpha_f J_f, lj : J \| V \| \alpha J \rangle|^2 \tag{3.59}
$$

The result (3.59) includes only a square modulus of the decay matrix element and is insensitive to the phase difference between the matrix elements. Naturally, the residual ion is characterized by the tensors of the rank k, which is not higher than the rank of the decaying system.

The statistical tensors (3.57)–(3.59) can be normalized in the standard way for the corresponding density matrices to have the unit trace. With the normalized statistical tensors of the decaying system, $\rho_{00}(\alpha J) = \hat{J}^{-1}$, one finds

$$
\rho_{kq}(\alpha_f J_f) = \left(\sum_{lj} |\langle \alpha_f J_f, lj : J \| V \| \alpha J \rangle|^2 \right)^{-1}
$$
$$
\times \rho_{kq}(\alpha J) \sum_{lj} (-1)^{J+J_f+j+k} \begin{Bmatrix} J & J & k \\ J_f & J_f & j \end{Bmatrix} |\langle \alpha_f J_f, lj : J \| V \| \alpha J \rangle|^2 \quad (3.60)
$$

We turn now to the radiation of a photon with a fixed wave vector $\mathbf{k}$ instead of ejection of an electron and derive equations similar to Eqs. (3.57)–(3.59). The starting point now is the relation

$$
\rho_{k_f q_f}(\alpha_f J_f; \mathbf{k}) = \sum_{\substack{M_f M_f' \\ \lambda = \pm 1}} (-1)^{J_f - M_f'} (J_f M_f, J_f - M_f' | k_f q_f)
$$
$$
\times \langle \alpha_f J_f M_f, \mathbf{k}\lambda |\rho| \alpha_f J_f M_f', \mathbf{k}\lambda \rangle
$$
$$
= \sum_{\substack{M_f M_f' \\ \lambda = \pm 1}} \sum_{\substack{pLM_L \\ p'L'M_L'}} (-1)^{J_f - M_f'} (J_f M_f, J_f - M_f' | k_f q_f) \frac{\hat{L}\hat{L}'}{8\pi} \lambda^p \lambda^{p'}
$$
$$
\times D^{L'}_{M_L' \lambda}(\varphi, \vartheta, 0) D^{L*}_{M_L \lambda}(\varphi, \vartheta, 0)
$$
$$
\times \langle \alpha_f J_f M_f, pLM_L |\rho| \alpha_f J_f M_f', p'L'M_L' \rangle \quad (3.61)
$$

The order of the principal steps of further transformations can be as follows: coupling the angular momentum of the residual atom J_f and angular momentum of the photon L into the total angular momentum J_t with the projection M_t; transformation from the density matrix of the total angular momentum J_t to the statistical tensors; and summation over M_f, M_f', M_t, M_t' [Eq. (A.91)]. Then the product of two D-functions in (3.61) is expanded in terms of D-functions according to Eq. (A.48) and the summation over M_L, M_L' is performed. The result of these rearrangements is the following expression:

$$
\rho_{k_f q_f}(\alpha_f J_f; \mathbf{k}) = \frac{1}{2\sqrt{4\pi}} \sum_{\lambda = \pm 1} \sum_{\substack{pLp'L' \\ kk_L}} (-1)^{1+L'} \hat{L}\hat{L}' \hat{k} \lambda^{p+p'} (L' - \lambda, L\lambda | k_L 0)
$$
$$
\times \begin{Bmatrix} J_f & L' & J \\ J_f & L & J \\ k_f & k_L & k \end{Bmatrix} \sum_{qq_L} (kq, k_L q_L | k_f q_f) \rho_{kq}(\alpha J) Y^*_{k_L q_L}(\vartheta, \varphi)
$$

$$\times \langle \alpha_f J_f, pL : J \| H_{\text{int}} \| \alpha J \rangle \langle \alpha_f J_f, p'L' : J \| H_{\text{int}} \| \alpha J \rangle^* \quad (3.62)$$

In the dipole approximation ($L = L' = 1$, $p = p' = 0$) this expression simplifies to the following:

$$\rho_{k_f q_f}(\alpha_f J_f; \mathbf{k}) = \frac{3}{2\sqrt{4\pi}} \left| \langle \alpha_f J_f \| D \| \alpha J \rangle \right|^2$$

$$\times \sum_{k k_L} \hat{k}(1 + (-1)^{k_L})(1 1, 1 - 1 | k_L 0) \begin{Bmatrix} J_f & 1 & J \\ J_f & 1 & J \\ k_f & k_L & k \end{Bmatrix}$$

$$\times \sum_{q q_L} (kq, k_L q_L | k_f q_f) \rho_{kq}(\alpha J) Y^*_{k_L q_L}(\vartheta, \varphi) \quad (3.63)$$

Note that due to the factor $(1 + (-1)^{k_L})$, k_L can be only even, i.e., $k_L = 0$ and 2. In addition, the $9j$-symbol in (3.63) has two identical rows and therefore it is nonzero only if $k_f + k_L + k$ is even [see (A.121)]. Thus $k_f + k$ is even and hence k_f and k are of the same parity. As a result of this analysis, we conclude that after emission of a dipole photon, an aligned atom can be only aligned, not oriented. Similarly, an oriented atom produces also an oriented residual atom.

If the initial atom is unpolarized ($k = q = 0$), then $k_L = k_f =$ even and the statistical tensors of the residual atom after emission of a dipole radiation are

$$\rho_{k_f q_f}(\alpha_f J_f; \mathbf{k}) = \frac{3}{\sqrt{4\pi}} \left| \langle \alpha_f J_f \| D \| \alpha J \rangle \right|^2 \hat{J}^{-1} \rho_{00}(\alpha J)$$

$$\times (-1)^{J_f + J + 1}(1 1, 1 - 1 | k_f 0) \begin{Bmatrix} J_f & 1 & J \\ 1 & J_f & k_f \end{Bmatrix} Y^*_{k_f q_f}(\vartheta, \varphi) \quad (3.64)$$

We see that in this case the residual atom can only be aligned along the direction of photon emission, which is similar to the case of electron emission.

Finally, integrating Eq. (3.63) over all emission angles gives

$$\rho_{kq}(\alpha_f J_f) = \left| \langle \alpha_f J_f \| D \| \alpha J \rangle \right|^2 \rho_{kq}(\alpha J)$$

$$\times (-1)^{J_f + J + k + 1} \begin{Bmatrix} J & J_f & 1 \\ J_f & J & k \end{Bmatrix} \quad (3.65)$$

This expression illustrates the obvious physical results: If the direction of the emitted photon is not detected, the residual atom can have only those statistical tensors that were present in the initial state. The absolute value of the final state-normalized tensor component cannot be larger than the corresponding component of the initial tensor.

3.4.   Polarization in Cascade Decay

When a highly excited atomic or ionic state is produced, such as a vacancy in a deep inner shell of a many-electron atom, the following rearrangement of the excited electronic system usually proceeds through a stepwise series of radiative and nonradiative decay processes known as a *transition cascade*. This complex process results in the emission of X-ray (fluorescence) photons and Auger (autoionization) electrons. The multitude of decay pathways and the variety of processes that can occur in a highly correlated many-electron system contribute to the complexity of the Auger electron and fluorescence spectra.

To overcome the difficulties in the analysis of the cascade and the resulting spectra, various coincidence techniques have been exploited, including coincidence measurements of any two particles (photons) emitted from the cascade.   The highly excited atomic and ionic states formed in various atomic processes are generally polarized (oriented and/or aligned) according to the dynamics and symmetry conditions of the corresponding processes.   Therefore, electron and photon emission during the cascade evolution is, as a rule, anisotropic.   A study of angular correlations between emitted particles in coincidence experiments shows great promise as a way of unraveling the complex spectra and investigating the cascade dynamics in detail.

In addition, the evolution of the polarization characteristics of atoms and ions in cascades of transitions from the initial states to lower-lying levels is important for modern studies based on recording characteristic X-ray radiation or Auger electrons at the lowest stages of the cascade.   In particular, one needs this information to obtain total cross sections of the processes under investigation by measuring the yield of the radiations in some fixed directions, taking into account an anisotropy in their angular distribution caused by possible alignment of the emitting atomic or ionic states.   In this section we consider two problems: the angular correlations between two successive particles and the evolution of the polarization characteristics of atoms in cascades.

3.4.1.   Angular Correlation between Two Particles Emitted Consecutively

We consider first the most general case of two successive particles emitted from an excited atom (ion). Here and below "particle" means either an Auger (autoionization) electron or a (fluorescent) photon. We assume that the atomic transitions occur in steps and the atomic states involved at each stage have well-defined total angular momentum and parity. Then the process can be represented as follows:

$$A_0(\alpha_0 J_0) \rightarrow A_1(\alpha_1 J_1) + a_1(j_1)$$
$$\quad\quad \hookrightarrow A_2(\alpha_2 J_2) + a_2(j_2) \tag{3.66}$$

Here $A_i(\alpha_i J_i), i = 0, 1, 2$ is an atomic (ionic) state characterized by total angular momentum J_i, and the set of all other quantum numbers denoted as α_i, $a_k(j_k), k = 1, 2$ applies to the particle, electron or photon, with total angular momentum j_k. The initial state with total angular momentum J_0 may be oriented and/or aligned so that it can have a nonstatistical population of magnetic substates which is described by the statistical tensors $\rho_{k_0 q_0}(\alpha_0 J_0)$. In general, all components of the statistical tensors with $k_0 \leq 2J_0$, $-k_0 \leq q_0 \leq k_0$ can be nonzero.

The general expression for the angular correlation function for two particles emitted in succession from a polarized initial state can be easily obtained by combining the results obtained in Sections 3.1 and 3.3. Owing to the assumption of the stepwise character of the process, we can consider each decay in the cascade independently. We assume that the detectors are insensitive to the polarization (spin) of the particles. The statistical tensors of the residual atom (ion) in the first decay corresponding to the emission of the particle a_1 in the direction $\mathbf{n}_1$ are given by Eq. (3.57). These statistical tensors determine the angular distribution of the second particle a_2 according to Eq. (3.10). The resulting expression, which gives the angular correlation function for two particles, can be presented in the following form:

$$W(\mathbf{n}_1, \mathbf{n}_2) = \sum_{k_0 q_0 k_1 k_2} G_{k_1 k_2 k_0} \rho_{k_0 q_0}(\alpha_0 J_0) \{Y_{k_1}(\mathbf{n}_1) \otimes Y_{k_2}(\mathbf{n}_2)\}_{k_0 q_0} \tag{3.67}$$

where $\{Y_{k_1}(\mathbf{n}_1) \otimes Y_{k_2}(\mathbf{n}_2)\}_{k_0 q_0}$ is the bipolar spherical harmonic (A.36).

Each term in the sum (3.67) consists of three factors. The last two factors, the statistical tensor and the spherical functions, are determined by the initial conditions and the geometry of the experiment. The dynamics of the decay process determine the first factor, the generalized angular correlation coefficient $G_{k_1 k_2 k_0}$. It contains the amplitudes of the decay processes and angular momentum coupling coefficients. Owing to the assumption of the stepwise character of the process, each of the coefficients $G_{k_1 k_2 k_0}$ may be further split into two factors, one of which depends entirely on the first transition and the other on the second:

$$G_{k_1 k_2 k_0} = B_{k_1 k_2 k_0}(J_0, J_1) A_{k_2}(J_1, J_2) \tag{3.68}$$

These factors are given by

$$B_{k_1 k_2 k_0}(J_0, J_1) = (-1)^{k_1 - k_0} \hat{J}_0^{-1} \hat{J}_1 \sum_{j_1 j_1'} C_{k_1 0}^*(j_1, j_1')$$

$$\times \begin{Bmatrix} J_1 \; j_1 \; J_0 \\ J_1 \; j_1' \; J_0 \\ k_2 \; k_1 \; k_0 \end{Bmatrix} \langle J_1, j_1 \| T \| J_0 \rangle \langle J_1, j_1' \| T \| J_0 \rangle^* \qquad (3.69)$$

$$A_{k_2}(J_1, J_2) = (-1)^{J_1+J_2} \hat{f}_1^{-1} \sum_{j_2 j_2'} C_{k_2 0}^*(j_2, j_2')$$

$$\times (-1)^{j_2'} \begin{Bmatrix} J_1 \; j_2 \; J_2 \\ j_2' \; J_1 \; k_2 \end{Bmatrix} \langle J_2, j_2 \| T \| J_1 \rangle \langle J_2, j_2' \| T \| J_1 \rangle^* \quad (3.70)$$

Here $\langle J_b, j \| T \| J_a \rangle$ denotes the decay amplitude which describes a transition from state J_a to state J_b by emission of a particle with total angular momentum j; $C_{k0}(j, j')$ are the corresponding radiation parameters (see Section 1.3). For Auger electrons, the radiation parameters are given by Eq. (1.105) and for dipole photons they are given by Eq. (1.125) (after substitution $L = L' = 1$, $p = p' = 0$).

Since the atomic states involved are assumed to have well-defined parities, and parity is conserved in the decay processes, it is easy to show that only even values of k_1 and k_2 can occur in (3.67). The values of k_0 and k_2 are limited by the conditions $0 \leq k_{0(2)} \leq 2J_{0(1)}$. In addition, the summation in (3.67) is limited by the triangle inequality $|k_0 - k_2| \leq k_1 \leq k_0 + k_2$, which follows from the properties of the Clebsch–Gordan coefficient in the bipolar spherical harmonics.

In general, the angular correlation pattern (3.67) is complex and difficult to analyze. However, by selecting the initial conditions and the geometry of the experiment, it is possible to simplify the pattern and therefore its analysis. For example, if the initial state of the cascade is aligned along some direction, which we choose to be the z-axis of the reference frame, and if one of the cascade particles is detected along this direction, then detection of the second particle will be axially symmetrical with respect to this axis. Therefore the angular correlation will depend on only one polar angle, ϑ. This result can be easily obtained from the general expression (3.67). In this case the angular correlation can be expressed as

$$W(\vartheta) = \sum_{k=\text{even}} a_k P_k(\cos \vartheta) \qquad (3.71)$$

where

$$a_k = \frac{1}{4\pi} \sum_{k_0 k_1} \hat{k} \hat{k}_1 \, (k_1 \, 0, k \, 0 | k_0 \, 0) \, \rho_{k_0 0}(\alpha_0 J_0) \, G_{k_1 k k_0} \qquad (3.72)$$

Another case is the geometry used in practically all experiments where both detectors are placed in a plane perpendicular to the symmetry axis of the initial state (z-axis). In this case, the angular correlation function can be expressed in the form:

$$W(\vartheta_1, \vartheta_2) = \frac{1}{2\pi} \sum_{k_0 k_1 k_2} \sum_{q=0,2,\dots} G_{k_1 k_2 k_0} \rho_{k_0 0}(\alpha_0 J_0) (2 - \delta_{0q})$$

$$\times (k_1 q, k_2 - q \,|\, k_0 0) \, \bar{P}_{k_1}^q (\cos \vartheta_1) (-1)^{q/2} \, \bar{P}_{k_2}^q (\cos \vartheta_2) \qquad (3.73)$$

Here ϑ_1 and ϑ_2 are polar angles of the particles; $\bar{P}_{k_2}^q (\cos \vartheta_2)$ are normalized associated Legendre polynomials defined by Eq. (A.9).

If the initial atomic state is unpolarized, the general expression (3.67) simplifies. Then $k_0 = q_0 = 0$ and $k_1 = k_2$; thus Eq. (3.67) takes the form:

$$W(\bar{\vartheta}) = \sum_{k=\text{even}} \bar{A}_k(J_0, J_1) A_k(J_1, J_2) P_k(\cos \bar{\vartheta}) \qquad (3.74)$$

where factor $A_k(J_1, J_2)$ is defined by Eq. (3.70) while for $\bar{A}_k(J_0, J_1)$ one obtains

$$\bar{A}_k(J_0, J_1) = (-1)^{J_1 + J_0} \hat{J}_1 \sum_{j_1 j_1'} C_{k0}^*(j_1, j_1')$$

$$\times (-1)^{j_1'} \begin{Bmatrix} J_1 & j_1 & J_0 \\ j_1' & J_1 & k \end{Bmatrix} \langle J_1, j_1 \,\|\, T \,\|\, J_0 \rangle \, \langle J_1, j_1' \,\|\, T \,\|\, J_0 \rangle^* \qquad (3.75)$$

Now the angular correlation function depends only on one angle, $\bar{\vartheta}$, the relative angle between two directions of particle emission. This fact has a simple physical explanation. The initial state is unpolarized; therefore the first particle is emitted isotropically. However, if we fix the direction of its emission and if it takes a definite angular momentum from the atom, then the intermediate atomic state will be aligned along the direction of emission of the first particle. The angular distribution of the second particle can be anisotropic, but it will be axially symmetrical about the direction of the first particle. k is even; therefore the distribution is also symmetric with respect to reflection through the $\bar{\vartheta} = 90°$ plane. The degree of complexity of the angular correlation function is determined by the condition $k \leq (2j_1, 2j_2, 2J_1)$. From this result it is clear that if any of the particles involved or the intermediate state have angular momenta equal to 0 or $\frac{1}{2}$, the angular correlation disappears, $W = \text{const}$.

If one of the particles is a dipole photon, then $k \leq 2$ and the angular correlation takes a familiar form [see Eq. (3.36)]:

$$W(\bar{\vartheta}) \sim 1 + \alpha_2^{J_1 \to J_2} \mathcal{A}_{20}(\alpha_1 J_1) P_2(\cos \bar{\vartheta}), \qquad (3.76)$$

where $\alpha_2^{J_1 \to J_2} = A_2(J_1, J_2)/A_0(J_1, J_2)$ is the intrinsic anisotropy parameter for the second decay and $\mathcal{A}_{20}(\alpha_1 J_1) = \bar{A}_2(J_0, J_1)/\bar{A}_0(J_0, J_1)$ is the alignment of the intermediate state along the direction of the first particle emission.

3.4.2. Evolution of the Spin Density Matrix of Excited Atoms in a Cascade of Electromagnetic Transitions

As was pointed out in the introduction to this section, the problem of the evolution of the polarization characteristics of an atom (ion) in a cascade is important for the interpretation of many modern experiments. Here we discuss this problem using as an example the cascade of electromagnetic transitions. The usual cascade equations for the total population of the levels involved in a cascade summed over all corresponding magnetic sublevels cannot be used for our purpose. The same approach extended to the space of magnetic sublevels of all levels in the cascade is very cumbersome and difficult for practical use because of the large dimension of this space. Moreover, it could lead to wrong results if it is applied to situations where the phase relations between the population amplitudes of different magnetic sublevels play a role. This is, for example, the case for the angular correlation between two emitted characteristic X-rays, which is used in modern experimental studies. The problem of the alignment transfer from the initial states of a highly excited atom or ion to its lower-lying states in the course of an electromagnetic cascade can be solved in an elegant way by using the method of recurrence relations for the statistical tensors of the states involved in the cascade.

Let us consider the cascade of radiative transitions $J_n \rightarrow J_{n'} + \gamma$ in the de-excitation process of a highly excited atom or ion with the initial distribution C_{J_i} for the total (summed over the magnetic sublevels) probabilities of populating the atomic levels that are proportional to the total cross sections of their formation:

$$C_{J_i} \sim \sigma_{J_i} \tag{3.77}$$

Let $\mathcal{A}_{kq}^0(J_i)$ be the normalized statistical tensors for these states at the same initial moment of the cascade. Then the statistical tensors $\mathcal{A}_{kq}(J_n)$ for all lower-lying states during the cascade can be obtained from the recurrence equations

$$\mathcal{A}_{kq}(J_n) = \frac{C_{J_n}}{P_{J_n}} \mathcal{A}_{kq}^0(J_n) + \sum_{\varepsilon_n' > \varepsilon_n} \frac{\lambda_R(J_n' \rightarrow J_n)}{\lambda_{\text{tot}}(J_n')} \frac{P_{J_n'}}{P_{J_n}} \mathcal{A}_{kq}^{J_n'}(J_n) \tag{3.78}$$

Here, $\mathcal{A}_{kq}^{J_n'}(J_n)$ is a normalized *partial statistical tensor* of the level J_n that corresponds to an unobserved spontaneous radiative transition from the level J_n' and that can be expressed in terms of the normalized statistical tensors $\mathcal{A}_{kq}(J_n')$ of that level according to Eq. (1.69):

$$\mathcal{A}_{kq}^{J_n'}(J_n) = (-1)^{J_n' + J_n + 1 + k} \hat{J}_n' \hat{J}_n \begin{Bmatrix} J_n' & J_n & 1 \\ J_n & J_n' & k \end{Bmatrix} \mathcal{A}_{kq}(J_n') \tag{3.79}$$

$\lambda_R(J_n' \rightarrow J_n)$ is the rate of radiative transition from the level J_n' to J_n and $\lambda_{\text{tot}}(J_n')$ is the total rate of transition (including the rate of Auger transitions if they are possible) from the level J'_n. The quantity P_{J_n} is a population of the level J_n, which is found from the well-known cascade equation:

$$P_{J_n} = C_{J_n} + \sum_{\varepsilon'_n > \varepsilon_n} \frac{\lambda_R(J'_n \to J_n)}{\lambda_{\text{tot}}(J'_n)} P_{J'_n} \tag{3.80}$$

Summation in Eqs. (3.78) and (3.80) expands over all levels of J'_n taking part in the cascade and having energies ε'_n greater than ε_n.

An important particular case of the process considered is a cascade of deexcitation of highly excited atoms or ions initially formed under conditions of axial symmetry. Here only the statistical tensors $\mathcal{A}^0_{k0}$ and hence only $\mathcal{A}_{k0}$ with $q = 0$ (the axial symmetry axis being taken as the quantization one) enter the problem at any stage of the cascade. However, the situation changes and the symmetry of the system is modified as soon as the photon corresponding to some intermediate stage of the cascade $J_a \to J_b + \gamma_{ab}$ is detected and the location of the detector $(\vartheta_\gamma, \varphi_\gamma)$ is fixed. The system, as a rule, loses its axial symmetry and is characterized by a new set of parameters. These can be obtained by using the same equations (3.78) and (3.80) with

$$C_{J_b} = 1; \qquad C_{J_{n \neq b}} = 0 \tag{3.81}$$

which contain now, for $\mathcal{A}^0_{kq}(J_i)$, the statistical tensors $\tilde{\mathcal{A}}_{kq}(J_b)$ for the state $|b\rangle$, which plays the role of the initial state for the next stage of the cascade after the photon γ_{ab} is detected. The tensors can be calculated using Eq. (3.63):

$$\tilde{\mathcal{A}}_{k_b q}(J_b) \sim \sum_{k_a} \tilde{\mathcal{A}}_{k_a 0}(J_a) \sum_{k_\gamma} [1 + (-1)^{k_\gamma}] \, (11, 1 - 1 | k_\gamma q)$$

$$\times \begin{Bmatrix} 1 & J_b & J_a \\ 1 & J_b & J_a \\ k_\gamma & k_b & k_a \end{Bmatrix} (k_a 0, k_\gamma q | k_b q) \, Y^*_{k_\gamma q}(\vartheta_\gamma, \varphi_\gamma) \tag{3.82}$$

Here $\tilde{\mathcal{A}}_{k_a 0}(J_a)$ are the statistical tensors of the upper state $|J_a\rangle$ in the detected transition $J_a \to J_b + \gamma_{ab}$.

Resonant and Two-Step Processes

4.1. Two-Step Reactions of Excitation and Radiation Decay of Discrete Atomic Levels

4.1.1. Resonance Scattering of Photons

Consider the angular distribution of photons emitted from an atomic discrete or autoionizing state excited by photon absorption (see Figure 4.1 for the scheme of the process under consideration):

$$A(\alpha_0 J_0) + \gamma_0 \longrightarrow A^*(\alpha J) \longrightarrow A(\alpha_f J_f) + \gamma \qquad (4.1)$$

The processes in Figure 4.1 are often lumped together as *resonance scattering of photons (resonance fluorescence)*. The second and the third ones, displayed in Figures 4.1(b) and 4.1(c), respectively, are processes of resonance inelastic scattering. In molecular optics, resonance inelastic scattering is known as *Raman scattering*: Figure 4.1(b) corresponds to emission of the red component, while

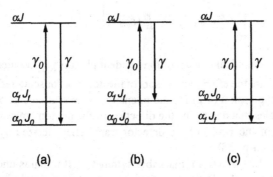

(a)   (b)   (c)

FIGURE 4.1. Schemes of transitions in resonant scattering of photons.

Figure 4.1(c) corresponds to emission of the violet component in the Raman spectra. A rigorous treatment of the resonance scattering of photons is performed in the second-order perturbation theory of quantum electrodynamics. In our approach, the phenomenon is treated as a pure two-step process. Within the dipole approximation the tensor structure of the transition operator for both steps, excitation and decay, is the simplest, and all angular momenta involved in the phenomenon are fixed in an ideal case. As a result, the angular distributions and polarization can be exhaustively described only by methods of the angular momentum algebra without any dynamical evaluations.

Assume that the atom in the final state $A(\alpha_f J_f)$ in Eq. (4.1) is not detected. It is assumed that the statistical tensors of the initial photons γ_0, the initial state of the atom $A(\alpha_0 J_0)$, and the efficiency tensors of the detector are given in the coordinate systems S^{ph}, S^{lab}, and S^{det}, respectively. Combining Eqs. (3.32), (2.12), (1.41), and (1.50) to calculate the parameters corresponding to the absorption and emission stages of the process, one can obtain the general expression for the intensity of the photon emission in an arbitrary geometry, including polarization features of the atomic target $A(\alpha_0 J_0)$, the incident photon γ_0, and the detector of the emitting photons γ. We will not present this rather cumbersome but straightforward general formula. Instead we will consider some typical cases.

4.1.1.1. Unpolarized Target

For unpolarized targets, the photon frame S^{ph} is convenient to take as the laboratory frame S^{lab}. The reduced statistical tensors of the photoexcited state follow from Eqs. (2.15) and (1.50). Substituting these tensors in Eq. (3.32), we obtain the intensity of scattered photons:

$$W(\vartheta, \varphi, \psi) = \frac{W_0}{4\pi} \left[1 + \frac{9}{2}(-1)^{J_0 - J_f}(2J+1) \sum_{k=1,2} \begin{Bmatrix} 1 & 1 & k \\ J & J & J_f \end{Bmatrix} \begin{Bmatrix} 1 & 1 & k \\ J & J & J_0 \end{Bmatrix} \right.$$

$$\left. \times \sum_{qq'} D_{qq'}^{k*}(\varphi, \vartheta, \psi)\, \rho_{kq}^{\gamma_0} (\varepsilon_{kq'}^{det})^* \right] \tag{4.2}$$

Here $\rho_{kq}^{\gamma_0}$ are the statistical tensors of the incident photon γ_0 presented in Table 1.1 on p. 25 and $\varepsilon_{kq'}^{det}$ are the efficiency tensors of the detector for scattered photons [see Eq. (1.176)]. Expression (4.2) for the intensity of the scattered photons depends on the three angles that determine the direction of the photon emission (ϑ, φ) and the orientation of the x-axis of the detector frame characterized by the angle ψ (see Figure 1.2 on p. 38).

Equation (4.2) shows a remarkable symmetry: It remains unchanged after the permutation $J_0 \leftrightarrow J_f$. This means that all correlation and polarization features of the red and violet components in the Raman spectra are identical regardless of

Table 4.1. Parameters β and ξ of Angular Distribution and Polarization in the Resonance Scattering of Photons [Eqs. (4.4) and (4.12)]

J_0	J	J_f	ξ	β	J_0	J	J_f	ξ	β
0	1	0	1	-1	3	4	4	1/8	11/40
0	1	1	1/2	1/2	3	4	5	$-1/2$	$-1/10$
0	1	2	$-1/2$	$-1/10$	1/2	1/2	1/2	2/3	0
1	0	1	0	0	1/2	1/2	3/2	$-1/3$	0
1	1	1	1/4	$-1/4$	1/2	3/2	1/2	5/6	$-1/2$
1	1	2	$-1/4$	1/20	1/2	3/2	3/2	1/3	2/5
1	2	1	3/4	$-7/20$	1/2	3/2	5/2	$-1/2$	$-1/10$
1	2	2	1/4	7/20	3/2	1/2	3/2	1/6	0
1	2	3	$-1/2$	$-1/10$	3/2	3/2	3/2	2/15	$-8/25$
2	1	2	1/4	$-1/100$	3/2	3/2	5/2	$-1/5$	2/25
2	2	2	1/12	$-7/20$	3/2	5/2	3/2	7/10	$-7/25$
2	2	3	$-1/6$	1/10	3/2	5/2	5/2	1/5	8/25
2	3	2	2/3	$-6/25$	3/2	5/2	7/2	$-1/2$	$-1/10$
2	3	3	1/6	3/10	5/2	3/2	5/2	3/10	$-1/50$
2	3	4	$-1/2$	$-1/10$	5/2	5/2	5/2	2/35	$-64/175$
3	2	3	1/3	$-1/35$	5/2	5/2	7/2	$-1/7$	4/35
3	3	3	1/24	$-3/8$	5/2	7/2	5/2	9/14	$-3/14$
3	3	4	$-1/8$	1/8	5/2	7/2	7/2	1/7	2/7
3	4	3	5/8	$-11/56$	5/2	7/2	9/2	$-1/2$	$-1/10$

the polarization of the incident beam. This symmetry is broken, in general, for polarized targets.

The angular distribution of the scattered photons follows from Eq. (4.2) after substituting the efficiency tensors of the detector, which is insensitive to polarization, $\varepsilon_{20}^{\text{det}} = \sqrt{2/3}$, and using Eq. (A.51):

$$W(\vartheta, \varphi) = \frac{W_0}{4\pi} \left[1 - 3\sqrt{\frac{2\pi}{15}} \beta \sum_q \rho_{2q}^{\gamma_0} Y_{2q}(\vartheta, \varphi) \right] \tag{4.3}$$

where

$$\beta = 3(-1)^{1+J_0-J_f}(2J+1) \begin{Bmatrix} 1 & 1 & 2 \\ J & J & J_f \end{Bmatrix} \begin{Bmatrix} 1 & 1 & 2 \\ J & J & J_0 \end{Bmatrix} \tag{4.4}$$

The anisotropy coefficient β determines the angular distribution of the scattered photons for an arbitrarily polarized initial photon beam. The coefficient β depends only on the values of the total angular momenta of the three atomic states involved in the resonance scattering (see Figure 4.1): J_0, J, and J_f. Table 4.1 presents the values of β for various combinations of these quantum numbers. The table includes only the cases $J_f \geq J_0$ because the coefficient β is symmetric with respect to the permutation $J_0 \leftrightarrow J_f$.

The circular component in the polarization of the incident photons has no effect on the angular distribution (4.3) because only the statistical tensors of the second rank $\rho_{2q}^{\gamma_0}$ determine the anisotropy of the scattered photons γ. For an unpolarized initial photon beam ($P_1 = P_2 = P_3 = 0$), we may choose the z^{ph}-axis along the beam. Substituting the value of the statistical tensor for this case, $\rho_{20}(\gamma_0) = 1/\sqrt{6}$ (see column S in Table 1.1 on p. 25) in Eq. (4.3), one obtains

$$W(\vartheta, \varphi) = \frac{W_0}{4\pi}\left(1 - \frac{\beta}{2}P_2(\cos\vartheta)\right) \tag{4.5}$$

For linearly polarized incident photons ($P_1 = 1$, $P_2 = P_3 = 0$), we choose the z^{ph}-axis along the direction of the polarization vector. Substituting now $\rho_{20}^{\gamma_0} = -2/\sqrt{6}$ (see column S' in Table 1.1) in Eq. (4.3), one obtains

$$W(\vartheta, \varphi) = \frac{W_0}{4\pi}(1 + \beta P_2(\cos\vartheta)) \tag{4.6}$$

Note that to get the simplest expressions for the angular distributions in the cases of unpolarized and linearly polarized initial photons, we choose different coordinate frames, and the angle ϑ in Eqs. (4.5) and (4.6) is counted from the direction of the initial photon beam and from the direction of its linear polarization, respectively. The angular distribution for arbitrarily polarized initial photons can be easily written using Eq. (4.3) and Table 1.1. For example, in the coordinate frame with the z^{ph}-axis along the incident beam, one finds for arbitrarily polarized initial photons with the Stokes parameters P_1, P_2 (the Stokes parameter P_3 does not matter):

$$\begin{aligned} W(\vartheta, \varphi) &= \frac{W_0}{4\pi}\left[1 - \frac{\beta}{4}\left(3\cos^2\vartheta - 1 - 3\sin^2\vartheta\,(P_1\cos 2\varphi + P_2\sin 2\varphi)\right)\right] \\ &= \frac{W_0}{4\pi}\left[1 - \frac{\beta}{4}\left(3\cos^2\vartheta - 1 - 3P_l\cos 2\Delta\sin^2\vartheta\right)\right] \end{aligned} \tag{4.7}$$

where

$$\Delta = \varphi - \varphi_0 \tag{4.8}$$

and P_l, φ_0 are given by Eqs. (1.117) and (1.118). Equation (4.7) shows that the angular distribution in the azimuthal angle φ depends only on Δ, i.e., on the azimuthal angle of the detector position relative to the principal axis of the polarization ellipse of the incident photon beam.

The degree of polarization of the scattered photons follows from the combination of Eqs. (3.48) and (3.49) with Eq. (4.2). In the case of an unpolarized target, the corresponding expression for the degree of linear polarization P_L (relative to the direction x^{det}, see Figure 1.2 on p. 38) as a function of the detector position takes the form (the z^{ph}-axis is along the incident photon beam):

$$P_L(\vartheta,\varphi,\psi) = \frac{3}{4}\beta$$

$$\times \frac{\sin^2\vartheta\cos 2\psi - P_l\cos 2\Delta(1+\cos^2\vartheta)\cos 2\psi + 2P_l\sin 2\Delta\cos\vartheta\sin 2\psi}{1 - \frac{\beta}{4}\left(3\cos^2\vartheta - 1 - 3P_l\cos 2\Delta\sin^2\vartheta\right)}$$

$$(4.9)$$

The degree of linear polarization (4.9) is expressed in terms of the anisotropy parameter β, defined by Eq. (4.4). The direction ψ_0 of the principal axis of the polarization ellipse of the scattered photon is defined by the requirement $P_2(\vartheta,\varphi,\psi) = 0$ with the result

$$\tan 2\psi_0 = P_l \frac{2\sin 2\Delta\cos\vartheta}{\sin^2\vartheta - P_l\cos 2\Delta(1+\cos^2\vartheta)} \tag{4.10}$$

The degree of circular polarization, P_c, is of the form

$$P_c(\vartheta,\varphi) = \frac{3}{2}\xi \frac{P_3\cos\vartheta}{1 - \frac{\beta}{4}\left(3\cos^2\vartheta - 1 - 3P_l\cos 2\Delta\sin^2\vartheta\right)} \tag{4.11}$$

Equation (4.11) shows that the degree of circular polarization needs one more parameter:

$$\xi = 3(-1)^{J_0-J_f}(2J+1)\begin{Bmatrix} 1 & 1 & 1 \\ J & J & J_f \end{Bmatrix}\begin{Bmatrix} 1 & 1 & 1 \\ J & J & J_0 \end{Bmatrix} \tag{4.12}$$

Values of ξ for different transitions are presented in Table 4.1. The parameter ξ is symmetric with respect to the permutation $J_0 \leftrightarrow J_f$, which is similar to the coefficient β [Eq. (4.4)]. A regularity is seen in Table 4.1: For transitions with increasing or decreasing angular momentum ($J-1 \to J \to J+1$ or $J+1 \to J \to J-1$), the angular distribution and polarization of scattered photons are the same and independent of J. This follows from the relation

$$(2J+1)\begin{Bmatrix} 1 & 1 & k \\ J & J & J+1 \end{Bmatrix}\begin{Bmatrix} 1 & 1 & k \\ J & J & J-1 \end{Bmatrix} = (-1)^k \frac{4}{(2-k)!(3+k)!} \tag{4.13}$$

which can be derived from expressions for $6j$-symbols with two pairs of equal momenta [see (A.111)–(A.116)]. The corresponding values of the coefficients are $\beta = -1/10$ and $\xi = -1/2$.

Equations (4.11) and (4.9) also define the Stokes parameters of the photon emitted in the direction ϑ,φ [see Eqs. (1.109), (1.114), and (1.115) and the discussion in Section 3.2.2] in the coordinate frame S^{det}:

$$P_1(\vartheta,\varphi,\psi) = P_L(\vartheta,\varphi,\psi)$$
$$P_2(\vartheta,\varphi,\psi) = P_L(\vartheta,\varphi,\psi+45°)$$
$$P_3(\vartheta,\varphi) = P_c(\vartheta,\varphi) \tag{4.14}$$

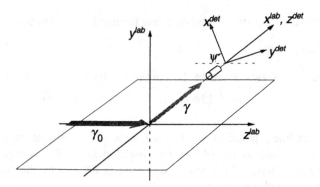

FIGURE 4.2. The geometry of an experiment on resonance photon scattering (see text).

It follows from (4.11) that the circular component of polarization can exist for a scattered photon only if the initial photon has nonzero circular polarization, $P_3 \neq 0$. The circular component of the initial photon does not affect the degree of linear polarization of the scattered photon (4.9). Recall that these conclusions were derived for the case of an unpolarized target. For an unpolarized initial photon beam ($P_1 = P_2 = P_3 = 0$), only linear polarization of the scattered photon (4.9) is nonzero:

$$P_L(\vartheta, \varphi, \psi) = \frac{3}{4} \beta \, \frac{\sin^2 \vartheta \cos 2\psi}{1 - \frac{\beta}{4}(3\cos^2 \vartheta - 1)} \tag{4.15}$$

Naturally, due to the axial symmetry of the process, Eq. (4.15) does not depend on the azimuthal angle φ.

To illustrate the application of the above formulas, let us consider in more detail the geometry of an experiment (Figure 4.2) where the detector is positioned perpendicular to the initial photon beam ($\vartheta = 90°$). This geometry is often used in practice. From Eq. (4.11) one finds that the circular polarization of the fluorescence vanishes at this angle. Nonzero P_c in measurements can reflect, for example, the initial polarization of the target. Equation (4.9) simplifies to the form

$$P_L(\psi) = \frac{3}{4} \beta \cos 2\psi \, \frac{1 - P_l \cos 2\Delta}{1 + \frac{\beta}{4}(1 + 3P_l \cos 2\Delta)} \tag{4.16}$$

The degree of linear polarization (4.16) depends on the angle ψ, which characterizes the direction of the polarimeter axis x^{det} with respect to z^{lab}, as shown in Figure 4.2. One observes the largest degree of linear polarization for $\psi = \frac{\pi}{2}n$ ($n = 0, 1, 2, 3$), that is, for measurements of the difference between intensities of the photon flux and the axis of the polarimeter along and perpendicular to the initial photon beam. The degree of linear polarization vanishes for the angles

$\psi = \frac{\pi}{4}n$ ($n = 1, 3, 5, 7$). This means that the intensities are the same for angles of the polarimeter axis that are 45° and 135° with respect to the incident beam. Therefore, the direction of the principal axis of the polarization ellipse coincides with the direction of the incident beam. The same result follows directly from Eq. (4.10) after putting $\vartheta = 90°$. Note that the linear polarization (4.16) vanishes for the incident beam, which is purely linearly polarized in the direction of the detector ($P_l = 1$, $\Delta = 0$). For an unpolarized target and a linearly polarized initial beam, the photons emitted in the direction of the initial polarization are not polarized ($P_1 = P_2 = P_3 = 0$).

The modulation of the intensity $W(\psi)$ as a function of the angle ψ of the polarimeter for the geometry shown in Figure 4.2 follows from Eq. (4.2) after substituting particular parameters and angles:

$$W(\psi) = \frac{W_0}{4\pi}(a + b\cos 2\psi) \qquad (4.17)$$

where the coefficients a and b are given by the relations

$$a = 1 + \frac{\beta}{4}(1 + 3P_l\cos 2\Delta), \quad b = \frac{3}{4}\beta(1 - P_l\cos 2\Delta) \qquad (4.18)$$

The intensity $W(\psi)$ shows the largest modulation when the incident beam is linearly polarized perpendicular to the direction of the detector ($P_l = 1, \Delta = 90°$), while there is no modulation when the incident beam is linearly polarized in the direction of the detector ($P_l = 1$, $\Delta = 0$).

4.1.1.2. Polarized Target

We assume that the statistical tensors of the target and the incident photon, $\rho_{k_0q_0}(\alpha_0 J_0)$ and $\rho_{kq}^{\gamma_0}$, are both given in the laboratory frame. For this choice, a generalization of Eq. (4.2) can be obtained directly from Eqs. (3.32) and (2.12). Owing to polarization of the target, an additional term appears in the intensity of photons in comparison with Eq. (4.2):

$$W(\vartheta, \varphi, \psi) = W^{(\mathrm{iso})}(\vartheta, \varphi, \psi) + \Delta W(\vartheta, \varphi, \psi) \qquad (4.19)$$

where $W^{(\mathrm{iso})}(\vartheta, \varphi, \psi)$ is the scattered photon flux for an unpolarized target given by Eq. (4.2) and

$$\Delta W(\vartheta, \varphi, \psi) = \frac{W^{(\mathrm{iso})}}{4\pi}\frac{9}{2}f^2\hat{J}_0\sum_{\substack{k_0>0 \\ q_0}}\mathscr{A}_{k_0q_0}(\alpha_0 J_0)\sum_{k_\gamma k}(-1)^{J+J_f+k+1}\hat{k}_0\hat{k}_\gamma \begin{Bmatrix} 1 & 1 & k \\ J & J & J_f \end{Bmatrix}$$

$$\times \begin{Bmatrix} J_0 & 1 & J \\ J_0 & 1 & J \\ k_0 & k_\gamma & k \end{Bmatrix} \sum_{qq'q_\gamma}(k_0 q_0, k_\gamma q_\gamma | kq)\, D_{qq'}^{k*}(\varphi, \vartheta, \psi)\rho_{k_\gamma q_\gamma}^{\gamma_0}(\varepsilon_{kq}^{\mathrm{det}})^* \qquad (4.20)$$

where $\mathcal{A}_{k_0 q_0}(\alpha_0 J_0)$ are the reduced statistical tensors of the initial atomic state. Owing to the property (A.121) of $9j$-symbols, $k_0 + k_\gamma + k =$ even. This relation is useful for predicting general features of photon scattering from polarized targets. For example, if the target is aligned ($k_0 =$ even) and the incident photon has no circular component ($k_\gamma =$ even) the scattered photons also do not have a circular component, regardless of the geometry of the experiment.

Consider a simple example of application of Eq. (4.19) to the forward scattering ($\vartheta = 0°$) of unpolarized photons from a polarized target with $J_0 = 1/2$. The value of k_0 in the additional term (4.20) is then fixed: $k_0 = 1$. Using values of the statistical tensors for unpolarized photons from Table 1.1 on p. 25 and Eqs. (A.44) and (A.54) for the Wigner D-functions, one obtains

$$\Delta W(\vartheta, \varphi, \psi) = C \, \mathcal{A}_{10}(\alpha_0 J_0) \, (\varepsilon_{10}^{\text{det}})^* \tag{4.21}$$

where C depends only on the total angular momenta J and J_f. Equation (4.21) shows that $\Delta W(\vartheta, \varphi, \psi)$ is nonvanishing only if the target polarization possesses a component P_z along the incident beam and the detector is sensitive to the circular polarization component of the radiation. Higher ranks of $k_0 > 1$ are possible for polarized targets with $J_0 > 1/2$. Then an additional photon flux (4.20) will occur as a result of the alignment of the target, even for a polarization-insensitive detector.

Integration of Eq. (4.19) over the scattered photon angles (ϑ, φ) gives the total scattered photon flux:

$$W = W_0 \left[1 + 3 \hat{J}_0 (-1)^{J+J_0+1} \sum_{\substack{k=1,2 \\ q}} \left\{ \begin{matrix} 1 & 1 & k \\ J_0 & J_0 & J \end{matrix} \right\} \mathcal{A}_{kq}(\alpha_0 J_0) \, (\rho_{kq}^{\gamma_0})^* \right] \tag{4.22}$$

Owing to the polarization of the target, the full expression for the scattered photon flux includes three incoherent terms corresponding to an isotropic target ($k = 0$), the orientation of the target ($k = 1$), and the alignment of the target ($k = 2$).

4.1.2. Electron–Photon Correlations in Coincidence $(e, e'\gamma)$ Experiments

Since the early 1970s [26] the most detailed results on atomic excitation by electron impact have been provided by investigations of the $(e, e'\gamma)$ process: measurements of coincidences between a scattered electron and a photon emitted in the process:

$$
\begin{array}{ccc}
e_0 \quad + \quad A(\alpha_0 J_0) & \longrightarrow & e \quad + \quad A^*(\alpha J) \\
 & & \downarrow \\
 & & A(\alpha_f J_f) \quad + \quad \gamma
\end{array}
\tag{4.23}
$$

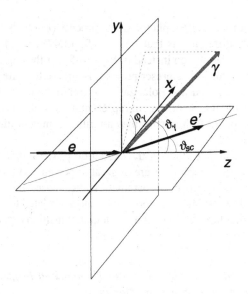

FIGURE 4.3. Kinematic diagram of the $(e, e'\gamma)$ experiment.

Detecting the scattered electron e fixes the scattering angle and separates, within the energy resolution, the excited atomic level αJ, while the angular distribution and/or the polarization of the emitted photon γ carry information on the amplitudes of the transition $A(\alpha_0 J_0) \longrightarrow A^*(\alpha J)$ between different magnetic sublevels of the target atom A induced by electron impact.

The geometry of the process $(e, e'\gamma)$ is displayed in Figure 4.3. Consider the excitation of a well-separated atomic state αJ with a sharp value of the total angular momentum. The angular distribution and polarization of photons emitted from polarized atoms were considered in Section 3.2. To obtain the angular correlation function between a scattered electron and an emitted photon, one has to put into Eq. (3.35) an expression for the reduced statistical tensors $\mathcal{A}_{2q}(\alpha J)$ of the atomic state αJ excited by electron impact. The tensors $\mathcal{A}_{2q}(\alpha J)$ depend on the particular conditions of the excitation [the first line of Eq. (4.23)]. These quantities can be expressed in terms of the transition amplitudes in the representation of projections, $T_{M_0 \mu_0 \to M \mu}(\mathbf{p}_0, \mathbf{p})$, or in terms of the reduced amplitudes, $\langle \alpha J, lj : \tilde{J} \| T \| \alpha_0 J_0, l_0 j_0 : \tilde{J} \rangle$, according to the general prescriptions of Section 2.2.1. For example, using Eqs. (2.53) or (2.62), the reduced statistical tensors $\mathcal{A}_{kq}(\alpha J)$ can be found for the case of an unpolarized electron beam and atomic target. More general expressions (2.39), (2.41) or (2.60), (2.61) should be used when considering reactions with polarized electrons and/or polarized target atoms. As a result, Eq. (3.35), which now includes the excitation amplitudes that depend on scattering angle, gives the electron–photon angular correlation function.

Polarization of the emitted photon in the process $(e, e'\gamma)$ is considered quite similarly. The only difference is that Eqs. (3.45)–(3.47) should be used for the Stokes parameters of the photon instead of Eq. (3.35) for the angular distribution.

Although using the above prescriptions it is simple to write general expressions for the $(e, e'\gamma)$ angular and polarization correlations in terms of the amplitudes, a general problem is a convenient representation of the angular correlation function and the choice of physically meaningful parameters directly related to the quantities observed.

Here the symmetry features of the excitation process and/or the symmetry features of the excitation amplitudes are of great importance. In the above discussion we used a general approach to parameterization based on the formalism of statistical tensors. The statistical tensors can be further related to other parameters convenient for particular reactions and particular transitions (see, for example, Ref. 9).

4.1.2.1. Unpolarized Initial Electron Beam, Unpolarized Target, Electron Detector Insensitive to Spin Polarization

In this simplest case, the atomic excitation process is symmetric with respect to the scattering plane. Taking x- and z-axes in this plane of symmetry and using the results of Section 1.4.3, we find that the polarization of the angular momentum J of the excited atomic state (with definite parity) is characterized by the independent statistical tensors presented in the second column of Table 1.2 on p. 30. As follows from Eq. (3.35) for an arbitrary value of $J > 1/2$, the angular distribution of emitted photons is characterized only by the statistical tensors of the rank $k = 2$. Therefore, only three independent real parameters, $\mathcal{A}_{20}(\alpha J)$, $\mathcal{A}_{21}(\alpha J)$, and $\mathcal{A}_{22}(\alpha J)$, or some equivalent set of three parameters, describe the $(e, e'\gamma)$ angular correlation function for arbitrary $J > 1/2$ and arbitrary J_0. If the z-axis is chosen along the incident beam, the angular correlation function can be written using the explicit expressions for the spherical harmonics from Table A.3 on p. 204:

$$\frac{d^2\sigma}{d\Omega_{sc}d\Omega_\gamma}(\vartheta, \varphi) = \frac{1}{4\pi}\frac{d\sigma}{d\Omega_{sc}}\left[1 + \alpha_2^\gamma\left(\frac{1}{2}\mathcal{A}_{20}(\alpha J)(3\cos^2\vartheta - 1)\right.\right.$$
$$\left.\left. -\sqrt{6}\mathcal{A}_{21}(\alpha J)\sin\vartheta\cos\vartheta\cos\varphi + \sqrt{\frac{3}{2}}\mathcal{A}_{22}(\alpha J)\sin^2\vartheta\cos 2\varphi\right)\right] \quad (4.24)$$

where coefficients α_2^γ are given by Eq. (3.33). The right side of Eq. (4.24) is normalized in such a way that after integration over angles (ϑ, φ), the differential excitation cross section $d\sigma/d\Omega_{sc}$ is obtained. The angular correlation function (4.24) is symmetric with respect to the scattering plane and its shape is described by only two parameters when the detector of photons moves in only one plane.

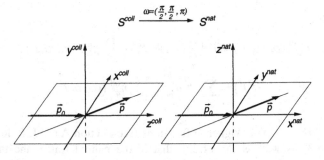

FIGURE 4.4. Collision and natural coordinate frames used to describe the $(e, e'\gamma)$ process.

For example, in the coplanar geometry, i.e., when the detector of photons moves in the scattering plane $(\varphi = 0, \pi)$, Eq. (4.24) takes the form

$$\frac{d^2\sigma}{d\Omega_{sc}d\Omega_\gamma}(\vartheta, \varphi) = \frac{1}{4\pi}\frac{d\sigma}{d\Omega_{sc}}A\left[1 + C\cos 2(\vartheta \mp \gamma)\right] \qquad (4.25)$$

with

$$A = 1 + \frac{1}{4}\alpha_2^\gamma\left(\mathcal{A}_{20}(\alpha J) + \sqrt{6}\mathcal{A}_{22}(\alpha J)\right) \qquad (4.26)$$

$$C = \frac{\sqrt{6}}{4A}\alpha_2^\gamma\left[4\mathcal{A}_{21}^2(\alpha J) + \left(\sqrt{\frac{3}{2}}\mathcal{A}_{20}(\alpha J) - \mathcal{A}_{22}(\alpha J)\right)^2\right]^{1/2} \qquad (4.27)$$

$$\tan 2\gamma = -\frac{2\mathcal{A}_{21}(\alpha J)}{\sqrt{\frac{3}{2}}\mathcal{A}_{20}(\alpha J) - \mathcal{A}_{22}(\alpha J)} \qquad (4.28)$$

The minus and plus signs in Eq. (4.25) correspond to $\varphi = 0°$ and $\varphi = 180°$, respectively, and angle γ is defined in the interval $(0°, 180°)$. The angle γ is the angle of tilting of the symmetry axis of the angular correlation function with respect to the incident electron beam.

In the above considerations we used the collision frame displayed in Figure 2.1. Another coordinate frame convenient for analyzing collisions is a *natural frame* with the z-axis perpendicular to the scattering plane and the x-axis along the incident beam. Both frames are shown in Figure 4.4. The collision frame is usually better for calculations of the excitation amplitudes, while the natural frame is often more convenient for a general analysis of parameterization. The collision frame transforms into the natural frame by means of the rotation $\omega = (\pi/2, \pi/2, \pi)$. Statistical tensors in collision and natural frames are related by the expressions

$$\rho_{kq}^{coll}\left(\alpha J,\alpha'J'\right)=i^{-q}\sum_{q'}(-1)^{q'}d_{qq'}^{k}\left(\frac{\pi}{2}\right)\rho_{kq'}^{nat}\left(\alpha J,\alpha'J'\right)$$

$$\rho_{kq}^{nat}\left(\alpha J,\alpha'J'\right)=\sum_{q'}i^{-q'}d_{qq'}^{k}\left(\frac{\pi}{2}\right)\rho_{kq'}^{coll}\left(\alpha J,\alpha'J'\right) \qquad (4.29)$$

where $d_{qq'}^{k}(\vartheta)$ are small Wigner D-functions (see Section A.6). It follows from Section 1.4.3, as well as Eq. (4.29), that in the natural frame, the three real independent parameters describing the $(e,e'\gamma)$ angular correlation function are $\mathcal{A}_{20}(\alpha J)$, $\mathrm{Re}\,\mathcal{A}_{22}(\alpha J)$, and $\mathrm{Im}\,\mathcal{A}_{22}(\alpha J)$. A general form of the angular correlation function in the natural frame may be derived similar to Eq. (4.24) and is of the form

$$\frac{d^2\sigma}{d\Omega_{sc}d\Omega_{\gamma}}(\vartheta,\varphi)=\frac{1}{4\pi}\frac{d\sigma}{d\Omega_{sc}}\left[1+\alpha_2^{\gamma}\left(\frac{1}{2}\mathcal{A}_{20}(\alpha J)(3\cos^2\vartheta-1)\right.\right.$$

$$\left.\left.+\sqrt{\frac{3}{2}}\sin^2\vartheta\,(\mathrm{Re}\,\mathcal{A}_{22}(\alpha J)\cos 2\varphi-\mathrm{Im}\,\mathcal{A}_{22}(\alpha J)\sin 2\varphi)\right)\right] \qquad (4.30)$$

where the angles ϑ and φ and the reduced statistical tensors are now given in the natural frame.

Consider the polarization of photons in the process $(e,e'\gamma)$. From Eqs. (3.45)–(3.47) and from the plane symmetry of the excitation process, it follows that not more than four independent real parameters describe the $(e,e'\gamma)$ polarization correlation function for arbitrary $J>1/2$. In the collision frame, the corresponding statistical tensors are $i\mathcal{A}_{11}(\alpha J)$, $\mathcal{A}_{20}(\alpha J)$, $\mathcal{A}_{21}(\alpha J)$, and $\mathcal{A}_{22}(\alpha J)$; while in the natural frame, an equivalent set is $\mathcal{A}_{10}(\alpha J)$, $\mathcal{A}_{20}(\alpha J)$, $\mathrm{Re}\,\mathcal{A}_{22}(\alpha J)$, and $\mathrm{Im}\,\mathcal{A}_{22}(\alpha J)$. Equations (3.45)–(3.47) show that independent of the geometry of the process, one should investigate the circular polarization of the emitted photon in order to obtain information on the statistical tensor of the first rank, $\mathcal{A}_{1q}(\alpha J)$.

The usual choice is observation of the Stokes parameters of the radiation emitted normal to the scattering plane. In this case the detector coordinate system coincides with the natural frame (Figure 4.4). Hence, there is no need for any rotation in Eqs. (3.45)–(3.47), provided the reduced statistical tensors $\mathcal{A}_{kq}(\alpha J)$ of the excited atom are taken in the natural frame. Using Eq. (A.53) we immediately obtain

$$P_1=-\sqrt{\frac{3}{2}}\frac{\alpha_2^{\gamma}\left(\mathcal{A}_{22}(\alpha J)+\mathcal{A}_{2-2}(\alpha J)\right)}{1+\alpha_2^{\gamma}\mathcal{A}_{20}(\alpha J)}=-\sqrt{6}\frac{\alpha_2^{\gamma}\,\mathrm{Re}\,\mathcal{A}_{22}(\alpha J)}{1+\alpha_2^{\gamma}\mathcal{A}_{20}(\alpha J)} \qquad (4.31)$$

$$P_2=-i\sqrt{\frac{3}{2}}\frac{\alpha_2^{\gamma}\left(\mathcal{A}_{22}(\alpha J)-\mathcal{A}_{2-2}(\alpha J)\right)}{1+\alpha_2^{\gamma}\mathcal{A}_{20}(\alpha J)}=\sqrt{6}\frac{\alpha_2^{\gamma}\,\mathrm{Im}\,\mathcal{A}_{22}(\alpha J)}{1+\alpha_2^{\gamma}\mathcal{A}_{20}(\alpha J)} \qquad (4.32)$$

$$P_3 = \sqrt{3} \frac{\alpha_1^\gamma \, \mathcal{A}_{10}(\alpha J)}{1 + \alpha_2^\gamma \, \mathcal{A}_{20}(\alpha J)} \tag{4.33}$$

A general relation holds

$$\tan 2\gamma = \frac{P_2}{P_1} \tag{4.34}$$

where γ is the tilt angle in the $(e, e'\gamma)$ angular correlation function (4.25). To obtain Eq. (4.34) one can transform the reduced statistical tensors in Eqs. (4.31) and (4.32) into the collision frame by using Eq. (4.29), apply Table A.6 on p. 211 for the D-functions, use the relations (1.133), (1.134), and compare the result with Eq. (4.28). Within the LS-coupling approximation, the angle γ for the excited atomic P state may be interpreted as orientation of the electron cloud with respect to the incident electron beam [9].

In general, the three Stokes parameters (4.31)–(4.33) are not enough to extract all information contained in the $(e, e'\gamma)$ process with an unpolarized incident electron beam and an unpolarized target atom. To obtain the fourth independent parameter (corresponding to a statistical tensor of the second rank), one has to invoke data on the angular correlation function, or move the detector of polarized photons to another position to measure the Stokes parameter P_1 and/or P_2 in this new position. For example, the Stokes parameter P_1 in the scattering plane (a P_4 parameter) is measured for this purpose by a photon detector positioned $90°$ with respect to the incident beam when the angle $0°$ for the polarimeter in Eq. (1.114) corresponds to the initial beam direction. Taking the relevant angles $\varphi = 3\pi/2$, $\vartheta = \pi/2$, $\psi = \pi/2$ in Eq. (3.45), we obtain

$$P_4 = \frac{\alpha_2^\gamma \left(3 \mathcal{A}_{20}(\alpha J) - \sqrt{6} \, \mathrm{Re} \, \mathcal{A}_{22}(\alpha J) \right)}{2 - \left(\alpha_2^\gamma \, \mathcal{A}_{20}(\alpha J) + \sqrt{6} \, \mathrm{Re} \, \mathcal{A}_{22}(\alpha J) \right)} \tag{4.35}$$

where the statistical tensors of the excited atom are taken in the natural frame as in Eqs. (4.31)–(4.33).

The above analysis of the process $(e, e'\gamma)$ with unpolarized beams with an electron detector insensitive to the spin polarization is general for all values of initial and final total angular momenta of the target J_0 and $J > 1/2$, when four independent dynamic parameters exist. Further restrictions on the number of independent parameters can follow from model approximations for scattering amplitudes, e.g., LS-coupling, plane-wave Born approximation, and so on.

4.1.2.2. General Case

The same equations [(3.35) and (3.45)–(3.47)] are valid for the angular distribution and the Stokes parameters of an emitted photon in more general cases when the incident electron beam and/or target are polarized, including the case when the detector of scattered electrons is sensitive to spin polarization. However, the plane symmetry of the excitation part in Eq. (4.23) is generally broken by additional directions related to the polarization states of the electrons and/or the target. As a result, the symmetry relations between components of the statistical tensors $\mathcal{A}_{kq}(\alpha J)$ of the excited atomic state $A^*(\alpha J)$ which were discussed in Section 1.4.3 are not valid. The angular distribution of the emitted photon (3.35) is therefore determined (for $J > 1/2$) by five real parameters related to the polarization state of the excited atom, for example, $\mathrm{Re}\,\mathcal{A}_{22}(\alpha J)$, $\mathrm{Im}\,\mathcal{A}_{22}(\alpha J)$, $\mathrm{Re}\,\mathcal{A}_{21}(\alpha J)$, $\mathrm{Im}\,\mathcal{A}_{21}(\alpha J)$, and $\mathcal{A}_{20}(\alpha J)$. The above components can be taken in both collision and natural frames.

The Stokes parameters (3.45)–(3.47) of the emitted photon are determined generally (for $J > 1/2$) by eight real parameters related to the polarization state of the excited atom. In addition to the five parameters listed above, three parameters from the tensor of the first rank exist: $\mathrm{Re}\,\mathcal{A}_{11}(\alpha J)$, $\mathrm{Im}\,\mathcal{A}_{11}(\alpha J)$, and $\mathcal{A}_{10}(\alpha J)$. The latter three parameters govern only the circular polarization of the emitted light [see Eq. (3.47)].

The statistical tensors of the excited atom are expressed in terms of the excitation amplitudes, but particular expressions now depend also on the polarization states of the incident electron and the target and/or on the efficiency tensors of the detector of scattered electrons. This provides broad possibilities for using the $(e, e'\gamma)$ experiments with polarized electrons and atoms in very detailed (at a level of *complete experiment* or *perfect experiment*) investigations of electron excitation amplitudes. Prescriptions on how to calculate the tensors $\mathcal{A}_{kq}(\alpha J)$ in particular situations were provided in Section 2.2.1.

4.1.2.3. Plane-Wave Born Approximation

For an unpolarized initial atom, one can substitute the reduced statistical tensors (2.71) into Eq. (3.35) and with the use of Eq. (A.32) obtain the angular correlation function in the PWBA for excitation with a single contributing multipole λ:

$$W_{\alpha_f J_f}(\vartheta, \varphi) = \frac{W_0}{4\pi}\left(1 + \beta^B(\lambda)P_2\left(\cos\vartheta_{\gamma Q}\right)\right) \tag{4.36}$$

where $\vartheta_{\gamma Q}$ is the angle between the direction of the momentum transfer $\mathbf{Q}$ and the direction of radiation emission:

$$\beta^B(\lambda) = \sqrt{\frac{3}{2}}(-1)^{J_0 - J_f + 1}f^2\hat{\lambda}^2(\lambda\,0, \lambda\,0|20)\begin{Bmatrix} J & J & 2 \\ 1 & 1 & J_f \end{Bmatrix}\begin{Bmatrix} J & J & 2 \\ \lambda & \lambda & J_0 \end{Bmatrix} \tag{4.37}$$

is the anisotropy parameter in the PWBA. The angular correlation function (4.36) is axially symmetric with respect to the direction of $\mathbf{Q}$ and symmetric with respect to

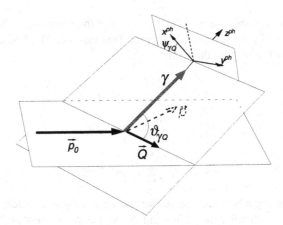

FIGURE 4.5. Kinematic diagram for the description of the $(e, e'\gamma)$ process in the PWBA.

to the plane normal to $\mathbf{Q}$. For the dipole excitation $(\lambda = 1)$, Eq. (4.36) exactly coincides with the angular distribution (4.6) and (4.4) of fluorescence after photoexcitation of an unpolarized atom by linearly polarized light: $\beta^B(\lambda = 1) = \beta$. The only difference is that in the PWBA, the emission angle is counted from the direction of the momentum transfer $\mathbf{Q}$, while for photoexcitation it is counted from the direction of linear polarization of the incident light.

In particular, for transitions with $J_0 = 0$, $J = 1$, $J_f = 0$ (a typical case for atoms with closed shells), Eq. (4.36) yields

$$W(\vartheta, \varphi) = \frac{W_0}{4\pi} \frac{3}{2} \sin^2 \vartheta_{\gamma Q} \qquad (4.38)$$

Any deviation from the shape (4.38) indicates a breakdown of the PWBA.

The formula for Stokes parameters of the fluorescence in the $(e, e'\gamma)$ process can be derived in the PWBA from Eqs. (2.71) and (3.45)–(3.47) with the use of an additional formula (A.46) and other properties of the D-functions (see Section A.6):

$$P_1 = -\frac{3}{2}\beta^B(\lambda)\frac{\sin^2 \vartheta_{\gamma Q} \cos 2\psi_{\gamma Q}}{1 + \beta^B(\lambda)P_2\left(\cos \vartheta_{\gamma Q}\right)} \qquad (4.39)$$

$$P_2 = -\frac{3}{2}\beta^B(\lambda)\frac{\sin^2 \vartheta_{\gamma Q} \sin 2\psi_{\gamma Q}}{1 + \beta^B(\lambda)P_2\left(\cos \vartheta_{\gamma Q}\right)} \qquad (4.40)$$

Here $\psi_{\gamma Q}$ is an angle between the axis x^{ph} and a plane defined by the direction of the momentum transfer $\mathbf{Q}$ and the direction of the radiation emission, shown in Figure 4.5. The Stokes parameter P_3 [Eq. (3.47)] is zero due to $\mathcal{A}_{1q}(\alpha J) = 0$, because in the PWBA, the atomic state being excited from an unpolarized atom is aligned but not oriented (see Section 2.2.1.3). For the usual set of Stokes parameters, P_1, P_2, and P_4 of the emitted radiation [see Eqs. (4.31), (4.32), and (4.35)] Eqs. (4.39) and (4.40) take the form

$$P_1 = -\frac{3\beta^B(\lambda)}{2 - \beta^B(\lambda)} \cos 2\vartheta_Q \qquad (4.41)$$

$$P_2 = -\frac{3\beta^B(\lambda)}{2 - \beta^B(\lambda)} \sin 2\vartheta_Q \qquad (4.42)$$

$$P_4 = -\frac{3\beta^B(\lambda)}{2 - \beta^B(\lambda)} \cos^2 \vartheta_Q \qquad (4.43)$$

where ϑ_Q is the angle between the incident beam and the momentum transfer $\mathbf{Q}$.

The angular distribution (4.36) and the Stokes parameters (4.39)–(4.43) of the photons emitted in the reaction $(e, e'\gamma)$ are independent of an atomic model and are defined only by the geometry of the process and the values of the total angular momenta of the initial atomic state J_0, the excited atomic state J, and the final atomic state after radiative decay, J_f. This statement is general within a PWBA, provided only a single multipole λ contributes to the excitation. Equations (4.36) and (4.39)–(4.43) remain valid for the case of several contributing multipoles, but the more general anisotropy parameter β^B should be used instead of $\beta^B(\lambda)$:

$$\beta^B = \sqrt{\frac{3}{2}}(-1)^{J_0-J_f+1}\hat{f}^2 \begin{Bmatrix} J & J & 2 \\ 1 & 1 & J_f \end{Bmatrix} \left[\sum_\lambda |M^B_\lambda(J_0 \to J)|^2\right]^{-1}$$

$$\times \sum_{\lambda\lambda'} \hat{\lambda}\hat{\lambda}' (\lambda 0, \lambda' 0 | 2 0) \begin{Bmatrix} J & J & 2 \\ \lambda & \lambda' & J_0 \end{Bmatrix} i^{\lambda'-\lambda} M^B_\lambda(J_0 \to J) M^{B*}_{\lambda'}(J_0 \to J) \quad (4.44)$$

The reduced multipole PWBA amplitude $M^B_\lambda(J_0 \to J)$ is defined by Eq. (2.67). Equation (4.44) can be obtained in a way similar to $\beta^B(\lambda)$, Eq. (4.37). One need only find the reduced statistical tensors $\mathcal{A}_{kq}(\alpha J; \mathbf{Q})$ using the general expression (2.66) instead of simplified expression (2.71) and substitute $k_0 = 0$ for the unpolarized target atom.

4.1.3. Angular Anisotropy and Polarization of Fluorescence in Electron–Atom and Ion–Atom Collisions

The measurement of photon emissions from collision-excited atoms is one of the oldest methods of investigating excitation cross sections. This process is represented by Eq. (4.23), but it differs from the reaction $(e, e'\gamma)$ considered in Section 4.1.2, since the scattered electrons are not detected. The energy loss in the collision is not fixed and in general the excited atomic level can be populated, not only directly from the ground state, but also indirectly via radiation cascades from the higher-lying levels excited by the same incident beam. This cascade vanishes near the excitation threshold when the incident energy is not enough to excite higher-lying levels. In this section we neglect the cascades and consider a well-isolated atomic level αJ excited by a particle impact. The angular distribution and polarization of photons from the level αJ can be found from Eqs. (3.35) and (3.45)–(3.47) after substituting the statistical tensors of the excited atom from Section 2.2.1 in a form convenient for the particular calculations.

The angular distribution and polarization of photons for an unpolarized target and an unpolarized incident particle beam take on especially simple form. With the quantization axis along the incident beam, statistical tensors of the excited state with only zero projection exist [see Eqs. (2.55) and (2.63)]. Then it follows from Eq. (3.35) that

$$W_{\alpha_f J_f}(\vartheta, \varphi) = \frac{W_0}{4\pi} \left(1 + \beta_2 P_2 (\cos \vartheta) \right) \tag{4.45}$$

where β_2 is the angular anisotropy coefficient:

$$\beta_2 = \alpha_2^{\gamma} \mathcal{A}_{20}(\alpha J) \tag{4.46}$$

The anisotropy coefficient (4.46) is a product of the reduced statistical tensor of the excited atom $\mathcal{A}_{20}(\alpha J)$, which depends on the excitation dynamics, and of the anisotropy parameter α_2^{γ}, a factor that represents the radiative decay part [in the dipole approximation, see Eq. (3.33) and Table 3.1 on p. 117].

Owing to the general relation between the threshold and high-energy limits of the reduced statistical tensors [see Eq. (2.86)], a similar relation holds for the anisotropy coefficient (4.46) (within the LS-coupling approximation and neglecting the effects of a cascade of radiation, as discussed in Section 2.2.1.4):

$$\beta_2(E \gg E_{\text{th}}) = -\frac{1}{2} \beta_2(E = E_{\text{th}}) \tag{4.47}$$

It is implied that the PWBA is valid for high energies of the incident electron. It follows from Eq. (4.46) that the anisotropy coefficient as a function of the incident electron energy changes sign at least once.

The Stokes parameters of the fluorescence follow from Eqs. (3.45)–(3.47):

$$P_1 = -\frac{3}{2} \beta_2 \frac{\sin^2 \vartheta \cos 2\psi}{1 + \beta_2 P_2 (\cos \vartheta)} \tag{4.48}$$

$$P_2 = -\frac{3}{2} \beta_2 \frac{\sin^2 \vartheta \sin 2\psi}{1 + \beta_2 P_2 (\cos \vartheta)} \tag{4.49}$$

$$P_3 = 0 \tag{4.50}$$

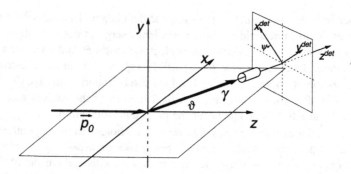

FIGURE 4.6. Kinematic diagram for measurements of fluorescence polarization.

where β_2 is the angular anisotropy coefficient (4.46). The angle ψ defines the direction of the axis x^{det} with respect to the reaction plane determined by the directions of the incident beam and emitted radiation, as shown in Figure 4.6.

Equations (4.45), (4.48), and (4.49) are very similar to Eqs. (4.36), (4.39), and (4.40), respectively. This fact illustrates the common symmetry properties of the processes, the coincidence experiment on the $(e, e'\gamma)$ reaction treated within the framework of the PWBA, and the noncoincident process with detection of fluorescence regardless of any approximation. The decaying state in both cases has a symmetry axis. In the first case, the decaying state is aligned along the momentum transfer vector $\mathbf{Q}$; in the second case it is aligned along the incident beam. As a result, all relations between correlation and polarization characteristics of the decay products are identical, as well as the general parameterizations themselves.

When the emitted photon is detected perpendicular to the incident particle beam ($\vartheta = \pi/2$ and for $x^{det} \parallel z$, one has to put $\psi = \pi$), only the Stokes parameter P_1 [Eq. (4.48)] is nonzero:

$$P_1 = \frac{3\beta_2}{\beta_2 - 2} \tag{4.51}$$

Equations (4.45), (4.48) and (4.49) show that the polarization and angular distribution of fluorescence are directly related to each other. To stress this fact, one can combine Eqs. (4.45) and (4.51), for example, in the form

$$W(\vartheta) = W(90°)(1 - P_1 \cos^2 \vartheta) \tag{4.52}$$

Relations similar to (4.51) and (4.52) are valid within the PWBA for the coincidence $(e, e'\gamma)$ process (see Section 4.1.2.3). In this case, however, the angle ϑ is counted from the direction of the momentum transfer $\mathbf{Q}$ instead of the direction of the incident beam.

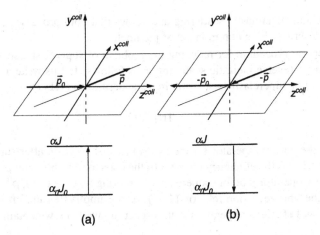

FIGURE 4.7. Kinematic diagrams and schemes of the transitions for inelastic (a) and corresponding superelastic (b) scattering.

4.1.4. Superelastic Scattering on Laser-Excited Polarized Atoms

A process

$$e + A^*(\alpha J) \longrightarrow e_0 + A(\alpha_0 J_0) \tag{4.53}$$

in which an incident electron gains energy from an excited target atom, deexciting the atom into its lower-lying state, is called *superelastic scattering*. Superelastic scattering can be considered as a process reverse to inelastic scattering:

$$e_0 + A(\alpha_0 J_0) \longrightarrow e + A^*(\alpha J) \tag{4.54}$$

Figure 4.7 shows a scheme for both processes, (4.53) and (4.54), together with the collision coordinate frame. In practice, the excited polarized atomic targets for superelastic scattering experiments are usually prepared by laser optical pumping [27] (see Section 5.1).

Since the Hamiltonian of the system atom + electron is invariant with respect to time reversion, the amplitudes of inelastic scattering (4.54) and superelastic scattering (4.53) are rigorously connected

$$\langle \alpha_0 J_0 M_0, -\mathbf{p}_0 \mu_0 \mid T \mid \alpha J M, -\mathbf{p}\mu \rangle$$
$$= (-1)^{J_0 + M_0 + 1/2 + \mu_0 + J + M + 1/2 + \mu} \langle \alpha J - M, \mathbf{p} - \mu \mid T \mid \alpha_0 J_0 - M_0, \mathbf{p}_0 - \mu_0 \rangle \tag{4.55}$$

Here $\mathbf{p}_0$ and $\mathbf{p}$ are linear momenta of the incoming and scattered electrons in the process (4.54), while $-\mathbf{p}$ and $-\mathbf{p}_0$ are those in the reverse process (4.53). Equation (4.55) provides a basis for all further connections between the characteristics of inelastic and superelastic scattering and shows that both processes can be, in principle, described by a common set of parameters.

A general procedure for reducing the problem of superelastic scattering to the problem of reverse inelastic scattering is as follows: Let us write a detection probability after superelastic scattering in a standard form (1.141):

$$W = \text{Tr}(\rho^0 \varepsilon^0) \tag{4.56}$$

Here ρ^0 is the density operator for the system atom + electron after superelastic scattering and ε^0 is the efficiency operator of the detectors. Expressing ρ^0 in terms of the density operator ρ before superelastic scattering by Eq. (1.24), $\rho^0 = T\rho T^+$, and using the time reversion relation (4.55) for the amplitudes and invariance of the density and efficiency operators with respect to inversion, we obtain

$$W = \sum_{\substack{\alpha J M \mu \\ \alpha' J' M' \mu'}} \sum_{\substack{\alpha_0 J_0 M_0 \mu_0 \\ \alpha'_0 J'_0 M'_0 \mu'_0}} (-1)^{J_0 + J'_0 + J - J' + M_0 - M'_0 + M - M' + \mu_0 - \mu'_0 + \mu - \mu'}$$

$$\times \eta \langle \alpha J - M, \mathbf{p} - \mu \,|\, T \,|\, \alpha_0 J_0 - M_0, \mathbf{p}_0 - \mu_0 \rangle$$

$$\times \langle \alpha' J' - M', \mathbf{p} - \mu' \,|\, T^+ \,|\, \alpha'_0 J'_0 - M'_0, \mathbf{p}_0 - \mu'_0 \rangle^*$$

$$\times \langle \alpha'_0 J'_0 M'_0, \mathbf{p}_0 \mu'_0 \,|\, \varepsilon^0 \,|\, \alpha_0 J_0 M_0, \mathbf{p}_0 \mu_0 \rangle \tag{4.57}$$

where $\eta = \pi_0 \pi'_0 \pi \pi'$ is the product of internal parities of the corresponding atomic states. Normally $\eta = 1$. All factors in Eq. (4.57) should be taken in the common coordinate frame. General Eqs. (4.57) and (4.58) provide a basis for further simplifications.

Consider, for example, the probability of the superelastic scattering of unpolarized electrons from a laser-excited polarized atom $A^*(\alpha J)$ when the detector of the scattered electrons e_0 in Eq. (4.53) is not sensitive to their spin state and the polarization of the final atomic state $A(\alpha_0 J_0)$ is not detected. Substituting into Eq. (4.57) the density matrices (1.40) for unpolarized electrons and unpolarized atoms, we obtain

$$W = \sum_{\substack{\alpha \alpha' J J' \\ M M'}} (-1)^{J - J' + M - M'} \left[\sum_{\substack{M_0 \mu \mu_0 \mu \\ \alpha_0 J_0}} T_{M_0 \mu_0 \rightarrow M \mu}(\mathbf{p}_0, \mathbf{p}) \, T^*_{M_0 \mu_0 \rightarrow M' \mu}(\mathbf{p}_0, \mathbf{p}) \right]$$

$$\times \langle \alpha J - M \,|\, \rho^{\text{las}} \,|\, \alpha' J' - M' \rangle$$

$$= \sum_{\substack{\alpha \alpha' J J' \\ M M'}} (-1)^{J - J' + M - M'} \langle \alpha J M \,|\, \rho \,|\, \alpha' J' M' \rangle \langle \alpha J - M \,|\, \rho^{\text{las}} \,|\, \alpha' J' - M' \rangle$$

$$\tag{4.58}$$

or, equivalently,

$$W = \sum_{\substack{\alpha\alpha'JJ' \\ kq}} (-1)^k \rho^*_{kq}(\alpha J, \alpha'J') \rho^{las}_{kq}(\alpha J, \alpha'J') \quad (4.59)$$

Here we used the short notation (2.40) for the transition matrix elements and applied Eq. (2.44). The density matrix $\langle \alpha JM | \rho^{las} | \alpha'J'M' \rangle$ and the corresponding statistical tensors $\rho^{las}_{kq}(\alpha J, \alpha'J')$ describe the polarization of the laser-excited state $A^*(\alpha J)$ in the process (4.53) under the condition that neither the incoming electron nor the target atom $A(\alpha_0 J_0)$ are polarized and the detector of the inelastically-scattered electrons is not sensitive to their polarization (this case was discussed in Section 2.2.1). We introduce the superscript "las" to distinguish the density matrix and the statistical tensors of the laser-excited state from the density matrix $\langle \alpha JM | \rho | \alpha'J'M' \rangle$ and the statistical tensors $\rho_{kq}(\alpha J, \alpha'J')$ of the same state $A^*(\alpha J)$ excited by electron impact in inelastic scattering (4.54). Introducing the reduced statistical tensors (1.50) and using the normalization of the probability (4.56) to the cross section, we obtain

$$\frac{d\sigma}{d\Omega} = \frac{d\sigma^{(iso)}}{d\Omega} \sum_{\substack{\alpha\alpha'JJ' \\ kq}} (-1)^k \mathcal{A}^{las}_{kq}(\alpha J, \alpha'J') \mathcal{A}^*_{kq}(\alpha J, \alpha'J') \quad (4.60)$$

Here the cross section $d\sigma^{(iso)}/d\Omega$ corresponds to the superelastic differential cross section from an unpolarized (isotropic) atom, i.e., when only the monopole term with $k = 0$ contributes to the sum at the right side of Eq. (4.60). Equation (4.60) expresses the cross section of superelastic scattering in terms of the amplitudes $T_{M_0\mu_0 \to M\mu}(\mathbf{p}_0, \mathbf{p})$ of the reverse inelastic scattering.

From Eq. (4.60) it follows that only those tensor combinations kq of the excitation amplitudes contribute to the superelastic cross section which are contained in the laser-excited, polarized, initial atomic state. This opens up the possibility of measuring different bilinear combinations of excitation amplitudes by changing the conditions of the laser optical pumping before the superelastic scattering. Equation (4.60) relates the differential cross section of the superelastic scattering from a polarized target to the reduced statistical tensors $\mathcal{A}_{kq}(\alpha J, \alpha'J')$ observed in the coincidence $(e, e'\gamma)$ experiment (see Section 4.1.2), but with two important complements. First, owing to the dipole character of the emitted photon, the $(e, e'\gamma)$ correlation experiment yields information for only a limited set of statistical tensors $\mathcal{A}_{kq}(\alpha J, \alpha'J')$ of the excited atom up to the rank $k = 2$ (see Section 4.1.2.2). Often this is not complete information on the polarization state of the collisionally excited atom and part of the information on the excitation amplitudes can be missed. In contrast, laser optical pumping with a large number of absorption and radiation events overcomes this restriction, and tensors $\mathcal{A}_{kq}(\alpha J)$ with much higher ranks of k can be in principle investigated. Second, the electron–atom excitation cross sections usually decrease rapidly with an increase in the scattering angle. Thus it is difficult to perform the coincidence $(e, e'\gamma)$ experiment in this kinematic region because of poor statistics. Superelastic scattering, which does not need the coincidence techniques, is a good alternative for this case.

Although we implied that the angular momentum J in Eqs. (4.59) and (4.60) is the angular momentum of the electronic shell, it may be of an arbitrary nature. For example, for atoms with nonzero nuclear spin, laser optical pumping can select hyperfine levels. Then Eq. (4.60) is valid for the statistical tensors of the total angular momentum $\mathbf{F} = \mathbf{J} + \mathbf{I}$, where I is the nuclear spin. Nevertheless, Eq. (4.60) still holds for the angular momentum of the electronic shell J if the transition operator T does not operate on the nuclear variables. [This can be shown rigorously, starting from the general Eq. (4.56) (where the trace over nuclear quantum numbers is included) and using the conservation of the nuclear spin I and its projection during the excitation.] In the latter case, the statistical tensors $\rho_{kq}^{\text{las}}(\alpha J, \alpha' J')$ of the angular momentum of the electronic shell should be calculated from the known statistical tensors of the hyperfine levels with the help of the trace procedure (1.69):

$$\rho_{kq}^{\text{las}}(\alpha J, \alpha' J') = \sum_{FF'}(-1)^{F'+I+J+k}\hat{F}\hat{F}' \begin{Bmatrix} J & J' & k \\ F' & F & I \end{Bmatrix} \rho_{kq}^{\text{las}}(JIF, J'IF') \qquad (4.61)$$

(we assumed that a certain isotope with nuclear spin I is excited by the laser optical pumping).

For the case when the total spin S_t of the system atom + scattering electron is conserved during the collision (*Percival–Seaton hypothesis*) [28], one can make further simplifications of Eq. (4.60). Consider transitions between multiplet atomic states with fixed orbital angular momenta L_0 and L and spins S_0 and S. Transforming the transition matrix elements in Eq. (4.57) to the representation of the total spin S_t and total orbital angular momentum L_t, assuming that the T-operator conserves the total spin S_t and its projection, and performing the necessary summations, we arrive at the expression

$$\frac{d\sigma}{d\Omega} = \frac{d\sigma^{(\text{iso})}}{d\Omega} \sum_{kq}(-1)^k \mathcal{A}_{kq}^{\text{las}}(\alpha L) \, \mathcal{A}_{kq}^*(\alpha L) \qquad (4.62)$$

Here the statistical tensors of the orbital angular momentum L of a laser-excited atom can be found from the statistical tensors (4.61) of the angular momentum J by Eq. (1.69).

Note that although the general form of the cross section for superelastic scattering looks similar in different cases [see Eqs. (4.60) and (4.62)], in fact, it can show very different behavior as a function of the polarization state of the laser-excited target. This is because the largest rank k in the statistical tensors is determined by different triangle inequalities.

Using Eqs. (4.56) or (4.57) and proceeding in analogy with the above derivations, one can obtain more general relations for the cross section of superelastic scattering when the incident electron beam is polarized and/or a spin-sensitive detector is used. For example, within the Percival–Seaton hypothesis, the superelastic scattering intensity in the case of an incident polarized electron beam can be written in the form

$$
W = \sum_{JJ'} \sum_{kk_s t} (-1)^{t+S_t+S+1/2} ff' \begin{Bmatrix} \frac{1}{2} & \frac{1}{2} & k_s \\ S & S & S_t \end{Bmatrix} \begin{Bmatrix} S & S & k_s \\ J & J' & k \\ L & L & t \end{Bmatrix}
$$
$$
\times \sum_{qq_s \tau} (kq, k_s q_s \,|\, t\tau) \rho_{kq}^{las} \left(\alpha J, \alpha' J' \right) \rho_{k_s q_s} \left(\frac{1}{2} \right) \left[\rho_{t\tau}^{S_t} (L) \right]^* \tag{4.63}
$$

where the statistical tensor $\rho_{k_s q_s} \left(\frac{1}{2} \right)$ characterizes polarization of the incident electron e in the process (4.53). The statistical tensor $\rho_{t\tau}^{S_t} (L)$ describes polarization of the orbital angular momentum L of the atomic state $A^*(LS)$ excited by electron impact due to the scattering channel with the total spin S_t [compare with Eqs. (2.46) and (2.44)]:

$$
\rho_{t\tau}^{S_t} (L) = \hat{S}_t^2 \sum_{M_L M_L'} (-1)^{L-M_L'} \left(LM_L, L - M_L' \,|\, t\tau \right) \sum_{M_{L_0}} T_{M_{L_0} \to M_L}^{S_t} T_{M_{L_0} \to M_L'}^{S_t *} \tag{4.64}
$$

Here a brief notation for the amplitudes is introduced:

$$
T_{M_{L_0} \to M_L}^{S_t} = \left\langle \alpha LM_L, \left(s\frac{1}{2} \right) S_t M_{S_t}, \mathbf{p} \,|\, T \,|\, \alpha_0 L_0 M_{L_0}, \left(S_0 \frac{1}{2} \right) S_t M_{S_t}, \mathbf{p}_0 \right\rangle \tag{4.65}
$$

4.2. Two-Step Reactions of Ionization and Decay

4.2.1. Electromagnetic Radiation from Ions Produced in Direct Ionization Processes

When an ion is produced in an atomic collision or by photoionization, it is usually excited and can subsequently radiate a photon. In this section we discuss the angular distributions and polarization of emitted photons. If the ion is produced by particle impact, the entire process can be presented as

$$
a_i + A(\alpha_0 J_0) \to a_f + A^+(\alpha J) + e_1
$$
$$
\hookrightarrow A^+(\alpha_f J_f) + \gamma \tag{4.66}
$$

If the primary beam is a photon beam, then the process is

$$\gamma_i + A(\alpha_0 J_0) \rightarrow A^+(\alpha J) + e_{ph}$$
$$\hookrightarrow A^+(\alpha_f J_f) + \gamma \qquad (4.67)$$

A typical example of such processes is generation of X-rays in inner-shell ionization; another example is photon emission in ionization with excitation of the outer atomic shell. Usually the collision time is much shorter than the lifetime of the excited ionic state. Consequently, a two-step approach can often be used to describe the whole process. Within the framework of the conventional two-step model, the ion produced in the ionization step is in a well-defined quantum state that may be characterized by the density matrix or statistical tensors. The polarization state of the ion is determined by the preceding ionization process. In turn it determines the angular distribution and polarization of the decay products. Thus we can expect that the coefficients that determine the anisotropy of the angular distribution and polarization will depend on the characteristics of both the ionization and the decay. On the other hand, in the spirit of the two-step model, we can consider each step independently. In the following discussion we always assume that the target atom is unpolarized.

4.2.1.1.   Angular Distribution and Polarization of Photons Generated by Particle Impact

Consider the photon production in the ionization process (4.66) induced by an unpolarized particle beam. If only photons (for example, X-rays) are detected and the detector is not sensitive to their polarization, then the angular distribution is axially symmetric with respect to the initial beam (which we choose as the z-axis) and is determined by Eq. (3.36):

$$W(\vartheta) = \frac{W_0}{4\pi}[1 + \beta P_2(\cos\vartheta)] = \frac{W_0}{4\pi}[1 + \alpha_2^\gamma \mathcal{A}_{20}(\alpha J) P_2(\cos\vartheta)] \qquad (4.68)$$

where the intrinsic anisotropy parameters α_2^γ are determined by Eq. (3.33). The alignment parameter $\mathcal{A}_{20}(\alpha J)$ can be calculated as outlined in Section 2.4.3, Eqs. (2.194)–(2.196) or equivalently, Eqs. (2.199) and (2.200). Combining Eqs. (2.194)–(2.196) with the definition (1.50), we obtain

$$\mathcal{A}_{20}(\alpha J) = \sum_{MM'} (-1)^{J-M'} \, (JM, J - M' | 20)$$
$$\times \int d\Omega_f \int d\Omega_1 \sum_{M_0 \mu_i \mu_f \mu_1} T_{M_0\mu_i \rightarrow M\mu_f\mu_1}(\mathbf{p}_i, \mathbf{p}_f, \mathbf{p}_1)$$
$$\times T^*_{M_0\mu_i \rightarrow M'\mu_f\mu_1}(\mathbf{p}_i, \mathbf{p}_f, \mathbf{p}_1)$$

$$\times \hat{f} \left(\int d\Omega_f \int d\Omega_1 \sum_{M_0 \mu_i \mu_f \mu_1 M} |T_{M_0 \mu_i \to M \mu_f \mu_1}(\mathbf{p}_i, \mathbf{p}_f, \mathbf{p}_1)|^2 \right)^{-1} \tag{4.69}$$

(The notations are the same as in Section 2.4.3.) The factorization of the anisotropy coefficient β is a direct consequence of the two-step character of the process. We see that the alignment parameter is determined only by the ionization step of the process, while the anisotropy parameter depends only on characteristics of the second (decay) step.

From symmetry arguments (see the discussion in Section 3.2.2) it is clear that in the case considered the photons can be only linearly polarized. The general expression for linear polarization of the decay photons is given by Eq. (3.49). This expression simplifies considerably if one takes into account that in this case only a $q = 0$ projection of the alignment tensor can be nonzero. Then the degree of linear polarization measured in the direction perpendicular to the initial beam is determined by Eq. (3.50):

$$P \equiv \frac{P_{\parallel} - P_{\perp}}{P_{\parallel} + P_{\perp}} = \frac{3\alpha_2^{\gamma} \mathcal{A}_{20}(\alpha J)}{\alpha_2^{\gamma} \mathcal{A}_{20}(\alpha J) - 2} \tag{4.70}$$

Consider an experiment where the photons from the process (4.66) are detected in coincidence with either the scattered particle a_f or the ejected electron e_1. We assume that the initial particle beam is unpolarized. Then the experimental conditions are symmetric with respect to a reflection through the plane that contains the beam and the detected particle. It is convenient to choose the z-axis along the beam and the x-axis in the symmetry plane. Then the angular distribution of the emitted photons is given by Eq. (3.35). We note here that in coincidence measurements, the alignment tensor $\mathcal{A}_{2q}$ which appears in Eq. (3.35) is differential with respect to the direction of the detected particle, $\mathcal{A}_{2q}(\alpha J; \mathbf{n})$; it is different from the integral alignment, which describes noncoincidence experiments. If we take into account that from symmetry considerations the relations $\mathcal{A}_{21} = -\mathcal{A}_{2-1}$ and $\mathcal{A}_{22} = \mathcal{A}_{2-2}$ follow [see Eq. (1.134)], then Eq. (3.35) can be reduced to an equation similar to (4.24). This general expression has an especially simple form if the angular distribution is measured in the symmetry plane:

$$W_{\parallel}(\vartheta) = \frac{W_0}{4\pi} A[1 + C\cos 2(\vartheta \pm \gamma)] \tag{4.71}$$

or in the plane perpendicular to the beam:

$$W_{\perp}(\varphi) = \frac{W_0}{4\pi} A_{\perp} [1 + C_{\perp} \cos 2\varphi] \tag{4.72}$$

where parameters A, C, and γ in Eq. (4.71) are expressed in terms of the alignment parameters $\mathcal{A}_{20}$, $\mathcal{A}_{21}$, and $\mathcal{A}_{22}$ by Eqs. (4.26)–(4.28), while the parameters $A_{\perp}$ and

$C_\perp$ in Eq. (4.72) are given by

$$A_\perp = 1 - \frac{1}{2}\alpha_2^\gamma \mathcal{A}_{20}(\alpha J; \mathbf{n}) \tag{4.73}$$

$$C_\perp = \frac{\sqrt{6}\alpha_2^\gamma \mathcal{A}_{22}(\alpha J; \mathbf{n})}{1 - \frac{1}{2}\alpha_2^\gamma \mathcal{A}_{20}(\alpha J; \mathbf{n})} \tag{4.74}$$

The differential alignment parameters $\mathcal{A}_{2q}(\alpha J; \mathbf{n})$ in coincidence experiments are expressed by the equation

$$\mathcal{A}_{2q}(\alpha J; \mathbf{n}_f) = \sum_{MM'} (-1)^{J-M'} (JM, J-M' | 2q)$$

$$\times \int d\Omega_1 \sum_{M_0\mu_i\mu_f\mu_1} T_{M_0\mu_i \to M\mu_f\mu_1}(\mathbf{p}_i, \mathbf{p}_f, \mathbf{p}_1) T^*_{M_0\mu_i \to M'\mu_f\mu_1}(\mathbf{p}_i, \mathbf{p}_f, \mathbf{p}_1)$$

$$\times \hat{J} \left(\int d\Omega_1 \sum_{M_0\mu_i\mu_f\mu_1 M} |T_{M_0\mu_i \to M\mu_f\mu_1}(\mathbf{p}_i, \mathbf{p}_f, \mathbf{p}_1)|^2 \right)^{-1} \tag{4.75}$$

when the scattered particle is detected and by

$$\mathcal{A}_{2q}(\alpha J; \mathbf{n}_1) = \sum_{MM'} (-1)^{J-M'} (JM, J-M' | 2q)$$

$$\times \int d\Omega_f \sum_{M_0\mu_i\mu_f\mu_1} T_{M_0\mu_i \to M\mu_f\mu_1}(\mathbf{p}_i, \mathbf{p}_f, \mathbf{p}_1) T^*_{M_0\mu_i \to M'\mu_f\mu_1}(\mathbf{p}_i, \mathbf{p}_f, \mathbf{p}_1)$$

$$\times \hat{J} \left(\int d\Omega_f \sum_{M_0\mu_i\mu_f\mu_1 M} |T_{M_0\mu_i \to M\mu_f\mu_1}(\mathbf{p}_i, \mathbf{p}_f, \mathbf{p}_1)|^2 \right)^{-1} \tag{4.76}$$

when the ejected electron is detected in coincidence with the photon.

The polarization of the photons is expressed by the general expressions (3.45)–(3.47). The linear polarization is determined by the same three parameters ($\mathcal{A}_{20}$, $\mathcal{A}_{21}$, and $\mathcal{A}_{22}$) as the angular distribution. However, in contrast to the noncoincidence measurement, in this case the photon can have a circular component that is connected to the first-rank statistical tensor $\rho_{11}(\alpha J; \mathbf{n})$.

If we choose a reference frame with the z-axis along the incident beam and the x-axis in the symmetry plane, then it is easy to show from the general Eq. (3.49) that the following equation is valid for the linear polarization of photons measured in the plane perpendicular to the beam [$\vartheta = \frac{\pi}{2}$ and the definition (3.50) is used, i.e., $\psi = 0$]:

$$P_\perp(\theta_n) = \frac{\alpha_2^\gamma \left(3\mathcal{A}_{20}(\alpha J; \theta_n) + \sqrt{6}\mathcal{A}_{22}(\alpha J; \theta_n)\cos 2\varphi \right)}{\alpha_2^\gamma \left(\mathcal{A}_{20}(\alpha J; \theta_n) - \sqrt{6}\mathcal{A}_{22}(\alpha J; \theta_n)\cos 2\varphi \right) - 2} \tag{4.77}$$

where θ_n is a polar angle of vector $\mathbf{n}$. In particular, the polarization of photons emitted along the x-axis ($\varphi = 0$) is

$$P_x(\theta_n) = \frac{\alpha_2^\gamma \left(3\mathcal{A}_{20}(\alpha J; \theta_n) + \sqrt{6}\mathcal{A}_{22}(\alpha J; \theta_n) \right)}{\alpha_2^\gamma \left(\mathcal{A}_{20}(\alpha J; \theta_n) - \sqrt{6}\mathcal{A}_{22}(\alpha J; \theta_n) \right) - 2} \qquad (4.78)$$

while the polarization of photons emitted along the y-axis ($\varphi = \pi/2$) is

$$P_y(\theta_n) = \frac{\alpha_2^\gamma \left(3\mathcal{A}_{20}(\alpha J; \theta_n) - \sqrt{6}\mathcal{A}_{22}(\alpha J; \theta_n) \right)}{\alpha_2^\gamma \left(\mathcal{A}_{20}(\alpha J; \theta_n) + \sqrt{6}\mathcal{A}_{22}(\alpha J; \theta_n) \right) - 2} \qquad (4.79)$$

4.2.1.2. Angular Distribution and Polarization of Photons Following Photoionization

Owing to the relative simplicity of the photoionization process in comparison with ionization by particle impact, it is instructive to consider the process (4.67) separately. First we consider experiments in which only emitted photons are detected. The angular distribution of the photons measured by a polarization-insensitive detector is given by Eq. (3.35). We recall that only a second-rank statistical tensor of the ionic state $\mathcal{A}_{2q}(\alpha J)$ influences the angular distribution. The most general expression for $\mathcal{A}_{2q}(\alpha J)$ follows from Eq. (2.161) for an arbitrarily polarized target atom and arbitrary polarization of the primary photon beam. Thus the two equations solve the problem in the general case.

Consider a case of an unpolarized target atom and a linearly polarized primary beam. It is natural to choose the z-axis along the beam polarization, which is an axis of symmetry in this case. The angular distribution of emitted photons is described by the same equation (4.68) as in particle impact ionization, but for the alignment parameter we have from Eq. (2.163) and Table 1.1 on p. 25 the following expression:

$$\mathcal{A}_{20}(\alpha J) = -\sqrt{2}\frac{A_{022}}{A_{000}} = -\sqrt{6}\hat{f}_f \Big(\sum_{ljJ} |M_{ljJ}|^2 \Big)^{-1}$$

$$\times \sum_{ljJJ'} (-1)^{J+J'+j+J_f+J_0+1} \hat{f}\hat{f}' \begin{Bmatrix} J & J_f & j \\ J_f & J' & 2 \end{Bmatrix} \begin{Bmatrix} 1 & J & J_0 \\ J' & 1 & 2 \end{Bmatrix} M_{ljJ}M^*_{ljJ'} \quad (4.80)$$

For the photoionization of an atom with closed subshells ($J_0 = 0$), one has

$$\mathcal{A}_{20}(\alpha J) = -\sqrt{6}\hat{f}_f \Big(\sum_{lj} |M_{lj1}|^2\Big)^{-1} \sum_{lj} (-1)^{j+J_f+1} \begin{Bmatrix} 1 & J_f & j \\ J_f & 1 & 2 \end{Bmatrix} |M_{lj1}|^2 \quad (4.81)$$

This case is especially simple; the angular distribution of photons is determined by the ratios of the photoionization amplitude squared. The polarization of emitted radiation is described by Eq. (4.70) with the alignment parameters (4.80) and (4.81).

Now turn to a case of an unpolarized or circularly polarized initial beam. Here we choose the z-axis along the beam. It is easy to show from Eqs. (2.163) and (3.35), taking into account the symmetry conditions, that in this case the angular distribution can be written in the same form as earlier, Eq. (4.68):

$$W(\vartheta) = \frac{W_0}{4\pi}[1 + \alpha_2^{\gamma}\tilde{\mathcal{A}}_{20}(\alpha J)P_2(\cos\vartheta)] \tag{4.82}$$

but where the alignment parameter $\tilde{\mathcal{A}}_{20}(\alpha J)$ is different from the previous case and it should be calculated in the reference frame with the z-axis along the beam. Using Eq. (3.35) we can show that numerically

$$\tilde{\mathcal{A}}_{20}(\alpha J) = -\frac{1}{2}\mathcal{A}_{20}(\alpha J) \tag{4.83}$$

where $\mathcal{A}_{20}(\alpha J)$ is given by Eq. (4.80) or (4.81). Thus the angular distribution for an unpolarized beam is often presented as

$$W(\vartheta) = \frac{W_0}{4\pi}[1 - \frac{1}{2}\beta P_2(\cos\vartheta)] \tag{4.84}$$

where β is the same as for linearly polarized photons.

Now we discuss an angle-resolved experiment in which fluorescence photons are detected in coincidence with photoelectrons. The angular distribution of emitted photons is given by Eq. (3.35), which we rewrite here in terms of differential cross sections. The double differential cross section for the photoionization of the atom that emits a photoelectron e and a fluorescent photon γ (4.67) can be expressed as follows (we assume again that the detector is insensitive to the photon's polarization):

$$\frac{d^2\sigma}{d\Omega_e d\Omega_\gamma} = \frac{d\sigma}{d\Omega_e}(\mathbf{n}_e)\frac{\omega_\gamma}{4\pi}\left\{1 + \alpha_2^{\gamma}\sum_q\sqrt{\frac{4\pi}{5}}Y_{2q}(\mathbf{n}_\gamma)\mathcal{A}_{2q}(\alpha J;\mathbf{n}_e)\right\} \tag{4.85}$$

where $\mathcal{A}_{2q}(\alpha J;\mathbf{n}_e) = \rho_{2q}(\alpha J;\mathbf{n}_e)/\rho_{00}(\alpha J;\mathbf{n}_e)$ is the differential alignment tensor that describes the ionic state J produced by photoionization with the photoelectron detected in the direction $\mathbf{n}_e$. $\frac{d\sigma}{d\Omega_e}(\mathbf{n}_e) \sim \rho_{00}(\alpha J;\mathbf{n}_e)$ corresponds to the photoionization cross section, ω_γ is the fluorescence yield, $\mathbf{n}_\gamma$ determines the direction of the fluorescent photon emission, and α_2^{γ} is the anisotropy parameter (3.33).

When the target atom is unpolarized, the differential statistical tensors of

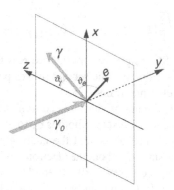

FIGURE 4.8. Kinematic diagram of the photoelectron–fluorescence photon coincidence measurements.

photoions are given by Eq. (2.169):

$$\rho_{kq}(\alpha J; \mathbf{n_e}) = \frac{1}{3(2J_0+1)} \sum_{k'q'\bar{k}\bar{q}} \frac{\sqrt{4\pi}}{2k'+1} (kq, k'q' | \bar{k}\bar{q}) Y_{k'q'}(\mathbf{n_e}) \rho_{\bar{k}\bar{q}}^{\gamma} B(k', k, \bar{k})$$

$$(4.86)$$

where coefficients $B(k', k, \bar{k})$, determined by Eq. (2.170), contain bilinear combinations of the dipole ionization amplitudes, and we recall that owing to parity conservation in photoionization, the value of k' is even. Thus, one can calculate the double differential cross section of Eq. (4.85) provided the dipole amplitudes are known, or alternatively, one can use Eqs. (4.85)–(4.86) to obtain information on the dipole amplitudes from the differential cross-section measurement.

The degree of linear polarization of the fluorescence detected in the direction $\mathbf{n}_\gamma \equiv (\vartheta_\gamma, \varphi_\gamma)$ is given by Eq. (3.49):

$$P(\varphi_\gamma, \vartheta_\gamma, \psi) = \frac{-\alpha_2^\gamma \sqrt{6} \sum_q \mathcal{A}_{2q}(\alpha J; \mathbf{n_e})[D_{q2}^{2*}(\varphi_\gamma, \vartheta_\gamma, \psi) + D_{q-2}^{2*}(\varphi_\gamma, \vartheta_\gamma, \psi)]}{2\left\{1 + \alpha_2^\gamma \sum_q \sqrt{\frac{4\pi}{5}} Y_{2q}(\mathbf{n}_\gamma) \mathcal{A}_{2q}(\alpha J; \mathbf{n_e})\right\}}$$

$$(4.87)$$

To illustrate the above equations, consider a particular geometry of the experiment presented in Figure 4.8. We choose the coordinate frame so that the y-axis is directed along the incident photon beam and the z-axis is along the principal axis of its linear polarization. Then the incident dipole photon statistical tensors $\rho_{\bar{k}\bar{q}}^{\gamma}$ in equation (4.86) can be expressed as

$$\rho_{00}^{\gamma} = \frac{1}{\sqrt{3}}, \ \rho_{10}^{\gamma} = 0, \ \rho_{1\pm1}^{\gamma} = -i\frac{P_3}{2}$$

$$\rho_{20}^\gamma = -\sqrt{\frac{2}{3}} + \sqrt{\frac{3}{8}}(1-P_1), \quad \rho_{2\pm1}^\gamma = 0, \quad \rho_{2\pm2}^\gamma = -\frac{1}{4}(1-P_1) \qquad (4.88)$$

where P_1 and P_3 are the Stokes parameters that express the degree of linear and circular polarizations, respectively. The second Stokes parameter, P_2, is zero because of our choice of reference frame.

From an experimental point of view, it is convenient to measure the angular distributions of electrons and fluorescent photons by rotating the analyzers around the incident photon beam direction so that the analyzers always view the same source volume. We thus assume that both the electrons and the fluorescent photons are detected in the xz-plane ($\varphi_e = \varphi_\gamma = 0$). Then $\mathcal{A}_{21}(\alpha J; \vartheta_e) = -\mathcal{A}_{2-1}(\alpha J; \vartheta_e)$ and $\mathcal{A}_{22}(\alpha J; \vartheta_e) = \mathcal{A}_{2-2}(\alpha J; \vartheta_e)$ hold and thus the double differential cross section (4.85) can be expressed as

$$\frac{d^2\sigma}{d\Omega_e d\Omega_\gamma} = A_0(\vartheta_e) + A_2(\vartheta_e)\cos 2\vartheta_\gamma + B_2(\vartheta_e)\sin 2\vartheta_\gamma \equiv I(\vartheta_e, \vartheta_\gamma) \qquad (4.89)$$

where

$$A_0(\vartheta_e) = \frac{d\sigma}{d\Omega_e}(\vartheta_e)\frac{\omega_\gamma}{4\pi}\left(1 + \frac{\alpha_2^\gamma}{4}\mathcal{A}_{20}(\alpha J; \vartheta_e) + \sqrt{\frac{3}{8}}\alpha_2^\gamma \mathcal{A}_{22}(\alpha J; \vartheta_e)\right)$$

$$A_2(\vartheta_e) = \frac{d\sigma}{d\Omega_e}(\vartheta_e)\frac{\omega_\gamma}{4\pi}\alpha_2^\gamma\left(\frac{3}{4}\mathcal{A}_{20}(\alpha J; \vartheta_e) - \sqrt{\frac{3}{8}}\mathcal{A}_{22}(\alpha J; \vartheta_e)\right)$$

$$B_2(\vartheta_e) = \frac{d\sigma}{d\Omega_e}(\vartheta_e)\frac{\omega_\gamma}{4\pi}\alpha_2^\gamma\left(-\sqrt{\frac{3}{2}}\mathcal{A}_{21}(\alpha J; \vartheta_e)\right) \qquad (4.90)$$

The statistical tensors $\rho_{kq}(\alpha J; \vartheta_e)$ and therefore the alignment tensor components $\mathcal{A}_{2q}(\alpha J; \theta_e)$ in (4.90) are expressed according to Eqs. (4.86) and (4.88) in terms of the dynamic parameters $B(k', k, \bar{k})$ and the Stokes parameters P_1 and P_3. Equation (4.89) can be regarded as the expression for the angular distribution of the fluorescent photons measured in coincidence with the photoelectron detected at a certain angle ϑ_e. The distribution has the simple form $1 + a\cos 2\vartheta'$ tilted with respect to the z-axis with the tilt angle $\psi = \frac{1}{2}\tan^{-1}B_2(\vartheta_e)/A_2(\vartheta_e)$. Measurement of the photon angular distribution gives two parameters a and ψ (or A_2/A_0 and B_2/A_0), which contain nontrivial information about the photoionization amplitudes.

Using expression (4.86) for the statistical tensors $\rho_{kq}(\alpha J; \vartheta_e)$, the same cross section can be presented in an alternative form that can be regarded as the expression for the photoelectron angular distribution measured in coincidence with the fluorescent photon detected at a certain angle ϑ_γ:

$$I(\vartheta_e, \vartheta_\gamma) = A_0'(\vartheta_\gamma) + A_2'(\vartheta_\gamma)\cos 2\vartheta_e + A_4'(\vartheta_\gamma)\cos 4\vartheta_e$$
$$+ B_2'(\vartheta_\gamma)\sin 2\vartheta_e + B_4'(\vartheta_\gamma)\sin 4\vartheta_e \qquad (4.91)$$

where $A'_k(\vartheta_\gamma)$ and $B'_k(\vartheta_\gamma)$ are functions of the coefficients $B(k',k,\bar{k})$ and the degree of linear P_1 and circular P_3 polarizations. In general, the photoelectron distribution (4.91) has an asymmetric shape. Its analysis can provide four independent parameters that may be used (together with others) for a complete determination of the photoionization amplitudes.

Consider the case when the fluorescent photons are detected in the x-axis direction in coincidence with the photoelectrons at an angle ϑ_e and the polarization is analyzed (see Figure 4.8). The degree of linear polarization is measured in the x direction and determined as $P_x(\vartheta_e) = (I_z - I_y)/(I_z + I_y)$. I_z and I_y are the yields of fluorescent photons polarized along the z- and y-axes, respectively, and in coincidence with photoelectrons detected at an angle ϑ_e. $P_x(\vartheta_e)$ can be obtained from the general expression (4.87) with $\varphi_\gamma = \psi = 0$, $\vartheta_\gamma = \pi/2$, and the result can be cast in the form

$$P_x(\vartheta_e) = \frac{\alpha_2^\gamma \left(3\mathcal{A}_{20}(\alpha J;\vartheta_e) + \sqrt{6}\mathcal{A}_{22}(\alpha J;\vartheta_e)\right)}{\alpha_2^\gamma \left(\mathcal{A}_{20}(\alpha J;\vartheta_e) - \sqrt{6}\mathcal{A}_{22}(\alpha J;\vartheta_e)\right) - 2} \qquad (4.92)$$

Note that the alignment tensor components $\mathcal{A}_{20}(\alpha J;\vartheta_e)$, $\mathcal{A}_{2\pm1}(\alpha J;\vartheta_e)$, and $\mathcal{A}_{2\pm2}(\alpha J;\vartheta_e)$ can be determined nontrivially from a combination of the measurements $I(\vartheta_e,\vartheta_\gamma)$ of Eq. (4.89) as a function of ϑ_γ and $P_x(\vartheta_e)$ of Eq. (4.92) at $\vartheta_\gamma = \pi/2$, provided α_2^γ is known.

We finally consider the angular distribution of photoelectrons measured in coincidence with fluorescent photons for the chosen geometry. Substitution of $\vartheta_\gamma = \pi/2$ simplifies Eq. (4.89) to the following:

$$\frac{d^2\sigma}{d\Omega_e d\Omega_\gamma} = \frac{d\sigma}{d\Omega_e}(\vartheta_e)\frac{\omega_\gamma}{4\pi}\left\{1 - \frac{1}{2}\alpha_2^\gamma\left(\mathcal{A}_{20}(\alpha J;\vartheta_e) - \sqrt{6}\mathcal{A}_{22}(\alpha J;\vartheta_e)\right)\right\} \quad (4.93)$$

This cross section can be presented in the form

$$\frac{d^2\sigma}{d\Omega_e d\Omega_\gamma} = A'_0 + A'_2\cos 2\vartheta_e + A'_4\cos 4\vartheta_e + B'_2\sin 2\vartheta_e \equiv I(\vartheta_e) \qquad (4.94)$$

where A'_0, A'_2, and A'_4 are functions of the coefficients $B(k',k,\bar{k})$ and the degree of linear polarization P_1, whereas B'_2 is determined by the product of $B(2,2,1)$, which is purely imaginary, and the degree of circular polarization, P_3.

Analyzing Eqs. (4.92) and (4.93), one immediately notices that both the angular correlation function (4.93) and the polarization of fluorescence (4.92) measured in coincidence with the photoelectron reveal circular and linear dichroism. For example, circular dichroism in the angular distribution of photoelectrons, which is determined as the difference in the intensities of photoelectrons ejected by right and left circularly polarized photons, can be expressed as

$$I(\vartheta_e, P_3 = 1) - I(\vartheta_e, P_3 = -1)$$

$$= \frac{d\sigma}{d\Omega_e}(\vartheta_e) \frac{\omega_\gamma}{4\pi} \alpha_2^\gamma \left(-\frac{3}{2\sqrt{5}}\right) iB(2,2,1)\sin 2\vartheta_e \quad (4.95)$$

Note that circular dichroism in the angular distribution disappears if the direction of photon emission is parallel (antiparallel) or perpendicular to that of the photoelectron emission. This is a simple consequence of the symmetry of the experiment and parity conservation in photoemission.

4.2.2. Angular Distribution and Polarization of Auger Electrons Produced in Direct Ionization Processes

An alternative method of deexcitation of a highly excited state of an ion formed in an atomic collision or by photoionization is Auger electron emission. The Auger process generated by particle impact can be presented as

$$a_i + A(\alpha_0 J_0) \rightarrow a_f + A^+(\alpha J) + e_1$$
$$\hookrightarrow A^{++}(\alpha_f J_f) + e_A \quad (4.96)$$

The photoinduced Auger process is presented as follows:

$$\gamma + A(\alpha_0 J_0) \rightarrow A^+(\alpha J) + e_{\text{ph}}$$
$$\hookrightarrow A^{++}(\alpha_f J_f) + e_A \quad (4.97)$$

In both cases the second row of Eqs. (4.96) and (4.97) represents the Auger decay of the ionic state with emission of the Auger electron, e_A. In a great number of cases the Auger process can be considered in the two-step model since as a rule the lifetime of the ionic Auger state is much longer than the collision time. Thus the ionization step and the decay can be considered independently. In the next section we discuss the angular distribution and spin polarization of Auger electrons within the framework of the two-step model.

4.2.2.1. Angular Distribution of Auger Electrons in a Noncoincidence Experiment

Consider first the process (4.96) when the projectiles and the target atoms are unpolarized. Suppose that only Auger electrons are detected and the detector is insensitive to their spin polarization. Then the angular distribution of the Auger electrons is axially symmetric with respect to the only direction that is fixed in the experiment for the ionization stage of the process (4.96): the direction of the initial beam, which we choose as the z-axis. According to Eq. (3.16), the angular distribution has the form

$$W_{\alpha_f J_f}(\vartheta) = \frac{W_0}{4\pi}\left[1 + \sum_{k=2,4,\ldots}^{k_{max}} \alpha_k \mathcal{A}_{k0}(\alpha J)P_k(\cos\vartheta)\right] \qquad (4.98)$$

where the anisotropy parameters α_k are given by Eq. (3.11), while the statistical tensors $\mathcal{A}_{k0}(\alpha J)$ are given by a simple generalization of Eq. (4.69) for arbitrary k or follow from Eq. (2.200) in the partial wave representation. The maximum value of k, $k_{max} \leq 2J$; therefore only Auger electrons from the states with $J \geq 1$ can be emitted anisotropically.

For a photoinduced Auger process (4.97), the angular distribution (4.98) simplifies to

$$W_{\alpha_f J_f}(\vartheta) = \frac{W_0}{4\pi}\left[1 + \alpha_2 \mathcal{A}_{20}(\alpha J)P_2(\cos\vartheta)\right] \qquad (4.99)$$

since in photoionization within the dipole approximation only the second-rank statistical tensors are nonzero. They are expressed in terms of photoionization amplitudes by Eqs. (4.80) or (4.81). (We recall that the z-axis is chosen along the linear polarization vector of the photon.)

If the target atom is polarized, the angular distribution of Auger electrons is expressed in general by Eq. (3.10). The statistical tensors of even rank that determine the angular distribution depend on the amplitudes of the ionization process and on the statistical tensors of the initial polarized state. For example, if the ion is produced by photoionization, the statistical tensors of the ionic state are given by Eqs. (2.161) and (2.162). Analysis of these expressions shows that the anisotropy of the Auger electrons can arise from the alignment of the target atom and/or can be induced by the linear polarized component of the photon beam. In addition, it can be induced by the orientation of the target, provided the photon beam is circularly polarized. The latter indicates that there may be circular dichroism in the angular distribution of Auger electrons from oriented targets, i.e., the difference in angular distributions of Auger electrons generated by right and left circularly polarized light.

4.2.2.2. Spin Polarization of Auger Electrons

In Section 3.1.3 we discussed the general problem of spin polarization of the electrons in nonradiative decay of an arbitrarily polarized state. Now we apply the results obtained to the process of Auger emission. We consider the case where only Auger electrons are detected by a spin-sensitive detector. As follows from the analysis in Section 3.1.3, spin polarization of Auger electrons appears when a decaying ionic state is aligned and/or oriented. An unpolarized ionic state gives rise to unpolarized Auger electrons. The general expressions for the polarization characteristics (statistical tensors) of ions produced in the process of ionization by particle impact (see Section 2.4.3) or by photoionization (see Section 2.3.5) provide the means for analyzing any particular situation.

Consider, for example, photoproduction of Auger electrons. In photoionization by unpolarized or linearly polarized light, the ion produced is aligned along the beam direction or photon polarization direction, respectively. As follows from the analysis of the general expression (3.27), the spin of the Auger electron in this case can only be oriented perpendicular to the plane containing the alignment axis and the electron emission direction. This is dynamic polarization (see Section 3.1.3). If the photoionization is produced by circularly polarized light, the ion produced is oriented. The oriented ion emits spin-polarized Auger electrons and the spin orientation direction depends on the emission angle. This is an example of the polarization transfer process.

Let us consider both cases on a more quantitative basis. We recall that the components of the vector of spin polarization for an electron are connected to the first-rank statistical tensors [see Eqs. (1.93)–(1.95)]. If the initial photon beam is linearly polarized, we choose the z-axis along the direction of its polarization, which is also the direction of the ion alignment (only tensors $\mathcal{A}_{20}, \mathcal{A}_{40}$, etc. exist in this reference frame). Choose the x-axis in such a way that the Auger electron momentum lies in the xz-plane. Combining Eqs. (1.93)–(1.95) with expression (3.25) for the statistical tensor of the Auger electron and using the known properties of spherical harmonics, we can easily show that in this case the spin polarization components $P_x = P_z = 0$. For the P_y component, we obtain from (3.25), properly normalized according to condition (1.93), the following expression:

$$P_y = \frac{\sum_{k=2,4,\ldots}^{k_{max}} \beta_k \mathcal{A}_{k0}(\alpha J) \bar{P}_k^1(\cos \vartheta)}{1 + \sum_{k=2,4,\ldots}^{k_{max}} \alpha_k \mathcal{A}_{k0}(\alpha J) P_k(\cos \vartheta)} \tag{4.100}$$

where the alignment parameters $\mathcal{A}_{k0}(\alpha J)$ can be obtained from Eq. (2.163); $\bar{P}_k^1(\cos \vartheta)$ are the normalized associated Legendre polynomials (A.10); ϑ is the Auger electron emission angle, α_k are the anisotropy parameters defined by Eq. (3.11); and parameters β_k are expressed in terms of Auger decay amplitudes as follows:

$$\beta_k = -4\sqrt{3}\hat{J} \sum_{lj \leq l'j'} (-1)^{l-J-J_f-j'} \hat{j}\hat{j}'\hat{l}\hat{l}' \, (l0,l'0|k0) \begin{Bmatrix} j & j' & k \\ J & J & J_f \end{Bmatrix} \begin{Bmatrix} l & \frac{1}{2} & j \\ l' & \frac{1}{2} & j' \\ k & 1 & k \end{Bmatrix}$$

$$\times \langle \alpha_f J_f, lj : J \| V \| \alpha J \rangle \langle \alpha_f J_f, l'j' : J \| V \| \alpha J \rangle^*$$

$$\times \left(\sum_{lj} |\langle \alpha_f J_f, lj : J \| V \| \alpha J \rangle|^2 \right)^{-1} \tag{4.101}$$

We note that the denominator of Eq. (4.100) is the angular distribution of Auger electrons.

We now point out the main properties of the dynamic polarization of Auger electrons, which follow from the general formulas (4.100) and (4.101).

1. Dynamic polarization arises only when the alignment is nonzero. It thus appears that the Auger electrons of the K spectrum are unpolarized because the $1s_{1/2}$ shell vacancy cannot be aligned. Of all the lines of the L spectrum, only the electrons corresponding to the L_3 vacancy decay can be polarized. For the latter case, $J = \frac{3}{2}$, expression (4.100) is simplified to:

$$\mathbf{P} = \mathbf{n} \frac{\xi \mathcal{A}_{20}(\alpha J) \sin 2\vartheta}{1 + \alpha_2 \mathcal{A}_{20}(\alpha J) P_2(\cos \vartheta)} \tag{4.102}$$

where $\xi = \sqrt{\frac{15}{16}} \beta_2$ and $\mathbf{n} = \frac{\mathbf{k}_i \times \mathbf{k}_A}{|\mathbf{k}_i \times \mathbf{k}_A|}$.

2. For polarization to arise, there must be interference by at least two decay channels (see the discussion in Section 3.1.3). If the selection rules admit only one channel, the dynamic polarization disappears.

3. Auger electrons produced by linearly polarized light are transversely polarized.

4. Polarization of Auger electrons vanishes if they are ejected at an angle of $\vartheta = 90°$ to the incident beam direction.

5. The net dynamic polarization (integrated over all angles of ejection) is zero.

If the ionizing photon beam is circularly polarized, the ion produced can be oriented [see Eq. (2.163)], i.e., the first rank statistical tensor of the ion can be nonzero. In this case it is convenient to choose the coordinate frame with the z-axis along the beam. (The Auger electron momentum is again in the xz-plane.) Using the definition (1.93)–(1.95), the following expressions can be obtained from Eq. (3.25):

$$P_x = \frac{\frac{3}{4} \gamma_1 \mathcal{A}_{10}(\alpha J) \sin 2\vartheta}{1 + \alpha_2 \mathcal{A}_{20}(\alpha J) P_2(\cos \vartheta)} \tag{4.103}$$

$$P_y = \frac{\xi_2 \mathcal{A}_{20}(\alpha J) \sin 2\vartheta}{1 + \alpha_2 \mathcal{A}_{20}(\alpha J) P_2(\cos \vartheta)} \tag{4.104}$$

$$P_z = \frac{\mathcal{A}_{10}(\alpha J)[\beta_1 + \gamma_1 P_2(\cos \vartheta)]}{1 + \alpha_2 \mathcal{A}_{20}(\alpha J) P_2(\cos \vartheta)} \tag{4.105}$$

where

$$\beta_1 = \frac{1}{N}\sqrt{2}\hat{f}\sum_{ljj'}(-1)^{l+j+j'}\hat{j}\hat{j'}\begin{Bmatrix} J & J & 1 \\ j & j' & J_f \end{Bmatrix}\begin{Bmatrix} j & j' & 1 \\ \frac{1}{2} & \frac{1}{2} & l \end{Bmatrix}$$

$$\times \langle \alpha_f J_f, lj : J \| V \| \alpha J \rangle \langle \alpha_f J_f, l'j' : J \| V \| \alpha J \rangle^* \qquad (4.106)$$

$$\gamma_1 = \frac{1}{N}(-1)^{J+J_f+1/2}2\sqrt{3}\hat{f}\sum_{ll'jj'}(-1)^{l'+j'-1/2}\hat{j}\hat{j'}\hat{l}\hat{l'}\,(l0,l'0|20)$$

$$\times \begin{Bmatrix} J & J & 1 \\ j & j' & J_f \end{Bmatrix}\begin{Bmatrix} l & \frac{1}{2} & j \\ l' & \frac{1}{2} & j' \\ 2 & 1 & 1 \end{Bmatrix}\langle \alpha_f J_f, lj : J \| V \| \alpha J \rangle$$

$$\times \langle \alpha_f J_f, l'j' : J \| V \| \alpha J \rangle^* \qquad (4.107)$$

$$N \equiv \sum_{lj}|\langle \alpha_f J_f, lj : J \| V \| \alpha J \rangle|^2 \qquad (4.108)$$

The component P_y is the same as earlier [see Eq. (4.102)]; it describes the dynamic polarization. However, the components P_x and P_z appear as a result of polarization transfer; they are proportional to the orientation of the decaying ion. Note that in this case the net (integrated) polarization is not zero, it is determined by the ionic orientation and parameter β_1.

4.2.2.3. Angular Correlations between Photoelectrons and Auger Electrons

Quite analogous to the photoelectron–fluorescence photon correlations considered in Section 4.2.1.2, there are angular correlations between photoelectrons and Auger electrons in the process (4.97) which are revealed by photoelectron–Auger electron coincidence experiments. Using Eq. (3.10) for the angular distribution of Auger electrons and Eq. (2.169) for the statistical tensors of the ion following photoelectron emission in a certain direction, one can write the following relations which determine the angular correlation function:

$$\frac{d^2\sigma}{d\Omega_{\text{ph}}d\Omega_A} = \frac{d\sigma}{d\Omega_{\text{ph}}}(\mathbf{n}_{\text{ph}})\frac{\omega_A}{4\pi}$$

$$\times \left(1 + \sum_{k_2=2,4,\ldots}\alpha_{k_2}\sum_{q_2}\sqrt{\frac{4\pi}{2k_2+1}}Y_{k_2q_2}(\mathbf{n}_A)\mathcal{A}_{k_2q_2}(\alpha J; \mathbf{n}_{\text{ph}})\right) \qquad (4.109)$$

where $\mathbf{n}_{\text{ph}}$ and $\mathbf{n}_A$ are the directions of the photoelectron and Auger electron emission, respectively, and ω_A is Auger electron yield. The statistical tensors of the ion after the photoelectron emission are determined by Eq. (2.169) (we recall that the target atom is assumed to be unpolarized):

$$\rho_{k_2 q_2}\left(\alpha J; \mathbf{n}_{\text{ph}}\right) = \sum_{k_1 k q_1 q} \sqrt{\frac{4\pi}{2k_1 + 1}} \left(k_2 q_2, k_1 q_1 \mid kq\right)$$
$$\times Y_{k_1 q_1}(\mathbf{n}_{\text{ph}}) \rho_{kq}^{\gamma} B(k_1, k_2, k) \qquad (4.110)$$

The coefficients $B(k_1, k_2, k)$ are defined by Eq. (2.170).

Using Eq. (4.110), the double differential cross section (4.109) can be rewritten in the alternative form:

$$\frac{d^2\sigma}{d\Omega_{\text{ph}} d\Omega_A} = \sum_{k_1 k_2 kq} \tilde{B}(k_1, k_2, k) 4\pi \{Y_{k_1}(\vartheta_1, \varphi_1) \otimes Y_{k_2}(\vartheta_2, \varphi_2)\}_{kq} \rho_{kq}^{\gamma} \qquad (4.111)$$

where $\tilde{B}(k_1, k_2, k) = (\hat{k}_1 \hat{k}_2)^{-1} \alpha_{k_2} B(k_1, k_2, k)$. The form of this expression coincides with that of Eq. (2.172), which describes the cross section of the direct double photoionization. This is natural since in both cases the angular correlations between two electrons emitted in one act of photoabsorption are considered.

However, the two-step model used in the derivation of Eq. (4.111) for the resonant process imposes some additional restrictions that are absent in Eq. (2.172). In fact, our assumption that in the resonant case the process proceeds through a well-defined intermediate state with definite angular momentum and parity limits the possible values of k_2: $k_2 \leq 2J$; while in the case of direct photoionization, the value of k_2 is not limited. In addition, in the resonant case, both values of k_1 and k_2 are even because of parity conservation in both steps of the process: photoionization and Auger decay. Therefore, the angular correlation pattern is always described by an even function with respect to the inversion of space coordinates. On the contrary, in direct double photoionization, it follows from parity conservation that only the sum of indices $k_1 + k_2$ is even. Therefore, the correlation pattern can contain odd components with respect to the inversion.

Similar to the case of photoelectron–fluorescence photon correlations, the general expression (4.109) simplifies for some particular conditions of experiment. For example, if the detectors of photoelectrons and Auger electrons are both in the plane perpendicular to the photon beam and the photons are linearly polarized, then choosing this plane as the xz-plane of the coordinate system (the z-axis is along the linear polarization of the photon), we notice that the conditions of the experiment are symmetric with respect to the reflection through the xz-plane and therefore the relations (1.134) holds. If the photoelectron detector is fixed at a certain angle and the Auger electron detector is rotated in the plane perpendicular to the beam, the angular correlation function can be presented as follows:

$$\frac{d^2\sigma}{d\Omega_{\text{ph}} d\Omega_A} = \sum_{k=0,2,\ldots} A_k \cos^k \vartheta_A + \sum_{k=2,4,\ldots} B_k \sin^k \vartheta_A \qquad (4.112)$$

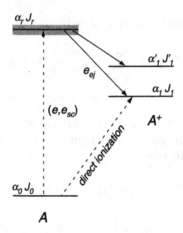

FIGURE 4.9. Scheme of autoionization process in electron-impact ionization.

where ϑ_A is the angle of the rotating detector. The coefficients A_k and B_k depend on the photon polarization and contain amplitudes of photoionization and Auger decay. The number of terms in the sums in Eq. (4.112) is restricted by the inequality $k \leq 2J$.

4.3. Polarization and Correlations in Autoionization Processes

The excitation and spontaneous decay of atomic autoionizing states is known as an important contribution to various ionization processes induced by photoabsorption or by electron–atom and ion–atom collisions. Taking into account that in these processes the excitation time of an autoionizing state is usually much shorter than its lifetime, one can consider the resonant ionization process as consisting of two independent steps. The first step is excitation of the target atom to an autoionizing state and the second one is its decay. If considered formally within this simplified picture, excitation mechanisms of a well-isolated autoionizing state do not differ from similar processes of excitation of discrete atomic levels. For this reason, the same theoretical approach as that outlined in Section 2.2.1 can be used to calculate polarization parameters of the autoionizing state and then to obtain the angular distribution and polarization characteristics of its decay products using the results of Chapter 3. However, in reality, the process is much more complicated because of an inherent interference between the two-step resonant ionization mechanism and the *direct ionization mechanism* when an atomic electron is ejected immediately to a continuum without the formation of any intermediate state of the target atom (Figure 4.9).

4.3.1. Parameterization of the Ionization Amplitudes in the Vicinity of an Isolated Autoionizing State

Consider an ionization process in the vicinity of an isolated autoionizing state $|\alpha_r J_r\rangle$ at excitation energy $\omega = E_r$

$$
A(\alpha_0 J_0) \longrightarrow A^+(\alpha_1 J_1) + e_{ej} \tag{4.113}
$$
$$
\searrow \qquad \nearrow
$$
$$
A^*(\alpha_r J_r)
$$

Let T be a transition operator connecting the ground state $|\alpha_0 J_0\rangle$ of the target atom with the autoionizing state $|\alpha_r J_r\rangle$ and with a number of open channels $|\alpha_1 J_1, \mathbf{p}_1 \mu_1\rangle$ in its vicinity where $\mathbf{p}_1$ and μ_1 denote the linear momentum and the spin component of the ejected electron e_{ej}. We do not specify here the transition operator T. It can be, for example, either an operator of electromagnetic interaction H_{int} in the photoionization case, or operator $T(\mathbf{Q}) = \sum_j^Z e^{i\mathbf{Q}\mathbf{r}_j}$ with the transferred momentum vector $\mathbf{Q} = \mathbf{p}_i - \mathbf{p}_f$ used in the PWBA calculations of the electron-impact ionization process, or operator $T(\mathbf{p}_i, \mathbf{p}_f) = \int \psi_{\mathbf{p}_f}^{(-)*}(\mathbf{r}) \sum_j^Z \frac{1}{|\mathbf{r}-\mathbf{r}_j|} \psi_{\mathbf{p}_i}^{(+)}(\mathbf{r}) d^3 r$ used instead of $T(\mathbf{Q})$ in the distorted-wave Born approximation (DWBA) calculations of the same process.

In the vicinity of the autoionizing state $|\alpha_r J_r\rangle$, the ionization amplitude $\langle \alpha_1 J_1, l_1 j_1 : JM \,|\, T \,|\, \alpha_0 J_0 M_0\rangle$ can be presented as a sum of the direct and resonant terms:

$$
\langle \alpha_1 J_1, l_1 j_1 : JM \,|\, T \,|\, \alpha_0 J_0 M_0\rangle
$$
$$
= \langle \alpha_1 J_1, l_1 j_1 : JM \,|\, T \,|\, \alpha_0 J_0 M_0\rangle_{dir}
$$
$$
+ \langle \alpha_1 J_1, l_1 j_1 : JM \,|\, T \,|\, \alpha_0 J_0 M_0\rangle_{res} \tag{4.114}
$$

where J_1, j_1, and J are the total angular momenta of the residual ion A^+, the ejected electron, and the target atom in the final state, respectively; l_1 is the orbital angular momentum of the ejected electron [additional quantum numbers can characterize the amplitudes in Eq. (4.114), for example, the spin projections of the incoming and scattered electrons in the case of electron–atom collisions].

The first term in Eq. (4.114) is calculated within the standard channel-coupling procedures by *prediagonalization* of the Hamiltonian in the subspace of open channels. As for the second term, one can construct it following the parameterization that was suggested by Fano [29] for a unified description of direct and resonant photoeffects and later extended to other ionization processes:

$$
\langle \alpha_1 J_1, l_1 j_1 : JM \,|\, T \,|\, \alpha_0 J_0 M_0\rangle_{res} = \delta_{JJ_r} \frac{V_r(\alpha_1 J_1, l_1 j_1 : J_r)}{V_r}
$$
$$
\times \left[\left(-\frac{i}{\varepsilon + i} \right) \langle v_r J_r M \,|\, T \,|\, \alpha_0 J_0 M_0\rangle_{dir} \right.
$$

$$+ \frac{\langle \alpha_r J_r M \,|\, T \,|\, \alpha_0 J_0 M_0 \rangle}{\pi V_r} \frac{1}{\varepsilon + 1} \Bigg] \tag{4.115}$$

Usually it is written in a more compact form:

$$\langle \alpha_1 J_1, l_1 j_1 : JM \,|\, T \,|\, \alpha_0 J_0 M_0 \rangle = \delta_{JJ_r} \frac{V_r(\alpha_1 J_1, l_1 j_1 : J_r)}{V_r} \langle v_r J_r M \,|\, T \,|\, \alpha_0 J_0 M_0 \rangle_{\mathrm{dir}}$$

$$\times \frac{q(\alpha_0 J_0 M_0 \to \alpha_r J_r M) - i}{\varepsilon + i} \tag{4.116}$$

Here the dimensionless parameters

$$q(\alpha_0 J_0 M_0 \to \alpha_r J_r M) = \frac{\langle \alpha_r J_r M \,|\, T \,|\, \gamma_0 J_0 M_0 \rangle}{\pi V_r \langle v_r J_r M \,|\, T \,|\, \alpha_0 J_0 M_0 \rangle_{\mathrm{dir}}} \tag{4.117}$$

depend generally on combinations of magnetic quantum numbers for the initial and final states. The number of parameters (4.117) can be reduced considerably with the use of particular approximations in calculating the transition amplitudes. For example, within the PWBA approach [see Eq. (2.64)], when the amplitudes do not depend on the spin quantum numbers of the fast scattering electron and owing to the specific axial symmetry inherent in this approximation, the isolated autoionizing state $|\alpha_r J_r \rangle$ can be specified by a single *profile index* $q_r(Q)$. (It is implied that in the PWBA the quantization axis is chosen along the vector of the transferred momentum Q.)

In Eqs. (4.116) and (4.117), $|v_r J_r M \rangle$ denotes a coherent superposition of the continuum state vectors $|\alpha_1 J_1, l_1 j_1 : JM \rangle$:

$$|v_r J_r M \rangle = \sum_{\alpha_1 J_1 l_1 j_1} \frac{V_r(\alpha_1 J_1, l_1 j_1 : J_r)}{V_r} |\alpha_1 J_1, l_1 j_1 : J_r M \rangle \tag{4.118}$$

which embodies the whole interaction V between these continuum states and the autoionizing state $|\alpha_r J_r M \rangle$ of the same magnetic quantum number M. The values $V_r(\alpha_1 J_1, l_1 j_1 : J_r)$ and V_r are connected to the matrix elements of this interaction and the partial and total decay widths of the autoionizing state:

$$V_r(\alpha_1 J_1, l_1 j_1 : J_r) = \langle \alpha_1 J_1, l_1 j_1 : J_r \,|\, V \,|\, \alpha_r J_r \rangle \tag{4.119}$$

$$V_r = \left(\sum_{\alpha_1 J_1 l_1 j_1} |V_r(\alpha_1 J_1, l_1 j_1 : J_r)|^2 \right)^{1/2} \tag{4.120}$$

$$\Gamma(\alpha_1 J_1, l_1 j_1 : J_r) = 2\pi |V_r(\alpha_1 J_1, l_1 j_1 : J_r)|^2 \tag{4.121}$$

$$\Gamma = \sum_{\alpha_1 J_1 l_1 j_1} \Gamma(\alpha_1 J_1, l_1 j_1 : J_r) = 2\pi |V_r|^2 \tag{4.122}$$

ε is the reduced deviation of the excitation energy E from the resonance energy

$$\varepsilon = \frac{E - (E_r + \Delta)}{\Gamma/2} \qquad (4.123)$$

where Δ is a shift of the resonance position caused by the interaction between the autoionizing state and open channels.

Being calculated according to Eqs. (4.114)–(4.116), the ionization amplitudes can be used to obtain any characteristics of corresponding ionization processes of interest in the same way as was done in Section 2 for direct ionization processes.

4.3.2. Alignment and Decay Characteristics in the Pure Two-Step Autoionization Process

There are a number of cases of practical importance where the direct ionization background for the autoionization resonances as well as the contribution of their interference to the cross sections and other characteristics of the ionization process turn out to be negligibly small. In such cases, the ionization can be treated as a pure two-step process in which the excitation of an autoionizing state is followed by the quite independent process of its spontaneous decay. Hence, all angular and polarization characteristics of the autoionization products are determined by polarization parameters of the autoionizing state formed at the first stage of the process, as well as by the corresponding decay amplitudes. A general theory of such a two-step process can be developed with the use of the results of Sections 2.1.1, 2.2, and Chapter 3.

4.3.2.1. Energy Dependence of the Alignment of Autoionizing States Excited in Electron–Atom Collisions

Consider the alignment of an isolated autoionizing state $|\alpha_r J_r\rangle$ excited in electron–atom collisions when neither the target atom nor the electron beam are polarized and when the scattered electron is not detected. Owing to symmetry arguments (see Section 1.4), the only nonzero reduced statistical tensors are $\mathcal{A}_{20}(\alpha_r J_r)$, $\mathcal{A}_{40}(\alpha_r J_r)$, ..., with the rank $k \leq 2J_r$, which contain complete information about the polarization properties of the excited state. To calculate them, one integrates over $d\Omega_{sc}$ the statistical tensors of the excited state $|\alpha_r J_r\rangle$ corresponding to fixed directions of the scattered electron [see Eqs. (2.46) and (2.44)]:

$$\rho_{k0}(\alpha_r J_r; \vartheta_{sc}, \varphi_{sc}) = \sum_{M_0 \mu_i \mu_f M_r M_r'} (-1)^{J_r - M_r'} (J_r M_r, J_r - M_r' | k0)$$

$$\times T_{M_0 \mu_i \to M_r \mu_f}(\mathbf{p}_i, \mathbf{p}_f) T^*_{M_0' \mu_i \to M_r' \mu_f}(\mathbf{p}_i, \mathbf{p}_f) \qquad (4.124)$$

The behavior of the integrated statistical tensors considered as a function of the energy E_0 of the incoming electron beam in the wide region from the excitation threshold $E_0 = E_{thr}$ to asymptotically high energies $E_0 \gg E_{thr}$ depends on the excitation mechanism of the state under consideration and the structural properties (wave functions) of both the ground and the excited states of the target atom. Nevertheless, it turns out that quite independently of these dynamic and structural factors, the curve $\mathcal{A}_{20}(E_0)$ of the alignment parameter [and hence the curve $\beta_2(E_0)$ of the angular anisotropy coefficient of the ejected autoionization electrons] crosses the abscissa axis at some value of E_0. This universal behavior is valid for any state $|\alpha_r J_r\rangle$ at any mechanism of its excitation. This is the same effect which we considered in Section 2.2.1.4 for the alignment of discrete levels of an atom excited by electron impact.

4.3.2.2. "Perfect Experiment" on Electron-Impact Excitation of Atomic Autoionizing States

Coincidence $(e, 2e)$ measurements [30] have become a powerful method for studying various aspects of the excitation and decay of atomic autoionizing states in electron–atom collisions. In this section we demonstrate the potential of this method used in combination with noncoincidence (e, e') measurements from the point of view of the concept of a *perfect experiment*. We consider the case of a pure two-step excitation–autoionization process and completely disregard the contribution from the direct ionization process and its interference with resonance ionization (Figure 4.9 on p. 170). The $^2S \rightarrow {}^2L$ transition in a "one-electron" atom (such as Na, K, and so on) can serve as a convenient example:

$$A(^2S) + e \rightarrow A^*(^2L) + e_{sc}$$
$$ \hookrightarrow A^+(^1S) + e_{ej} \qquad (4.125)$$

We consider a general case where both the target atom and the incoming beam can be polarized. No assumption is made concerning the electron-impact excitation mechanism besides conservation of the total spin of the $A + e$ system.

Here we use the LS-coupling approximation and denote the amplitude of electron-impact excitation of the target atom by $\langle LM_L SM_S \,|\, T(\mathbf{p}_i, \mathbf{p}_f) \,|\, S_0 M_{S_0} \rangle$ where $\mathbf{p}_i$ and $\mathbf{p}_f$ are electron linear momenta in the initial and final states and and LM_L, SM_S, and $S_0 M_{S_0}$ are the total orbital angular momentum and total spin quantum numbers with their projections of the whole system $(e + A)$ in the initial and final states. The spin density matrix ρ_f of the total system in the final state $A^* + e_{sc}$ is related to the density matrix ρ_i of the system $A + e$ in the initial state by Eq. (1.24):

$$\rho_f = T \rho_i T^+ \qquad (4.126)$$

If the Percival–Seaton hypothesis (conservation of the total spin $\mathbf{S} = \mathbf{S}_A(^2S) + \mathbf{s}_i = \mathbf{S}_A(^2L) + \mathbf{s}_f$) is accepted:

$$\langle LM_L SM_S \,|\, T(\mathbf{p}_i, \mathbf{p}_f) \,|\, S_0 M_{S_0} \rangle = \delta_{SS_0} \delta_{M_S M_{S_0}} F_{SM_L}(\mathbf{p}_i, \mathbf{p}_f) \quad (4.127)$$

the density matrix can be presented as follows:

$$\langle LM_L SM_S \,|\, \rho_f \,|\, LM'_L S'M'_S \rangle$$
$$= F_{SM_L}(\mathbf{p}_i, \mathbf{p}_f) \langle SM_S \,|\, \rho_i \,|\, S'M'_S \rangle F_{S'M'_L}(\mathbf{p}_i, \mathbf{p}_f)^* \quad (4.128)$$

where each of the amplitudes is a complex quantity:

$$F_{SM_L}(\mathbf{p}_i, \mathbf{p}_f) \equiv |F_{SM_L}| e^{i\delta_{SM_L}} \quad (4.129)$$

We denote as $F_{M_L}^{(s)}(\mathbf{p}_i, \mathbf{p}_f)$ and $F_{M_L}^{(t)}(\mathbf{p}_i, \mathbf{p}_f)$ the excitation amplitudes $F_{SM_L}(\mathbf{p}_i, \mathbf{p}_f)$ with $S = 0, 1$ in singlet (s) and triplet (t) states, respectively.

To consider how many independent parameters characterize these amplitudes, let us analyze the well known example of the transition $^2S \to {}^2P$. Owing to reflection symmetry [$F_{M_L=0}^{(s,t)} = 0$ in the natural frame], one has only seven parameters. Namely, these are two subsets of three parameters for the singlet and triplet cases:

$$|F_{+1}^{(s,t)}|; \qquad |F_{-1}^{(s,t)}|; \qquad \delta^{(s,t)} = \delta_{+1}^{(s,t)} - \delta_{-1}^{(s,t)} \quad (4.130)$$

and the phase difference between the singlet and triplet amplitudes:

$$\Delta_{+1} = \delta_{+1}^{(s)} - \delta_{+1}^{(t)} \quad (4.131)$$

Consider first an autoionization $(e, 2e)$ process when spin observables are not measured in the final state. The angular distribution of electrons ejected by the autoionizing atom and measured in coincidence with the scattered electrons is given by the general equation (3.10), with alignment parameters of the decaying state dependent on $\mathbf{p}_i$ and $\mathbf{p}_f$. The singlet and triplet interactions contribute additively to the statistical tensors $\rho_{kq}(L)$ of the decaying state:

$$\rho_{kq}(L; \vartheta, \varphi) = w^{(s)} \rho_{kq}^{(s)}(L; \vartheta, \varphi) + w^{(t)} \rho_{kq}^{(t)}(L; \vartheta, \varphi) \quad (4.132)$$

where

$$\rho_{00}^{(s,t)}(L; \vartheta, \varphi) = \frac{1}{\sqrt{3}} \left(|F_{+1}^{(s,t)}|^2 + |F_{-1}^{(s,t)}|^2 \right)$$

$$\rho_{10}^{(s,t)}(L; \vartheta, \varphi) = \frac{1}{\sqrt{2}} \left(|F_{+1}^{(s,t)}|^2 - |F_{-1}^{(s,t)}|^2 \right)$$

$$\rho_{20}^{(s,t)}(L; \vartheta, \varphi) = \frac{1}{\sqrt{6}} \left(|F_{+1}^{(s,t)}|^2 + |F_{-1}^{(s,t)}|^2 \right)$$

$$\rho_{22}^{(s,t)}(L; \vartheta, \varphi) = \left[\rho_{2-2}^{(s,t)}(L; \vartheta, \varphi) \right]^* = F_{+1}^{(s,t)} \left(F_{-1}^{(s,t)} \right)^* \quad (4.133)$$

Equations (4.132) and (4.133) can be derived by taking the trace of the double statistical tensor $\rho_{k_L q_L k_S q_S}(LS)$ over the total spin S according to Eq. (1.69), where $\rho_{k_L q_L k_S q_S}(LS)$ are expressed in terms of the amplitudes F_{SM_L} by Eqs. (4.128) and (1.62). Below we will consider the angular distribution of the elected electrons $\frac{d^2\sigma}{d\Omega_{sc}d\Omega_{ej}}$ as function of ejection angles $\vartheta_{ej}, \varphi_{ej}$ at fixed $\vartheta_{sc}, \varphi_{sc}$. Similar to Eqs. (4.132) and (4.133), the additivity of the singlet and triplet contributions takes place also for the distribution:

$$\frac{d^2\sigma}{d\Omega_{sc}d\Omega_{ej}}(\vartheta_{ej},\varphi_{ej}) = \frac{d^2\sigma^{(s)}}{d\Omega_{sc}d\Omega_{ej}}(\vartheta_{ej},\varphi_{ej}) + \frac{d^2\sigma^{(t)}}{d\Omega_{sc}d\Omega_{ej}}(\vartheta_{ej},\varphi_{ej}) \quad (4.134)$$

When either the target atom or the incoming electron beam is not polarized, then the weight factors $w^{(s,t)}$ are independent of the degree of polarization of the other partner; in such a case singlet and triplet contributions are weighted statistically:

$$w(S) = \frac{2S+1}{4} \quad (4.135)$$

In particular, the angular distribution of the autoionization electrons in the scattering plane, $W(\varphi_{ej}) = \frac{d^2\sigma}{d\Omega_{sc}d\Omega_{ej}}\left(\vartheta_{ej} = \frac{\pi}{2}, \varphi_{ej}\right)$, at fixed $\mathbf{p}_i, \mathbf{p}_f$ can be parameterized as

$$\begin{aligned} W(\varphi_{ej}) &= W^{(s)}(\varphi_{ej}) + W^{(t)}(\varphi_{ej}) \\ &= \frac{1}{4}\frac{d\sigma^{(s)}}{d\Omega_{sc}}\left[1 - P_l^{(s)}\cos 2(\varphi_{ej} - \gamma^{(s)})\right] \\ &\quad + \frac{3}{4}\frac{d\sigma^{(t)}}{d\Omega_{sc}}\left[1 - P_l^{(t)}\cos 2(\varphi_{ej} - \gamma^{(t)})\right] \end{aligned} \quad (4.136)$$

where

$$\frac{d\sigma^{(s,t)}}{d\Omega_{sc}} = |F_{+1}^{(s,t)}|^2 + |F_{-1}^{(s,t)}|^2 \quad (4.137)$$

are the differential cross sections of the $^2S \to {}^2P$ excitation due to the singlet and triplet interactions,

$$\gamma^{(s,t)} \equiv -\frac{1}{2}\delta^{(s,t)} \quad (4.138)$$

are the phase differences, Eq. (4.130) and the factors $P_l^{(s,t)}$ are

$$P_l^{(s,t)} \equiv \frac{2|F_{+1}^{(s,t)}||F_{-1}^{(s,t)}|}{|F_{+1}^{(s,t)}|^2 + |F_{-1}^{(s,t)}|^2} \quad (4.139)$$

Note that the latter are the same combinations of the excitation amplitudes as those that should determine the Stokes parameters P_1, P_2, and P_3 of the emitted light in the corresponding $(e, e'\gamma)$ case:

$$P_1 + iP_2 = P_l\, e^{2i\gamma} \tag{4.140}$$

The partial angular distributions $W^{(s)}(\varphi_{ej})$ and $W^{(t)}(\varphi_{ej})$ display the shape and orientation of the charge clouds collisionally induced in the excited atom by means of singlet and triplet interactions. Generally they differ from each other as a result of exchange scattering.

Using both a polarized target and a polarized beam (and combining such measurements with those performed with unpolarized particles), one can separate the singlet and triplet contributions in Eq. (4.136). In particular, one suppresses the contribution of the singlet interaction completely by using the collinear polarized target and beam:

$$W(\varphi_{ej})\,|_{\uparrow\uparrow} = \frac{d\sigma^{(t)}}{d\Omega_{sc}}\left[1 - P_l^{(t)}\cos 2(\varphi_{ej} - \gamma^{(t)})\right] \tag{4.141}$$

The measurements discussed here cannot provide the phase difference parameter Δ_{+1}. This can be obtained by spin-resolved measurements in the final state if, for example, one measures the depolarization of the incoming polarized electron beam after the scattering provided the other six parameters of the complete set (4.130) and (4.131) are obtained without spin-resolved measurements in the final state. As the depolarization factor one can use the ratio of the normal (to the scattering plane) components of the polarization vector $P_\perp$ of the scattered and incoming electrons:

$$T(\vartheta_{sc}, \varphi_{sc}) = \frac{\left(P_\perp\right)_f}{\left(P_\perp\right)_i} = \frac{\sum_M\left[|F_M^{(t)}|^2 + 2\cdot\mathrm{Re}\left(F_M^{(s)}F_M^{(t)*}\right)\right]}{\sum_M\left[|F_M^{(s)}|^2 + 3|F_M^{(t)}|^2\right]} \tag{4.142}$$

To summarize, three steps can be suggested that lead to the "perfect experiment" in electron-impact ionization of a one-electron ^2S-atom via excitation and decay of the ^2P autoionizing state and that can provide the total set of the excitation amplitudes in a model-independent way:

1. Measurements of the angular distribution of autoionization electrons in the $(e, 2e)$ experiment with an unpolarized target and an unpolarized beam by a detector insensitive to spin polarization. Here the shape and orientation of the electron cloud in the autoionizing state averaged over contributions from singlet and triplet interactions can be obtained.

2. The $(e, 2e)$ measurements when both the target atom and the incoming beam are polarized. Combined with step 1 this measurement provides information about separated charge cloud distributions induced by singlet and triplet interactions. In this case, an almost complete set of independent parameters of the excitation amplitudes $F_{\pm 1}^{(s,t)}(\vartheta_{sc}, \varphi_{sc})$ can be obtained, but without the phase difference between the singlet and triplet amplitudes Δ_+.

3. Spin-resolved measurements in the final state (combined with steps 1 and 2) complete the perfect experiment.

In this section we considered a special case of ^2P autoionizing states decaying to the ^1S state of the residual ion. Generally, beyond the $^2S \rightarrow {}^2P \rightarrow {}^1S$ case, comparing the excitation of an autoionizing level and of a discrete one from the point of view of the general possibilities of coincidence methods $(e, e'\gamma)$ and $(e, 2e)$, one notes a fundamental advantage of the latter method. In the $(e, e'\gamma)$ case, only the dipole transitions play a role in photon decay at any value of the angular momentum of the discrete level formed. In contrast, the electron emission from an autoionizing state is free of this kind of limitation, the multipolarity of the emission being governed only by angular momenta and parity of the decaying and final states. This means that any observables related to the photon emitted in the process $(e, e'\gamma)$ are absolutely insensitive to any orientation and alignment parameters of the emitting state if the rank of corresponding statistical tensors $\rho_{kq}(\alpha J)$ exceeds $k = 2$. This is not so in the autoionization $(e, 2e)$ case.

4.3.3.   Resonance Profile of Correlation and Polarization Parameters in the Vicinity of an Isolated Autoionization State

In this section we consider three examples of application of resonant parameterization of the ionization amplitudes given in Section 4.3.1.

4.3.3.1.   Resonances in Coincidence $(e, 2e)$ Spectra

Consider a general case of electron-impact ionization of an unpolarized atom A from its initial state $|\alpha_0 J_0\rangle$ in the vicinity of an isolated autoionizing state $|\alpha_r J_r\rangle$ at the excitation energy $E = E_r$ provided that the incoming electron beam is not polarized nor is the detector of ejected electrons sensitive to their polarization. We denote as $\alpha_1 J_1$ the total set of quantum numbers characterizing the ionic state that can be populated in the process under consideration:

$$A(\alpha_0 J_0) + e_0 \rightarrow \left(A^+(\alpha_1 J_1) + e_{ej}\right) + e_{sc} \qquad (4.143)$$

Similar to Section 4.3.1 the state vectors $|\alpha_1 J_1, l_1 j_1 : JM\rangle$ specify the ionization channels where $l_1 j_1$ denote the orbital and total angular momenta of the ejected electrons. Let $T_{M_0\mu_i \to M\mu_f}(\mathbf{p}_i, \mathbf{p}_f) = \langle \alpha_1 J_1 l_1 : JM, \mathbf{p}_f \mu_f \,|\, T \,|\, \alpha_0 J_0 M_0, \mathbf{p}_i \mu_i \rangle$ be the ionization amplitude of the electron-scattering process where $\mathbf{p}_i, \mu_i$ and $\mathbf{p}_f, \mu_f$ are the linear momenta and spin components of the scattered electron in the initial and final states, respectively; $\mathbf{p}_1$ will stand for the linear momentum of the ejected electron. No special approximation concerning the atom–electron interaction and therefore no special form of the transition operator T will be accepted in this section.

One can derive equations for all angular correlation and orientation characteristics of the ionization process in the standard manner starting from the density matrix (or, equivalently, statistical tensors) of the entire system $A^+ + e_{ej}$ in the final state:

$$
\begin{aligned}
\rho_{kq}&(\alpha_1 J_1 (l_1 j_1) J; \alpha_1' J_1' (l_1' j_1') J') \\
&= \frac{1}{2(2J_0 + 1)} \sum_{M_0 \mu_0 M \mu M'} (-1)^{J'-M'} (JM, J' - M' | kq) \\
&\quad \times T_{M_0 \mu_i \to M\mu_f}(\mathbf{p}_i, \mathbf{p}_f) T^*_{M_0 \mu_i \to M'\mu_f}(\mathbf{p}_i, \mathbf{p}_f)
\end{aligned} \tag{4.144}
$$

The angular distribution of electrons ejected to a particular ionization channel $A^+ + e_{ej}$ (with the excitation energy ω) and observed in coincidence with the scattered electrons is given by partial triple-differential cross sections (TDC) of the process, which can be calculated by convoluting statistical tensors (4.144) of the system $A^+ + e_{ej}$ in the final state and the efficiency tensors of the detector:

$$
\begin{aligned}
\frac{d^3\sigma}{d\Omega_{sc} d\Omega_{ej} d\omega} &= \frac{p_f}{p_i} \sum_{l_1 j_1 l_1' j_1' J J' kq} \rho_{kq}(\alpha_1 J_1, l_1 j_1 : J; \alpha_1 J_1, l_1' j_1' : J') \\
&\quad \times \varepsilon^*_{kq}(\alpha_1 J_1, l_1 j_1 : J; \alpha_1 J_1, l_1' j_1' : J')
\end{aligned} \tag{4.145}
$$

When the detector is not sensitive to spin polarization of the ejected electrons, the efficiency tensors take the form [see Eq. (1.171)]:

$$
\begin{aligned}
\varepsilon_{kq}(\alpha_1 J_1, l_1 j_1 : J; \alpha_1 J_1, l_1' j_1' : J') &= \frac{1}{\sqrt{4\pi}} (-1)^{J + J_1 + k - 1/2} \hat{l}_1 \hat{l}_1' \hat{j}_1 \hat{j}_1' \hat{J} \hat{J}' \frac{1}{\hat{k}} \\
&\quad \times (l_1 0, l_1' 0 | k0) \begin{Bmatrix} j_1 & l_1 & 1/2 \\ l_1' & j_1' & k \end{Bmatrix} \begin{Bmatrix} J' & j_1' & J_1 \\ j_1 & J & k \end{Bmatrix} Y^*_{kq}(\vartheta_{ej}, \varphi_{ej})
\end{aligned} \tag{4.146}
$$

Combining Eqs. (4.144)–(4.146), one obtains the TDC as a bilinear combination of the ionization amplitudes with its coefficients depending on discrete quantum numbers only. Note that contributions from the channels with different J are not additive here. Then applying the general equations (4.114)–(4.116) for the ionization amplitude one obtains the TDC as a sum of contributions from pure direct and pure two-step (via autoionizing states) processes plus an interference term. In the Shore parameterization [31], the TDC has the following form:

$$\frac{d^3\sigma}{d\Omega_{sc}d\Omega_{ej}d\omega} = f_{(\alpha_1 J_1)}(\mathbf{p}_i, \mathbf{p}_f, \mathbf{p}_1)$$
$$+ \frac{a_{(\alpha_1 J_1)}(\mathbf{p}_i, \mathbf{p}_f, \mathbf{p}_1)\varepsilon + b_{(\alpha_1 J_1)}(\mathbf{p}_i, \mathbf{p}_f, \mathbf{p}_1)}{\varepsilon^2 + 1} \qquad (4.147)$$

The coefficient a in this formula is the profile asymmetry parameter showing the asymmetry in the shape of the resonance. The asymmetry originates from interference between the direct and resonant ionization amplitudes and vanishes after averaging of the cross section over an excitation energy range ΔE much wider than the resonance width Γ_r. The other coefficient, b, can be considered as a resonance yield parameter showing an excess of the averaged resonance curve for the cross section over the direct ionization background.

The Shore parameterization (4.147) can be given an equivalent form of the *Fano formula*

$$\frac{d^3\sigma}{d\Omega_{sc}d\Omega_{ej}d\omega} = \left(\frac{d^3\sigma}{d\Omega_{sc}d\Omega_{ej}d\omega}\right)_1 + \left(\frac{d^3\sigma}{d\Omega_{sc}d\Omega_{ej}d\omega}\right)_2 \frac{(q_r + \varepsilon)^2}{\varepsilon^2 + 1} \qquad (4.148)$$

where q_r is the profile index of the resonance which is a function of the directions of both the scattered and the ejected electrons.

4.3.3.2. *Resonances in Energy Dependence of Polarization of Excited States of Residual Ions*

Consider the effects of the autoionizing resonances in coincidence experiments of the type $(e, e'\gamma)$ where the fluorescence radiation from an ionic state $|\alpha_1 J_1\rangle$ in the process (4.143) is detected in coincidence with the scattered electron. Here we assume that the ejected electrons are not observed and the directions of $\mathbf{p}_i$ and $\mathbf{p}_f$ are fixed. We ignore here the population of the state $|\alpha_1 J_1\rangle$ via radiative cascade from higher-lying levels.

The angular distribution of fluorescence emitted by an ion in the transition $|\alpha_1 J_1\rangle \rightarrow |\alpha_f J_f\rangle$ as well as its polarization is determined by the reduced statistical tensors $\mathcal{A}_{kq}(\alpha_1 J_1)$, $k \leq 2$ (see Sections 3.2.1 and 3.2.2). Corresponding statistical tensors $\rho_{kq}(\alpha_1 J_1)$ of the excited ionic state follow from the statistical tensors (4.144) of the entire system $A^+ + e_{ej}$ and are calculated according to Eq. (1.69):

$$\rho_{kq}(\alpha_1 J_1) = \sum_{JJ'l_1 j_1} (-1)^{J'+J_1+j_1+k} \hat{f}\hat{f}' \left\{ \begin{array}{ccc} J_1 & J_1 & k \\ J' & J & j_1 \end{array} \right\}$$
$$\times \rho_{kq}(\alpha_1 J_1, l_1 j_1 : J; \alpha_1 J_1, l_1 j_1 : J') \qquad (4.149)$$

With the statistical tensors [see Eq. (4.144)] and the ionization amplitudes $T_{M_0\mu_i \to M\mu_f}(\mathbf{p}_i, \mathbf{p}_f)$ parameterized according to Eqs. (4.114)–(4.116), the alignment parameters $\mathcal{A}_{kq}(\alpha_1 J_1)$ of the excited ionic state $|\alpha_1 J_1\rangle$ can be presented as a fraction, with both the numerator and the denominator written in the Shore parameterization:

$$\mathcal{A}_{kq}(\alpha_1 J_1) = \left(f_n + \frac{a_n \varepsilon + b_n}{\varepsilon^2 + 1}\right) \Big/ \left(f_d + \frac{a_d \varepsilon + b_d}{\varepsilon^2 + 1}\right) \qquad (4.150)$$

Here the parameters $f_{n,d}$, $a_{n,d}$, and $b_{n,d}$ are smooth functions of the electron energy loss in the vicinity of the resonance; their dependence on the electron scattering angle ϑ_{sc} is determined by the ionization dynamics.

4.3.3.3. Angular Distribution and Polarization of Photoelectrons in Resonant Photoionization Processes

In Section 2.3 we introduced the partial angular correlation and polarization characteristics of photoionization of an unpolarized atom $A(\alpha_0 J_0)$ corresponding to a definite final state $|\alpha_f J_f\rangle$ of the residual ion A^+. Consider now their energy dependence (*resonance profile*) in the vicinity of an isolated autoionization state $|\alpha_r J_r\rangle$ of the target atom at excitation energy E_r. We discuss the anisotropy coefficient $\beta_{\alpha_f J_f}(E)$ of the angular distribution of photoelectrons [see Eq. (2.133)]:

$$\frac{d\sigma_{\alpha_f J_f}}{d\Omega_e} = \frac{\sigma_{\alpha_f J_f}}{4\pi}\left(1 + \beta_{\alpha_f J_f}(E)P_2(\cos\vartheta_e)\right) \qquad (4.151)$$

and the parameter $\xi_{\alpha_f J_f}(E)$, which determines their spin polarization [see Eq. (2.154)]:

$$P_y(E) = \frac{\xi_{\alpha_f J_f}(E)\sin 2\vartheta_e}{1 + \beta(E)P_2(\cos\vartheta_e)} \qquad (4.152)$$

For the direct photoionization, their general expressions were given by Eqs. (2.135) and (2.155) where the symbols $M_{ljJ} \equiv \langle\alpha_f J_f, lj : J\|D\|\alpha_0 J_0\rangle$ were used for partial amplitudes of the photoionization process. We can use the same equations for our purpose, combining them with the general Fano parameterization (4.115)–(4.116) of the photoionization amplitude in the vicinity of an autoionizing state as a sum of the direct and resonant terms:

$$\langle\alpha_f J_f, lj : J\|D\|\alpha_0 J_0\rangle = \langle\alpha_f J_f, lj : J\|D\|\alpha_0 J_0\rangle_{\text{dir}}$$
$$+ \delta_{J,J_r} \frac{V_r(\alpha_f J_f, lj : J_r)}{V_r}\langle v_r J_r\|D\|\alpha_0 J_0\rangle_{\text{dir}}\frac{q_r - i}{\varepsilon + i} \qquad (4.153)$$

Here the continuum state vector $|v_r J_r\rangle$ is constructed according to the general formula (4.118) and

$$q_r = \frac{\langle \alpha_r J_r \| D \| \alpha_0 J_0 \rangle}{\pi V_r \langle v_r J_r \| D \| \alpha_0 J_0 \rangle_{\text{dir}}} \tag{4.154}$$

is the photoabsorption profile index, which is a common parameter for the whole set of the partial amplitudes of photoionization. It determines the profile of the total photoionization cross section in the vicinity of E_r:

$$\sigma(E) = \sum_{\alpha_f J_f J l j} |\langle \alpha_f J_f, l j : J \| D \| \alpha_0 J_0 \rangle_{\text{dir}}|^2$$

$$+ |\langle v_r J_r \| D \| \alpha_0 J_0 \rangle_{\text{dir}}|^2 \left[\frac{(\varepsilon + q_r)^2}{\varepsilon^2 + 1} - 1 \right] \tag{4.155}$$

Substituting Eq. (4.153) into (2.135), one can parameterize the energy dependence of the angular anisotropy coefficient for photoelectrons in the vicinity of an autoionizing state by the following form:

$$\beta(E) = \frac{X\varepsilon^2 + Y\varepsilon + Z}{A\varepsilon^2 + B\varepsilon + C} \tag{4.156}$$

where the quantities A, B, C, X, Y, and Z do not depend on the energy. Inserting Eq. (4.153) into (2.154), the similar form can be obtained for the energy dependence of the polarization parameter $\xi(E)$ with new coefficients X, Y, Z but the same coefficients A, B, and C.

4.3.4. Polarization Characteristics of Negative Ion Resonances in Electron–Atom Collisions

The formation and decay of negative-ion resonances in electron–atom collisions can be considered as a kind of autoionization process. Using the Feshbach approach, the amplitude of excitation of an atomic state $|n\rangle$

$$A + e \rightarrow A_n^* + e_{sc} \tag{4.157}$$

in the vicinity of an isolated negative-ion resonance with energy E_r and with an internal wave function Ψ_r can be presented as the sum of direct and resonance terms:

$$F_{n0}(\mathbf{p}_i, \mathbf{p}_f) = F_{n0}^{\text{dir}}(\mathbf{p}_i, \mathbf{p}_f) + F_{n0}^{\text{res}}(\mathbf{p}_i, \mathbf{p}_f) \tag{4.158}$$

The first term is calculated using standard methods of the theory of direct inelastic scattering in electron–atom collisions by either taking into account channel coupling in a corresponding subspace of open channels or, more approximately, applying the DWBA or even PWBA approaches if the incoming electron energy is high enough to rely on these approximations. The second term can be parameterized in the vicinity of E_r according to the general prescriptions of the unified theory of direct and resonant reactions:

$$F_{n0}^{res}(\mathbf{p}_i, \mathbf{p}_f) = -\frac{1}{2\pi} \frac{\left\langle \Psi_{n,\mathbf{p}_f}^{(-)} | V | \Psi_r \right\rangle \left\langle \Psi_r | V | \Psi_{0,\mathbf{p}_i}^{(+)} \right\rangle}{E - (E_r + \Delta) + \frac{i}{2}\Gamma_r} \tag{4.159}$$

where $\Psi_{n,\mathbf{p}_f}^{(-)}$ and $\Psi_{0,\mathbf{p}_i}^{(+)}$ are generally multichannel wave functions of the whole system $A + e$ in the initial and final states. The resonance width Γ_r and the resonance shift Δ are determined by the interaction V coupling the negative-ion resonance state with the continuum:

$$\Gamma_r = 2\pi \sum_n \int \left| \left\langle \Psi_{n,\mathbf{p}_f}^{(-)} | V | \Psi_r \right\rangle \right|^2 \rho_n(E) d\Omega_n \tag{4.160}$$

$$\Delta = -\sum_n P \int dE' \frac{\left| \left\langle \Psi_{n,\mathbf{p}_f}^{(-)} | V | \Psi_r \right\rangle \right|^2}{E - E'} \rho_n(E') d\Omega_n' \tag{4.161}$$

where, as usually, the density of the continuum states $\rho_n(E)$ depends on the choice of the normalization of the final state wave function $\Psi_{n,\mathbf{p}_f}^{(-)}$.

Here we give a schematic presentation of the theory of inelastic electron–atom scattering via formation of negative-ion resonances, concentrating on a pure two-step resonance process and disregarding the direct excitation mechanism. Our aim is to investigate the general features of the process concerning alignment and orientation characteristics of the negative-ion resonances formed, the mechanism of their transfer to discrete or autoionizing states of the target atom, and the angular anisotropy and polarization parameters of the decay products.

Consider a general case of the excitation of a polarized atom $A(\alpha_i J_i)$ to a state $A^*(\alpha_f J_f)$ induced by a polarized electron beam via formation of an isolated negative-ion resonance $A^-(\alpha_C J_C)$:

$$A(\alpha_i J_i) + e \rightarrow A^-(\alpha_C J_C)$$
$$\downarrow \tag{4.162}$$
$$A^*(\alpha_f J_f) + e_{sc}$$

The statistical tensors of the negative ion resonance $\rho_{k_C q_C}^C(\alpha_C J_C)$ are calculated according to Eq. (2.32) with $j_x = \frac{1}{2}$ for the electron.

Generally one can be interested in alignment and orientation parameters of the negative-ion resonances for different purposes. These parameters determine the angular distribution of the scattered electrons, the polarization parameters of the atomic state produced (angular distribution and polarization of the decay products), and various correlation characteristics important for coincidence measurements of the scattered electron and the decay product of the excited target atom in the final state.

The angular distribution of electrons scattered via formation and decay of negative-ion resonance to a channel $A^*(\alpha_f J_f) + e_{sc}$ and detected in a one-arm (noncoincidence) experiment (e, e') is determined by alignment parameters of the decaying state [i.e., by its statistical tensors $\rho^C_{k_C q_C}(\alpha_C J_C)$ of even rank] and the corresponding decay amplitudes $\langle \alpha_f J_f, l_f j_f : J_C \| V \| \alpha_C J_C \rangle$.

Alignment of the negative-ion resonance is partially induced by the incoming electron and is partially due to the initial alignment of the target atom. The angular distribution of the scattered electrons produced in the decay of the negative-ion resonance is not sensitive to statistical tensors $\rho^C_{k_C q_C}(\alpha_C J_C)$ of odd rank, as was discussed in Section 3.1.2. In the case of a one-arm (e, e') experiment, the odd-rank tensors $\rho^C_{k_C q_C}(\alpha_C J_C)$ determine the spin polarization of the scattered electrons. Taking into account that the state $|\alpha_C J_C\rangle$ is of definite parity and any spin-1/2 particle cannot be aligned, the negative-ion resonance state $A^-(\alpha_C J_C)$ can be vector polarized ($k_C = $ odd) only if at least one of the colliding objects (the target atom A and/or the incoming electron beam) is vector polarized.

The statistical tensors $\rho^C_{k_C q_C}(\alpha_C J_C)$ of both even and odd ranks determine the corresponding statistical tensors of the residual excited atom after the decay of the negative-ion resonance. This question was considered in a general form in Section 3.3. We give here the final equation connecting the reduced statistical tensors of the decaying state $|\alpha_C J_C\rangle$ and the final atomic state $|\alpha_f J_f\rangle$ [see Eq. (3.59)]:

$$\mathcal{A}_{kq}(\alpha_f J_f) = \mathcal{A}_{kq}(\alpha_C J_C) \hat{J}_f \hat{J}_C \left(\sum_{l_f j_f} \langle \alpha_f J_f, l_f j_f : J_C \| V \| \alpha_C J_C \rangle |^2 \right)^{-1}$$

$$\times \sum_{l_f j_f} (-1)^{J_C + J_f + j_f + k} \begin{Bmatrix} J_C & J_C & k \\ J_f & J_f & j_f \end{Bmatrix}$$

$$\times |\langle \alpha_f J_f, l_f j_f : J_C \| V \| \alpha_C J_C \rangle|^2 \qquad (4.163)$$

No interference between the decay amplitudes takes place in the alignment and orientation transfer from the negative-ion state to the final atomic state until the scattered electron is detected.

The polarization parameters of the atomic final state in the case of observed scattered electrons are given as a function of the scattering direction $(\vartheta_{sc}, \varphi_{sc})$ [see Eq. (3.57)]:

$$\rho_{k_f q_f}(\alpha_f J_f; \vartheta_{sc}, \varphi_{sc}) = \frac{1}{\sqrt{4\pi}} \sum_{k_C k_l} \sum_{l_f l'_f j_f j'_f} (-1)^{k_C + k_f + j'_f + 1/2} \hat{l}_f \hat{l}'_f \hat{j}_f \hat{j}'_d \hat{k}_C$$

$$
\times (l_f 0, l'_f 0 \mid k_l 0)
\begin{Bmatrix} j_f & j'_f & k_l \\ l'_f & l_f & 1/2 \end{Bmatrix}
\begin{Bmatrix} J_f & j_f & J_C \\ J_f & j'_f & J_C \\ k_f & k_l & k_C \end{Bmatrix}
$$

$$
\times \sum_{q_C q_l} (k_C q_C k_l q_l \mid k_f q_f) \rho_{k_C q_C}(\alpha_C J_C) Y^*_{k_l q_l}(\vartheta_{sc}, \varphi_{sc})
$$

$$
\times \langle \alpha_f J_f, l_f j_f : J_C \| V \| \alpha_C J_C \rangle \langle \alpha_f J_f, l'_f j'_f : J_C \| V \| \alpha_C J_C \rangle^* \qquad (4.164)
$$

Contrary to Eq. (4.163), the statistical tensors of the final atomic state related to a fixed direction of scattered electrons are sensitive, in principle, to the phase relations between the decay amplitudes $\langle \alpha_f J_f, l_f j_f : J_C \| V \| \alpha_C J_C \rangle$. The final atomic state $\mid \alpha_f J_f \rangle$ can be vector polarized even for an aligned negative-ion state provided the decay proceeds via several channels with the amplitudes $\langle \alpha_f J_f, l_f j_f : J_C \| V \| \alpha_C J_C \rangle$.

Complements

5.1. Polarization of Atoms by Laser Optical Pumping

The creation of lasers stimulated the rapid development of many new directions in atomic physics. In a number of them, the role of laser radiation is restricted to "preparation" of the polarized excited target. Consider the main features of the density matrix and the statistical tensors of the excited atomic state after optical pumping by a laser within the two-level approximation. Each of the two levels, the ground state $\alpha_0 J_0$ and the excited state $\alpha_1 J_1$, includes, generally, a set of sublevels with different projections of the total angular momentum: $M_0 = -J_0, \ldots, J_0$ and $M_1 = -J_1, \ldots, J_1$. Assume that the laser field is weak and does not split the sublevels. In practice, the laser radiation in experiments has a high degree of linear or circular polarization and we consider here laser pumping by linearly and circularly polarized light. For pumping by linearly polarized light, the stimulated transitions $\alpha_0 J_0 \longrightarrow \alpha_1 J_1$ and $\alpha_1 J_1 \longrightarrow \alpha_0 J_0$ proceed only between magnetic sublevels with $M_0 = M_1$ (the quantization axis along the polarization vector of the laser field), while for pumping by circularly polarized light, the stimulated transitions proceed between the sublevels with $M_1 = M_0 \pm 1$ (the quantization axis along the laser beam; the plus and minus signs are for right and left circular polarization, respectively). The quantization z-axis in this subsection will always be chosen as stated above. So the stimulated transitions take place only within definite pairs of magnetic sublevels, while transitions between sublevels from different pairs occur only as a result of spontaneous photoemission.

Let us write the rate equations for an arbitrary pair of magnetic sublevels $\tilde{M}_1$ and $\tilde{M}_0$ connected by the stimulated transitions:

$$\frac{dW_{\tilde{M}_1}}{dt} = -W_{\tilde{M}_1} \sum_{M_0} A_{\tilde{M}_1 M_0} - W_{\tilde{M}_1} B_{\tilde{M}_1 \tilde{M}_0} I_\nu + W_{\tilde{M}_0} B_{\tilde{M}_0 \tilde{M}_1} I_\nu$$

$$\frac{dW_{\tilde{M}_0}}{dt} = \sum_{M_1} A_{M_1\tilde{M}_0} W_{M_1} + W_{\tilde{M}_1} B_{\tilde{M}_1\tilde{M}_0} I_\nu - W_{\tilde{M}_0} B_{\tilde{M}_0\tilde{M}_1} I_\nu \qquad (5.1)$$

Here W_M is the number of atoms in the substate with the projection M, I_ν is the spatial density of the laser radiation with a given polarization ν, and $A_{M_1 M_0} \equiv A(M_1 \to M_0)$ and $B_{M_1 M_0} \equiv B(M_1 \to M_0)$ are the rates of spontaneous and stimulated transitions. The validity of (5.1) needs a special justification that is outside the scope of this book. Particular expressions for $A_{M_1 M_0}$ and $B_{M_1 M_0}$ will not be important here; all that is important is their proportionality to the absolute square of the matrix element of the dipole operator (2.10).

Consider the spin density matrix of the target. Suppose that it is diagonal at $t = 0$; for example, the atoms are unpolarized and all the atoms are in the ground state. During the stimulated transitions and the spontaneous photoemission, the density matrices of the atomic angular momentum for the ground and excited states keep the diagonal form (provided the emitted photon is unobserved). This is clear from the conservation of the projection of the statistical tensor of the total angular momentum [see Eq.(1.58)] of the system atom + photon and is due to our special choice of the z-axis. Therefore, the populations of the magnetic sublevels in the ground and laser-excited states, W_{M_0} ($M_0 = -J_0, \ldots, J_0$) and W_{M_1} ($M_1 = -J_1, \ldots, J_1$), respectively, present all nonvanishing elements of the density matrix of the ground and excited states. We will not analyze the time evolution of the density matrix, but find the density matrix of the excited state in a stationary regime when dynamic equilibrium is reached. Taking the time derivatives in Eq. (5.1) equal to zero and applying the Wigner–Eckart theorem (A.62) for factorizing magnetic quantum numbers in the coefficients $A_{M_1 M_0}$, one obtains a system of equations for the diagonal elements W_{M_1} of the density matrix of the excited state:

$$W_{\tilde{M}_1} = \sum_{M_1} \left(J_0 \tilde{M}_1 - \lambda, 1 M_1 - \tilde{M}_1 + \lambda \,\middle|\, J_1 M_1 \right)^2 W_{M_1}$$
$$\tilde{M}_1 = -J_1, \ldots, J_1 \qquad (5.2)$$

where $\lambda = +1, -1, 0$ for right circularly, left circularly, and linearly polarized laser beams, respectively. The system of equations (5.2) always has a nontrivial solution if $J_1 \neq J_0$. Solving it together with the normalization condition $\sum_{M_1} W_{M_1} = 1$, one finds the desired density matrix.

Consider the important particular case $J_1 = J_0 + 1$ in more detail. The solution of the system (5.2) for a circularly polarized beam ($\lambda = \pm 1$) gives

$$W_{M_1}^{(\pm)} = \left\langle \alpha_1 J_1 M_1 \,\middle|\, \hat{\rho}^{(\pm)} \,\middle|\, \alpha_1 J_1 M_1 \right\rangle = \delta_{\pm J_1 M_1} \qquad (5.3)$$

This equation shows that all excited atoms accumulate in the magnetic substate with the largest absolute value of the projection M_1 in the direction of the laser beam, while the sign of M_1 coincides with the sign of circular polarization (or the sign of the Stokes parameter P_3). From Eq. (5.3) one obtains for the statistical tensors (1.41) the following expression:

$$\rho_{kq}^{(\pm)}(\alpha_1 J_1) = \frac{\hat{k}}{\hat{J}_1}(J_1 \pm J_1, k0|J_1 \pm J_1)\delta_{q0}$$

$$= (\pm 1)^k \hat{k}(2J_1)! [(2J_1-k)!(2J_1+k+1)!]^{-1/2}\delta_{q0} \quad (5.4)$$

An explicit expression for the Clebsch–Gordan coefficient is used in Eq. (5.4). The relation follows:

$$\rho_{k0}^{(+)}(\alpha_1 J_1) = (-1)^k \rho_{k0}^{(-)}(\alpha_1 J_1) \quad (5.5)$$

which shows that switching from right to left circularly polarized laser light leads to a change in the sign of the statistical tensors with an odd rank, while those with an even rank remain unchanged. The changing of sign for tensors with an odd rank is a source of the "circular pumping dichroism," which can manifest itself in changes in different observables, such as the angular distribution and polarization of the products of atomic reactions with laser-excited targets.

The solution of Eq. (5.2) for a linearly polarized laser beam ($\lambda = 0$) can be found by using explicit expressions for the Clebsch–Gordan coefficients of the form

$$W_{M_1}^{(0)} = \langle \alpha_1 J_1 M_1 | \hat{\rho} | \alpha_1 J_1 M_1 \rangle = (J_0 M_1, J_1 - M_1 | J_0 + J_1 0)^2 \quad (5.6)$$

The sublevels with $M_1 = \pm J_1$ are not populated (this is a direct consequence of the equality $J_1 = J_0 + 1$). Substitution of Eq. (5.6) into (1.41) and summation over the projections with the use of Eq. (A.89) gives for the statistical tensors:

$$\rho_{kq}(\alpha_1 J_1) = (-1)^k \hat{k}\sqrt{2J_0 + 2J_1 + 1} \begin{Bmatrix} J_1 & J_0 & J_0 + J_1 \\ J_0 + J_1 & k & J_1 \end{Bmatrix}$$

$$\times (J_0 + J_1 0, k0 | J_0 + J_1 0)\delta_{q0} \quad (5.7)$$

From the property of the Clebsch–Gordan coefficients (A.94), it follows that the tensors (5.7) with odd rank vanish. Therefore, a state excited by pumping with linearly polarized light is aligned, but not oriented. For both types of pumping, circular and linear, only tensors with zero projection occur for the excited atom (recall the special choice of the coordinate frames). Note that statistical tensors of the excited atom of a rank higher than 2 have their origin in the multiphoton nature of the pumping process. This is impossible when only one photon is absorbed by an unpolarized target.

5.2.   Time Evolution of Statistical Tensors and Depolarization of Atomic Angular Momenta

Consider the following typical situation: At time $t = 0$, the polarization state of an atomic multiplet ^{2S+1}L is described by the double statistical tensors $\rho_{k_L q_L k_S q_S}(LS; t = 0)$. L and S are not conserved due to the spin–orbit interactions. The problem is to find the polarization of orbital angular momentum L, spin S and total angular momentum J at time t. This can be done by applying Eq. (1.56) and the results of Section 1.2.4. A new feature that should be taken into account is a finite lifetime, or, equivalently, nonzero decay widths of the excited atomic levels. The decay of a level αJ can be incorporated phenomenologically into the theory by introducing a decay width $\Gamma_{\alpha J}$ and multiplying the corresponding time-dependent eigenfunction by the factor $\exp\left(-\frac{1}{2}\Gamma_{\alpha J}t\right)$. As a result, Eq. (1.56) becomes

$$\rho_{kq}(J, J'; t) = \rho_{kq}(J, J'; t = 0)\exp(i\omega_{JJ'}t - \Gamma_{JJ'}t) \tag{5.8}$$

where $\omega_{JJ'} = E_J - E_{J'}$; $\Gamma_{JJ'} = (\Gamma_J + \Gamma_{J'})/2$; we omit the state index α for brevity. Applying Eqs. (1.64), (1.65), and (1.69) leads to the following expression for the time-dependent statistical tensor of the orbital angular momentum:

$$\rho_{kq}(L; t) = \sum_{\substack{JJ' \\ k_L q_L k_S q_S}} (-1)^{S+k+L+J'} \hat{J}^2 \hat{J}'^2 \hat{k}_L \hat{k}_S (k_L q_L, k_S q_S | kq)$$

$$\times \begin{Bmatrix} J J' k \\ L L S \end{Bmatrix} \begin{Bmatrix} L & S & J \\ L & S & J' \\ k_L & k_S & k \end{Bmatrix} \exp[i\omega_{JJ'}t - \Gamma_{JJ'}t]\rho_{k_L q_L k_S q_S}(LS; t = 0) \tag{5.9}$$

Equation (5.9) shows that the evolution of polarization of the orbital angular momentum L of the multiplet depends on fine-structure splitting, decay widths, and quantum numbers of levels, and on the initial polarization of orbital and spin angular momenta described by the double statistical tensor. An important particular case is an initially unpolarized spin of the excited state just after excitation. Substituting $k_S = q_S = 0$ into Eq. (5.9) [see Eq. (A.122)] gives the product form

$$\rho_{kq}(L; t) = G_k(L, t)\rho_{kq}(L; t = 0) \tag{5.10}$$

where

$$G_k(L, t) = \hat{S}^{-2}\sum_{JJ'} \hat{J}^2 \hat{J}'^2 \begin{Bmatrix} J J' k \\ L L S \end{Bmatrix}^2 \exp[i\omega_{JJ'}t - \Gamma_{JJ'}t]$$

$$= \hat{S}^{-2}\sum_{JJ'} \hat{J}^2 \hat{J}'^2 \begin{Bmatrix} J J' k \\ L L S \end{Bmatrix}^2 \exp(-\Gamma_{JJ'}t)\cos\omega_{JJ'}t \tag{5.11}$$

Equations (5.10) and (5.11) define a *depolarization factor* $G_k(L,t)$ for the orbital angular momentum. The factor $G_k(L,t)$ is always no larger than unity [see Eq. (A.131)]. The reduced statistical tensors of the orbital angular momentum take the form

$$\mathcal{A}_{kq}(L,t) = \frac{G_k(L;t)}{G_0(L;t)} \mathcal{A}_{kq}(L,t=0)$$

$$= \left[\sum_J \hat{J}^2 \exp(-\Gamma_J t) \right]^{-1} \hat{L}^2 \sum_{JJ'} \hat{J}^2 \hat{J}'^2 \left\{ \begin{matrix} J & J' & k \\ L & L & S \end{matrix} \right\}^2$$

$$\times \exp(-\Gamma_{JJ'} t) \cos \omega_{JJ'} t \, \mathcal{A}_{kq}(L,t=0) \qquad (5.12)$$

Equation (5.12) shows that observables that depend on the statistical tensors $\mathcal{A}_{kq}(L,t)$ in general oscillate with time due to spin–orbit interactions. These *quantum beats* originate from the interference terms with $J \neq J'$ in Eqs. (5.9) and (5.11). The quantum beats can be observed if the time resolution of the detector is high enough. Since the interference terms do not contribute to $\mathcal{A}_{00}(L,t)$, the quantum beats cannot exist for the observables described by the statistical tensors of zero rank, like cross sections. For a discussion of the theory of quantum beats, we refer the reader to Ref. 8.

An equation analogous to (5.9) can be obtained for statistical tensors of spin S. The depolarization factor for spin, $G_k(S,t)$, can be introduced in the same manner as for L, but since the orbital angular momentum L is almost always polarized after the collision, the factorized form (5.10) for spin S is practically useless. In general, analysis of the statistical tensors of the orbital angular momenta is of greater importance for studies of polarization and correlation phenomena because operators describing the decay of atomic states (for example, the dipole operator or the Coulomb operator) act only on the spatial variables.

The time evolution of the statistical tensors of the total angular momentum J of the electronic shell due to the hyperfine interactions can be considered in analogy with the previous case. The analogue of Eq. (5.9) has the form

$$\rho_{kq}(J;t) = \sum_{\substack{FF' \\ k_J q_J k_I q_I}} (-1)^{I+k+J+F'} \hat{F}^2 \hat{F}'^2 \hat{k}_J \hat{k}_I \, (k_J q_J, k_I q_I | kq)$$

$$\times \left\{ \begin{matrix} F & F' & k \\ J & J & I \end{matrix} \right\} \left\{ \begin{matrix} J & I & F \\ J & I & F' \\ k_J & k_I & k \end{matrix} \right\} \exp[i\omega_{FF'} t - \Gamma_{FF'} t] \rho_{k_J q_J k_I q_I}(JI;t=0) \qquad (5.13)$$

Here I is the nuclear spin and $\mathbf{F} = \mathbf{J} + \mathbf{I}$ is the total angular momentum of an atom, including the nucleus. The value of I is fixed. For a mixture of isotopes, the summation over I in (5.13) has to be incorporated. When the nuclear spin is initially unpolarized, one obtains

$$\rho_{kq}(J;t) = G_k(J,t)\rho_{kq}(J;t=0) \tag{5.14}$$

where the depolarization factor is of the form

$$G_k(J,t) = \hat{I}^{-2} \sum_{FF'} \hat{F}^2 \hat{F}'^2 \begin{Bmatrix} F & F' & k \\ J & J & I \end{Bmatrix}^2 \exp[i\omega_{FF'}t - \Gamma_{FF'}t] \tag{5.15}$$

In a general case when both fine-structure and hyperfine interactions are important, the statistical tensors of three angular momenta, L, S, and I, should be considered. Applying Eqs. (1.64), (1.65), and (1.69) twice, we arrive at the expression

$$\begin{aligned}
\rho_{kq}(L;t) = &\sum_{JJ'FF'} (-1)^{J+J'+L+S+I+F'} \hat{J}^2\hat{F}^2\hat{J}'^2\hat{F}'^2 \begin{Bmatrix} L & L & k \\ J' & J & S \end{Bmatrix} \begin{Bmatrix} J & J' & k \\ F' & F & I \end{Bmatrix} \\
&\times \exp(i\omega_{JF,J'F'}t - \Gamma_{JF,J'F'}t) \sum_{k_Lk_Sk_Jk_I} \hat{k}_L\hat{k}_S\hat{k}_J\hat{k}_I \begin{Bmatrix} L & S & J \\ L & S & J' \\ k_L & k_S & k_J \end{Bmatrix} \begin{Bmatrix} J & I & F \\ J' & I & F' \\ k_J & k_I & k \end{Bmatrix} \\
&\times \sum_{q_Lq_Sq_Jq_I} (k_L q_L, k_S q_S | k_J q_J)(k_J q_J, k_I q_I | kq)\rho_{k_Lq_Lk_Sq_Sk_Iq_I}(LSI;t=0)
\end{aligned}$$

$$\tag{5.16}$$

where $\omega_{JF,J'F'} = E_{JF} - E_{J'F'}$; $\Gamma_{JF,J'F'} = (\Gamma_{JF} + \Gamma_{J'F'})/2$. We consider the sharp values of the nuclear spin I and of the electronic spin S. In the derivation of Eq. (5.16) the following approximation was used: We considered the angular momentum J as being a good quantum number even though the hyperfine interaction was present. Equation (5.16) is a generalization of Eq. (5.9). Assuming that the nuclear spin I and the spin of electronic shell S at time $t = 0$ are unpolarized ($k_S = q_S = k_I = q_I = 0$), we obtain

$$\rho_{kq}(L;t) = G_k(L,t)\rho_{kq}(L;t=0) \tag{5.17}$$

where the depolarization factor now has the form

$$\begin{aligned}
G_k(L,t) = &(\hat{I}\hat{S})^{-2} \sum_{JFJ'F'} \hat{F}^2\hat{J}^2\hat{F}'^2\hat{J}'^2 \begin{Bmatrix} L & L & k \\ J' & J & S \end{Bmatrix}^2 \begin{Bmatrix} F & F' & k \\ J' & J & I \end{Bmatrix}^2 \\
&\times \exp[i\omega_{JF,J'F'}t - \Gamma_{JF,J'F'}t]
\end{aligned} \tag{5.18}$$

5.3. Influence of Time Evolution of Statistical Tensors on Angular Distribution and Polarization of Decay Products

Throughout this book situations are considered where an atomic level with the total angular momentum J is well isolated (it is implied that J is the total angular momentum of the electronic shell). Here we discuss a modification of the formalism developed, allowing for the influence of the hyperfine splitting of the decaying level αJ on characteristics of the decay products. Similarly, this modification can be used to treat the influence of the fine splitting of the multiplet.

Assume that a well-isolated atomic level αJ, where J is the angular momentum of the electronic shell, is a hyperfine structure multiplet and that in an experiment all the hyperfine structure levels are excited at time $t = 0$. In practice, this describes a coherent excitation, when the spectral width of the impact is much larger than the hyperfine splitting, or, equivalently, the time of the excitation is much shorter than the period of precession of the angular momentum J due to the hyperfine interaction. The collision time is almost always short enough for excitation of an atom by a particle impact to use an assumption of instant excitation. (The same is often true for excitation of an atomic beam in beam–foil experiments, but less often in laser-pulsed excitation.) Any polarization or correlation characteristic measured for the decay products can be expressed in terms of the probabilities (1.151) and (1.152), which for the case of hyperfine structure can be written in terms of the statistical and efficiency tensors of the total angular momentum F:

$$W = \sum_{\substack{\alpha\alpha'FF' \\ kq}} \rho_{kq}\left(\alpha JI : F, \alpha' JI : F'\right) \varepsilon_{kq}^*\left(\alpha JI : F, \alpha' JI : F'\right) \tag{5.19}$$

or in terms of double statistical tensors of the electronic J and nuclear I angular momenta:

$$W = \sum_{\substack{k_J q_J k_I q_I \\ \alpha\alpha'}} \rho_{k_J q_J k_I q_I}\left(\alpha JI, \alpha' JI\right) \varepsilon_{k_J q_J k_I q_I}^*\left(\alpha JI, \alpha' JI\right) \tag{5.20}$$

The statistical tensors in Eqs. (5.19) and (5.20) describe the decaying system in the final state. We consider sharp values of J and I (the latter corresponds to an isotopically pure target); otherwise one should introduce nondiagonal terms in Eqs. (5.19) and (5.20) with J,J' and I,I' and sum up over J,J',I,I'. Suppose that the polarization state of nuclei is not directly measured by the detector. Substituting Eq. (1.159) into Eq. (5.20) gives the relation:

$$W = \sum_{\alpha\alpha'kq} \rho_{kq}\left(\alpha J, \alpha' J\right) \varepsilon_{kq}^*\left(\alpha J, \alpha' J\right) \tag{5.21}$$

[which coincides with Eq. (1.151) for sharp j]. Now assume that the decay operator T does not act on the nuclear degrees of freedom. The vast majority of operators in atomic physics fulfill this condition. Then the statistical tensors $\rho_{kq}(\alpha J, \alpha' J)$ of the final state can be expressed in terms of the statistical tensors of the decaying excited state by Eq. (3.5) (with obvious extension to the decay of a state with unfixed quantum numbers), which evolves in time in accordance with Eq. (5.13). Consider a usual situation when the detector of the decay product has been switched on at time $t = 0$ and has a resolution time Δt. Collecting Eqs. (5.21), (3.5), and (5.13), and integrating over time from 0 to Δt gives

$$
W = \sum_{\substack{\alpha\alpha'\beta\beta' \\ kq}} \varepsilon_{kq}^*(\alpha J, \alpha' J)\, \hat{J}^{-2} \langle \alpha J \| T \| \beta J \rangle \langle \alpha' J \| T \| \beta' J \rangle
$$

$$
\times \sum_{\substack{k_J q_J k_I q_I \\ FF'}} (-1)^{I+k+J+F'} \hat{F}^2 \hat{F}'^2 \hat{k}_J \hat{k}_I \, (k_J q_J, k_I q_I | kq) \begin{Bmatrix} F & F' & k \\ J & J & I \end{Bmatrix} \begin{Bmatrix} J & I & F \\ J & I & F' \\ k_J & k_I & k \end{Bmatrix}
$$

$$
\times \rho_{k_J q_J k_I q_I}(\beta JI, \beta' JI; t = 0) \int_0^{\Delta t} \exp(i\omega_{FF'}t - \Gamma_{FF'}t)\, dt \tag{5.22}
$$

A simple result follows for the case when just after excitation the nuclear spin I is not polarized ($k_I = q_I = 0$). Usually this condition is fulfilled for targets with nuclei unpolarized before the excitation. Then Eq. (5.22) gives

$$
W = \sum_{\alpha\alpha'kq} G_k(J, \Delta t)\rho_{kq}(\alpha J, \alpha' J; t = 0)\, \varepsilon_{kq}^*(\alpha J, \alpha' J) \tag{5.23}
$$

where

$$
G_k(J, \Delta t) = \hat{I}^{-2} \sum_{FF'} \hat{F}^2 \hat{F}'^2 \begin{Bmatrix} F & F' & k \\ J & J & I \end{Bmatrix}^2 \frac{1}{\Gamma_{FF'}} \left(1 + \frac{\omega_{FF'}^2}{\Gamma_{FF'}^2}\right)^{-1}
$$

$$
\times \left[1 - \left(\cos(\omega_{FF'}\Delta t) - \frac{\omega_{FF'}}{\Gamma_{FF'}} \sin(\omega_{FF'}\Delta t)\right) \exp(-\Gamma_{FF'}\Delta t)\right] \tag{5.24}
$$

A comparison of Eqs. (5.23) and (1.151) (the latter is taken for fixed angular momentum) results in the conclusion that under rather general assumptions one can treat the two-step process of the excitation and decay of the fine structure level equivalently, whether or not it is a hyperfine structure multiplet. The only difference is a depolarization factor $G_k(J, \Delta t)$ before the statistical tensors of the excited state. This allows for the time evolution (precession of the angular momentum of the electronic shell) due to the hyperfine interactions. In particular, within the above assumptions, all expressions for correlation and polarization characteristics of the decay products containing the reduced statistical tensors $\mathcal{A}_{kq}(\alpha J)$ are valid also for the case when the level αJ is the hyperfine structure multiplet; one should only make a replacement:

$$\mathcal{A}_{kq}(\alpha J) \Longrightarrow \frac{G_k(J,\Delta t)}{G_0(J,\Delta t)} \mathcal{A}_{kq}(\alpha J) \qquad (5.25)$$

where $G_k(J,\Delta t)$ are given by Eq. (5.24).

Consider the extreme cases of Eq. (5.24). We assume that the decay widths of the hyperfine levels are equal: $\Gamma = \Gamma_F = \Gamma_{F'}$. When $\Delta t \gg \Gamma$ (i.e., the observation time is much longer than the lifetime of the decaying level), the factor (5.24) takes the form

$$G_k(J,\Delta t) = \hat{I}^{-2} \sum_{FF'} \hat{F}^2 \hat{F}'^2 \left\{ \begin{matrix} F & F' & k \\ J & J & I \end{matrix} \right\}^2 \frac{1}{\Gamma} \left(1 + \frac{\omega_{FF'}^2}{\Gamma^2} \right)^{-1} \qquad (5.26)$$

In particular, Eq. (5.26) is valid for the case when the time of the instant excitation is not fixed, as in noncoincidence experiments. When the hyperfine splitting is negligible ($\omega_{FF'} \ll \Gamma$), the summation over FF' in Eq. (5.26) [see Eq. (A.131)] gives $G_k(J,\Delta t) = 1/\Gamma$ and the reduced statistical tensors (5.25) do not change. This result is obvious, since for $\omega_{FF'} \ll \Gamma$ we again have the single isolated level αJ. When the hyperfine levels do not overlap ($\omega_{FF'} \gg \Gamma$), the diagonal terms with $F = F'$ dominate the sum in Eq. (5.26); interference terms are negligible, and Eq. (5.26) takes the form

$$G_k(J,\Delta t) = \frac{1}{\Gamma} \hat{I}^{-2} \sum_F \hat{F}^4 \left\{ \begin{matrix} F & F & k \\ J & J & I \end{matrix} \right\}^2 \qquad (5.27)$$

A similar approach can be used for treating the fine structure of the excited multiplet ^{2S+1}L within the LS-coupling approximation. Consider the decay of an isolated level with a total orbital angular momentum L. We assume first that spin–orbit interaction can be neglected and that the decay operator conserves spin (for example, the Coulomb operator or the dipole operator). If the polarization of the spin of the system after the decay is not directly observed, one can express observables related to the decay products of the state L in terms of probabilities:

$$W = \sum_{\alpha\alpha'kq} \rho_{kq}(\alpha L, \alpha' L) \, \varepsilon_{kq}^*(\alpha L, \alpha' L) \qquad (5.28)$$

Now consider the case of a decaying multiplet ^{2S+1}L accounting for the fine-structure splitting of the energy levels. Repeating the above derivation with replacements

$$J \to L, I \to S, F \to J \qquad (5.29)$$

we find that one can treat the case of an isolated level with orbital angular momentum L and the case of the LS-multiplet equivalently, provided the spin S of the excited decaying multiplet is not polarized just after the excitation at time $t = 0$ and depolarization factors are introduced:

$$W = \sum_{\alpha\alpha'kq} G_k(L,\Delta t)\rho_{kq}\left(\alpha L, \alpha'L; t = 0\right)\varepsilon_{kq}^*\left(\alpha L, \alpha'L\right) \tag{5.30}$$

where

$$G_k(L,\Delta t) = \hat{S}^{-2}\sum_{JJ'} \hat{J}^2\hat{J'}^2 \left\{\begin{matrix} J & J' & k \\ L & L & S \end{matrix}\right\}^2 \frac{1}{\Gamma_{JJ'}}\left(1 + \frac{\omega_{JJ'}^2}{\Gamma_{JJ'}^2}\right)^{-1}$$

$$\times \left[1 - \left(\cos(\omega_{JJ'}\Delta t) - \frac{\omega_{JJ'}}{\Gamma_{JJ'}}\sin(\omega_{JJ'}\Delta t)\right)\exp(-\Gamma_{JJ'}\Delta t)\right] \tag{5.31}$$

Equation (5.31) is a counterpart of Eq. (5.24) with the replacements (5.29). Equations (5.25)–(5.27) for the hyperfine structure are also transformed directly to the case of fine structure, for example, equation

$$G_k(L,\Delta t) = \frac{1}{\Gamma}\hat{S}^{-2}\sum_{J}\hat{J}^4 \left\{\begin{matrix} J & J & k \\ L & L & S \end{matrix}\right\}^2 \tag{5.32}$$

is valid for well-separated fine-structure levels of the multiplet ($\omega_{JJ'} \gg \Gamma$). Note that the assumption $\Gamma = \Gamma_J = \Gamma_{J'}$ is often not a good approximation and one has to use the more general $\Gamma_{JJ'}$ instead of Γ.

Appendix

There are many books in which the properties of spherical harmonics, nj-symbols, etc. are presented exhaustively. Here only those equations are given that are necessary and, we believe, sufficient to perform all the angular momentum algebra in the book.

A.1. Pauli Matrices

A.1.1. Explicit Form

Three Pauli matrices σ_i $(i = x, y, z)$ correspond to the operators of projections of the angular momentum $J = 1/2$:

$$J_i = \frac{1}{2}\sigma_i \tag{A.1}$$

$$\sigma_x = \begin{pmatrix} 0 & 1 \\ 1 & 0 \end{pmatrix}, \quad \sigma_y = \begin{pmatrix} 0 & -i \\ i & 0 \end{pmatrix}, \quad \sigma_z = \begin{pmatrix} 1 & 0 \\ 0 & -1 \end{pmatrix} \tag{A.2}$$

A.1.2. Main Properties

1. Hermiticity: $\sigma_i^+ = \sigma_i$.

2. Unitarity: $\sigma_i \sigma_i^+ = I$, where I is the unit matrix. Therefore,

$$\sigma_x^2 = \sigma_y^2 = \sigma_z^2 = I \tag{A.3}$$

3. Anticommutativity

$$\sigma_i \sigma_j = -\sigma_j \sigma_i, \quad (i \neq j) \tag{A.4}$$

4. Product of two Pauli matrices

$$\sigma_x \sigma_y = i\sigma_z;, \quad \sigma_y \sigma_z = i\sigma_x;, \quad \sigma_z \sigma_x = i\sigma_y \tag{A.5}$$

Equations (A.3)–(A.5) can be unified by the expression

$$\sigma_i \sigma_j = I\delta_{ij} + i\sum_k \varepsilon_{ijk}\sigma_k \tag{A.6}$$

where ε_{ijk} is the absolutely antisymmetric tensor with $\varepsilon_{xyz} = 1$ which change sign under permutation of any pair of indices. Thus:

$$\varepsilon_{ijk} = \begin{cases} 0, & \text{if at least two of the indices are equal} \\ +1 & \text{if the indices form a cyclic permutation of } x, y, z \\ -1 & \text{if the indices form a noncyclic permutation of } x, y, z. \end{cases} \tag{A.7}$$

5. Commutative relations follow from Eq. (A.6):

$$[\sigma_i \sigma_j] = 2i\sum_k \varepsilon_{ijk}\,\sigma_k \tag{A.8}$$

A.2.   *Legendre Polynomials and Associated Legendre Polynomials*

A.2.1.   *Definition of Associated Legendre Polynomials*

$$P_l^m(\cos\vartheta) = \sin^m\vartheta \left[\frac{d}{d(\cos\vartheta)} \right]^m P_l(\cos\vartheta) \tag{A.9}$$

where $P_l(\cos\vartheta)$ is the Legendre polynomial. Sometimes the associated Legendre polynomials with another normalization are used

$$\bar{P}_l^m = \sqrt{\frac{(2l+1)(l-m)!}{2(l+m)!}}\, P_l^m(\cos\vartheta) \tag{A.10}$$

A.2.2.   *Tables*

Associated Legendre polynomials (A.9) for small l are given in Table A.1 on p. 199 and some of their numerical values in Table A.2 on p. 199.

Table A.1. Associated Legendre Polynomials for $l = 0, 1, 2, 3, 4$

l	m		$P_m^l(\cos\vartheta)$
0	0		1
1	0		$\cos\vartheta$
1	1		$\sin\vartheta$
2	0	$\frac{1}{2}(3\cos^2\vartheta - 1)$	$\frac{1}{4}(1 + 3\cos 2\vartheta)$
2	1	$3\cos\vartheta\sin\vartheta$	$\frac{3}{2}\sin 2\vartheta$
2	2	$3\sin^2\vartheta$	$\frac{3}{2}(1 - \cos 2\vartheta)$
3	0	$\frac{1}{2}(5\cos^3\vartheta - 3\cos\vartheta)$	$\frac{1}{8}(3\cos\vartheta + 5\cos 3\vartheta)$
3	1	$\frac{3}{2}\sin\vartheta(5\cos^2\vartheta - 1)$	$\frac{3}{8}(\sin\vartheta + 5\sin 3\vartheta)$
3	2	$15\cos\vartheta\sin^2\vartheta$	$\frac{15}{4}(\cos\vartheta - \cos 3\vartheta)$
3	3	$15\sin^3\vartheta$	$\frac{15}{4}(3\sin\vartheta - \sin 3\vartheta)$
4	0	$\frac{1}{8}(35\cos^4\vartheta - 30\cos^2\vartheta + 3)$	$\frac{1}{64}(9 + 20\cos 2\vartheta + 35\cos 4\vartheta)$
4	1	$\frac{5}{2}\sin\vartheta(7\cos^3\vartheta - 3\cos\vartheta)$	$\frac{5}{16}(2\sin 2\vartheta + 7\sin 4\vartheta)$
4	2	$\frac{15}{2}\sin^2\vartheta(7\cos^2\vartheta - 1)$	$\frac{15}{16}(3 + 4\cos 2\vartheta - 7\cos 4\vartheta)$
4	3	$105\sin^3\vartheta\cos\vartheta$	$\frac{105}{8}(2\sin 2\vartheta - \sin 4\vartheta)$
4	4	$105\sin^4\vartheta$	$\frac{105}{8}(3 - 4\cos 2\vartheta + \cos 4\vartheta)$

Table A.2. Numerical Values of $P_m^l(\cos\vartheta)$ for Particular Angles

l	m	0°	90°	180°	45°	135°	magic angles (54.7°)	magic angles (125.3°)
0	0	1	1	1	1	1	1	1
1	0	1	0	-1	$\frac{1}{\sqrt{2}}$	$-\frac{1}{\sqrt{2}}$	$\frac{1}{\sqrt{3}}$	$-\frac{1}{\sqrt{3}}$
1	1	0	1	0	$\frac{1}{\sqrt{2}}$	$\frac{1}{\sqrt{2}}$	$\frac{\sqrt{2}}{\sqrt{3}}$	$\frac{\sqrt{2}}{\sqrt{3}}$
2	0	1	$-\frac{1}{2}$	1	$\frac{1}{4}$	$\frac{1}{4}$	0	0
2	1	0	0	0	$\frac{3}{2}$	$-\frac{3}{2}$	$\sqrt{2}$	$-\sqrt{2}$
2	2	0	3	0	$\frac{3}{2}$	$\frac{3}{2}$	2	2
3	0	1	0	-1	$-\frac{1}{4\sqrt{2}}$	$\frac{1}{4\sqrt{2}}$	$-\frac{2}{3\sqrt{3}}$	$\frac{2}{3\sqrt{3}}$
3	1	0	$-\frac{3}{2}$	0	$\frac{9}{4\sqrt{2}}$	$\frac{9}{4\sqrt{2}}$	$\frac{\sqrt{2}}{\sqrt{3}}$	$\frac{\sqrt{2}}{\sqrt{3}}$
3	2	0	0	0	$\frac{15}{2\sqrt{2}}$	$-\frac{15}{2\sqrt{2}}$	$\frac{10}{\sqrt{3}}$	$-\frac{10}{\sqrt{3}}$
3	3	0	15	0	$\frac{15}{2\sqrt{2}}$	$\frac{15}{2\sqrt{2}}$	$\frac{10\sqrt{2}}{\sqrt{3}}$	$\frac{10\sqrt{2}}{\sqrt{3}}$
4	0	1	$\frac{3}{8}$	1	$-\frac{13}{32}$	$-\frac{13}{32}$	$-\frac{7}{18}$	$-\frac{7}{18}$
4	1	0	0	0	$\frac{5}{8}$	$-\frac{5}{8}$	$-\frac{5\sqrt{2}}{9}$	$\frac{5\sqrt{2}}{9}$
4	2	0	$-\frac{15}{2}$	0	$\frac{75}{8}$	$\frac{75}{8}$	$\frac{20}{3}$	$\frac{20}{3}$
4	3	0	0	0	$\frac{105}{4}$	$-\frac{105}{4}$	$\frac{10\sqrt{2}}{3}$	$-\frac{10\sqrt{2}}{3}$
4	4	0	105	0	$\frac{105}{4}$	$\frac{105}{4}$	$\frac{140}{3}$	$\frac{140}{3}$

A.3.   Spherical Harmonics

A.3.1.   Definition

$$Y_{lm}(\vartheta,\varphi) = \Theta_{lm}(\vartheta)\,\Phi_m(\varphi) \tag{A.11}$$

$$\Phi_m(\varphi) = \frac{1}{\sqrt{2\pi}}\,e^{im\phi} \tag{A.12}$$

$$\Theta_{lm}(\vartheta) = (-1)^m \sqrt{\frac{(2l+1)(l-m)!}{2(l+m)!}}\,P_l^m(\cos\vartheta) \qquad (m \geq 0) \tag{A.13}$$

$$\Theta_{l-|m|}(\vartheta) = (-1)^m\,\Theta_{l|m|}(\vartheta) \tag{A.14}$$

where $P_l^m(\cos\vartheta)$ is the associated Legendre polynomial defined by Eq. (A.9).

A.3.2.   Useful Symmetries

$$Y_{lm}(\vartheta,\varphi) = (-1)^m\,Y_{l-m}^*(\vartheta,\varphi) \tag{A.15}$$

$$Y_{lm}(\pi-\vartheta,\varphi) = (-1)^{l+m}\,Y_{lm}(\vartheta,\varphi) \tag{A.16}$$

$$Y_{lm}(\vartheta,\pi+\varphi) = (-1)^m\,Y_{lm}(\vartheta,\varphi) \tag{A.17}$$

Inversion:

$$P_r Y_{lm}(\vartheta,\varphi) = Y_{lm}(\pi-\vartheta,\pi+\varphi) = (-1)^l\,Y_{lm}(\vartheta,\varphi) \tag{A.18}$$

Rotation:

Spherical harmonics are covariant components of the irreducible tensor operator and are transformed during the rotation of the coordinate system $S \xrightarrow{\alpha,\beta,\gamma} S'$ by the law

$$Y_{lm'}(\vartheta',\varphi') = \sum_m D_{mm'}^l(\alpha,\beta,\gamma)\,Y_{lm}(\vartheta,\varphi) \tag{A.19}$$

where $D_{mm'}^l(\alpha,\beta,\gamma)$ is the Wigner D-function (Section A.6), ϑ,φ and ϑ',φ' are the spherical angles of a given vector in the initial frame S and the new rotated frame S', respectively. The Euler angles α,β,γ characterize the rotation of the frame S to the frame S'.

A.3.3. Particular Angles

$$Y_{lm}(0, \varphi) = \delta_{m0} \sqrt{\frac{2l+1}{4\pi}} \tag{A.20}$$

$$Y_{lm}(\frac{\pi}{2}, \varphi) = 0 \qquad (l+m = \text{odd}) \tag{A.21}$$

$$Y_{lm}(\frac{\pi}{2}, \varphi) = (-1)^{(l+m)/2} e^{im\varphi}$$
$$\times \sqrt{\frac{2l+1}{4\pi} \frac{(l+m-1)!!}{(l+m)!!} \frac{(l-m-1)!!}{(l-m)!!}} \qquad (l+m = \text{even}) \tag{A.22}$$

A.3.4. Orthonormalization and Completeness

$$Y_{l_1 m_1}(\vartheta, \varphi) Y^*_{l_2 m_2}(\vartheta, \varphi) \, d\Omega$$
$$= \int_0^{2\pi} d\varphi \int_0^{\pi} d\vartheta \sin\vartheta \, Y_{l_1 m_1}(\vartheta, \varphi) Y^*_{l_2 m_2}(\vartheta, \varphi) = \delta_{l_1 l_2} \delta_{m_1 m_2} \tag{A.23}$$

$$\sum_{l=0}^{\infty} \sum_{m=-l}^{l} Y_{lm}(\vartheta_1, \varphi_1) Y^*_{lm}(\vartheta_2, \varphi_2) = \delta(\varphi_1 - \varphi_2) \delta(\cos\vartheta_1 - \cos\vartheta_2) \tag{A.24}$$

A.3.5. Selected Formulas

$$\langle \vartheta, \varphi \,|\, lm \rangle = Y_{lm}(\vartheta, \varphi) \tag{A.25}$$

$$\sum_{m=-l}^{l} |Y_{lm}(\vartheta, \varphi)|^2 = \frac{2l+1}{4\pi} \tag{A.26}$$

$$\int_0^{2\pi} d\varphi \, Y_{lm}(\vartheta, \varphi) = \frac{1}{2} \sqrt{4\pi(2l+1)} \, \delta_{m0} P_l(\cos\vartheta) \tag{A.27}$$

$$\int d\Omega Y_{l_1 m_1}(\vartheta, \varphi) Y_{l_2 m_2}(\vartheta, \varphi) Y^*_{l_3 m_3}(\vartheta, \varphi)$$

$$= \sqrt{\frac{(2l_1 + 1)(2l_2 + 1)}{4\pi(2l_3 + 1)}} (l_1 0, l_2 0 | l_3 0)(l_1 m_1, l_2 m_2 | l_3 m_3) \tag{A.28}$$

The Clebsch–Gordan expansion

$$Y_{l_1 m_1}(\vartheta, \varphi) Y_{l_2 m_2}(\vartheta, \varphi) = \sum_{LM} \sqrt{\frac{(2l_1 + 1)(2l_2 + 1)}{4\pi(2L + 1)}}$$

$$\times (l_1 0, l_2 0 | L 0)(l_1 m_1, l_2 m_2 | LM) Y_{LM}(\vartheta, \varphi) \tag{A.29}$$

A.3.6. Connection with Legendre Polynomial

$$Y_{l0}(\vartheta, \varphi) = \sqrt{\frac{2l + 1}{4\pi}} P_l(\cos \vartheta) \tag{A.30}$$

A.3.7. Scalar Product of Spherical Harmonics

Definition

$$(Y_l(\vartheta_1, \varphi_1) Y_l(\vartheta_2, \varphi_2)) = \sum_m Y^*_{lm}(\vartheta_1, \varphi_1) Y_{lm}(\vartheta_2, \varphi_2) \tag{A.31}$$

Relation to Legendre polynomials

$$(Y_l(\vartheta_1, \varphi_1) Y_l(\vartheta_2, \varphi_2)) = \frac{2l + 1}{4\pi} P_l(\cos \omega_{12}) \tag{A.32}$$

where ω_{12} is an angle between the directions ϑ_1, φ_1 and ϑ_2, φ_2:

$$\cos \omega_{12} = \cos \vartheta_1 \cos \vartheta_2 + \sin \vartheta_1 \sin \vartheta_2 \cos(\varphi_1 - \varphi_2)$$

A.3.8. Multipole Expansions

Plane wave

$$\exp(i\mathbf{kr}) = 4\pi \sum_{lm} i^l j_l(kr) Y^*_{lm}(\vartheta_k, \varphi_k) Y_{lm}(\vartheta_r, \varphi_r)$$

$$= 4\pi \sum_l i^l j_l(kr)(Y_l(\vartheta_k, \varphi_k) Y_l(\vartheta_r, \varphi_r))$$

$$= \sum_l i^l (2l + 1) j_l(kr) P_l(\cos \omega_{rk}) \tag{A.33}$$

where $j_l(x)$ is the spherical Bessel function;

Coulomb interaction

$$\frac{1}{|\mathbf{r}_1 - \mathbf{r}_2|} = \frac{4\pi}{r_>} \sum_l \frac{1}{(2l+1)} \left(\frac{r_<}{r_>}\right)^l \sum_m Y_{lm}^*(\vartheta_1, \varphi_1) Y_{lm}(\vartheta_2, \varphi_2)$$

$$= \frac{1}{r_>} \sum_l \left(\frac{r_<}{r_>}\right)^l P_l(\cos\omega_{12}) \qquad (A.34)$$

where $r_< = r_2, r_> = r_1$ if $r_1 > r_2$ and $r_< = r_1, r_> = r_2$ if $r_1 < r_2$;

Delta function

$$\delta(\mathbf{r}_1 - \mathbf{r}_2) = \frac{1}{r_1^2}\delta(r_1 - r_2)\delta(\cos\vartheta_1 - \cos\vartheta_2)\delta(\varphi_1 - \varphi_2)$$

$$= \delta(r_1 - r_2)\frac{1}{r_1^2}\sum_{lm} Y_{lm}^*(\vartheta_1, \varphi_1) Y_{lm}(\vartheta_2, \varphi_2) \qquad (A.35)$$

A.3.9. Table

Spherical harmonics for small l are given in Table A.3 on p. 204.

A.4. Bipolar spherical harmonics

A.4.1. Definition

$$\{Y_{l_1}(\vartheta_1, \varphi_1) \otimes Y_{l_2}(\vartheta_2, \varphi_2)\}_{LM}$$

$$= \sum_{m_1 m_2} (l_1 m_1, l_2 m_2 | LM) Y_{l_1 m_1}(\vartheta_1, \varphi_1) Y_{l_2 m_2}(\vartheta_2, \varphi_2) \qquad (A.36)$$

where $Y_{lm}(\vartheta, \varphi)$ are the spherical harmonics (see Section A.3). The bipolar spherical harmonics (A.36) form the irreducible tensor of rank L with components M.

A.4.2. Relation to Scalar Product of Spherical Harmonics

$$(Y_l(\vartheta_1, \varphi_1) Y_l(\vartheta_2, \varphi_2)) = (-1)^l \sqrt{2l+1} \{Y_l(\vartheta_1, \varphi_1) \otimes Y_l(\vartheta_2, \varphi_2)\}_{00} \qquad (A.37)$$

Table A.3. Spherical Harmonics for $l = 0, 1, 2, 3, 4$

l	m	$Y_{lm}(\vartheta, \varphi)$	Zeroes (deg)
0	0	$\frac{1}{2}\sqrt{\frac{1}{\pi}}$	
1	± 1	$\mp\frac{1}{2}\sqrt{\frac{3}{2\pi}}\sin\vartheta\, e^{\pm i\varphi}$	0; 180
1	0	$\frac{1}{2}\sqrt{\frac{3}{\pi}}\cos\vartheta$	90
2	± 2	$\frac{1}{4}\sqrt{\frac{15}{2\pi}}\sin^2\vartheta\, e^{\pm 2i\varphi}$ $\frac{1}{8}\sqrt{\frac{15}{2\pi}}(1-\cos 2\vartheta)e^{\pm 2i\varphi}$	0; 180
2	± 1	$\mp\frac{1}{2}\sqrt{\frac{15}{2\pi}}\cos\vartheta\sin\vartheta\, e^{\pm i\varphi}$ $\mp\frac{1}{4}\sqrt{\frac{15}{2\pi}}\sin 2\vartheta\, e^{\pm i\varphi}$	0; 90; 180
2	0	$\frac{1}{4}\sqrt{\frac{5}{\pi}}(3\cos^2\vartheta-1)$ $\frac{1}{8}\sqrt{\frac{5}{\pi}}(1+3\cos 2\vartheta)$	54.7; 125.3
3	± 3	$\mp\frac{1}{8}\sqrt{\frac{35}{\pi}}\sin^3\vartheta\, e^{\pm 3i\varphi}$ $\mp\frac{1}{32}\sqrt{\frac{35}{\pi}}(3\sin\vartheta-\sin 3\vartheta)e^{\pm 3i\varphi}$	0; 180
3	± 2	$\frac{1}{4}\sqrt{\frac{105}{2\pi}}\cos\vartheta\sin^2\vartheta\, e^{\pm 2i\varphi}$ $\frac{1}{16}\sqrt{\frac{105}{2\pi}}(\cos\vartheta-\cos 3\vartheta)e^{\pm 2i\varphi}$	0; 90; 180
3	± 1	$\mp\frac{1}{8}\sqrt{\frac{21}{\pi}}(5\cos^2\vartheta-1)\sin\vartheta\, e^{\pm i\varphi}$ $\mp\frac{1}{32}\sqrt{\frac{21}{\pi}}(\sin\vartheta+5\sin 3\vartheta)e^{\pm i\varphi}$	0; 63.4; 116.6; 180
3	0	$\frac{1}{4}\sqrt{\frac{7}{\pi}}(5\cos^2\vartheta-3)\cos\vartheta$ $-\frac{1}{16}\sqrt{\frac{7}{\pi}}(3\cos\vartheta+5\cos 3\vartheta)$	32.9; 90; 140.8
4	± 4	$\frac{3}{16}\sqrt{\frac{35}{2\pi}}\sin^4\vartheta\, e^{\pm 4i\varphi}$ $\frac{3}{128}\sqrt{\frac{35}{2\pi}}(3-4\cos 2\vartheta+\cos 4\vartheta)e^{\pm 4i\varphi}$	0; 180
4	± 3	$\mp\frac{3}{8}\sqrt{\frac{35}{\pi}}\sin^3\vartheta\cos\vartheta\, e^{\pm 3i\varphi}$ $\mp\frac{3}{64}\sqrt{\frac{35}{\pi}}(2\sin 2\vartheta-\sin 4\vartheta)e^{\pm 3i\varphi}$	0; 90; 180
4	± 2	$\frac{3}{8}\sqrt{\frac{5}{2\pi}}\sin^2\vartheta(7\cos^2\vartheta-1)e^{\pm 2i\varphi}$ $\frac{3}{64}\sqrt{\frac{5}{2\pi}}(3+4\cos 2\vartheta-7\cos 4\vartheta)e^{\pm 2i\varphi}$	0; 67.8; 112.2; 180
4	± 1	$\mp\frac{3}{8}\sqrt{\frac{5}{\pi}}\sin\vartheta(7\cos^3\vartheta-3\cos\vartheta)e^{\pm i\varphi}$ $\mp\frac{3}{64}\sqrt{\frac{5}{\pi}}(2\sin 2\vartheta+7\sin 4\vartheta)e^{\pm i\varphi}$	0; 49.1; 90; 130.9; 180
4	0	$\frac{3}{16}\sqrt{\frac{1}{\pi}}(35\cos^4\vartheta-30\cos^2\vartheta+3)$ $\frac{3}{128}\sqrt{\frac{1}{\pi}}(9+20\cos 2\vartheta+35\cos 4\vartheta)$	30.6; 70.1; 109.9; 149.4

Table A.4. Solid Spherical Harmonics for $l = 0,1,2,3,4$

l	m	$\mathcal{Y}_{lm}(x,y,z)$	l	m	$\mathcal{Y}_{lm}(x,y,z)$
0	0	$\frac{1}{2}\sqrt{\frac{1}{\pi}}$	3	± 1	$\mp\frac{1}{8}\sqrt{\frac{21}{\pi}}(x\pm iy)(5z^2-r^2)$
1	± 1	$\mp\frac{1}{2}\sqrt{\frac{3}{2\pi}}(x\pm iy)$	3	0	$\frac{1}{4}\sqrt{\frac{7}{\pi}}(5z^2-3r^2)z$
1	0	$\frac{1}{2}\sqrt{\frac{3}{\pi}}z$	4	± 4	$\frac{3}{32}\sqrt{\frac{70}{\pi}}(x\pm iy)^4$
2	± 2	$\frac{1}{4}\sqrt{\frac{15}{2\pi}}(x\pm iy)^2$	4	± 3	$\mp\frac{3}{8}\sqrt{\frac{35}{\pi}}(x\pm iy)^3z$
2	± 1	$\mp\frac{1}{2}\sqrt{\frac{15}{2\pi}}(x\pm iy)z$	4	± 2	$\frac{3}{16}\sqrt{\frac{10}{\pi}}(x\pm iy)^2(7z^2-r^2)$
2	0	$\frac{1}{4}\sqrt{\frac{5}{\pi}}(3z^2-r^2)$	4	± 1	$\mp\frac{3}{8}\sqrt{\frac{5}{\pi}}(x\pm iy)(7z^2-3r^2)z$
3	± 3	$\mp\frac{1}{8}\sqrt{\frac{35}{\pi}}(x\pm iy)^3$	4	0	$\frac{15}{16}\sqrt{\frac{1}{\pi}}\left(7z^4-6z^2r^2+\frac{3}{5}r^4\right)$
3	± 2	$\frac{1}{4}\sqrt{\frac{105}{2\pi}}(x\pm iy)^2z$			

A.4.3. Orthogonalization and Completeness

$$\int\int d\Omega_1 d\Omega_2 \left\{Y_{l_1}(\vartheta_1,\varphi_1)\otimes Y_{l_2}(\vartheta_2,\varphi_2)\right\}_{LM}\left\{Y_{l_1'}(\vartheta_1,\varphi_1)\otimes Y_{l_2'}(\vartheta_2,\varphi_2)\right\}^*_{L'M'}$$
$$= \delta_{l_1 l_1'}\delta_{l_2 l_2'}\delta_{LL'}\delta_{MM'} \tag{A.38}$$

$$\sum_{l_1 l_2 LM}\left\{Y_{l_1}(\vartheta_1,\varphi_1)\otimes Y_{l_2}(\vartheta_2,\varphi_2)\right\}_{LM}\left\{Y_{l_1}(\vartheta_1',\varphi_1')\otimes Y_{l_2}(\vartheta_2',\varphi_2')\right\}^*_{LM}$$
$$= \delta(\Omega_1-\Omega_1')\delta(\Omega_2-\Omega_2') \tag{A.39}$$

where $\delta(\Omega-\Omega') = \delta(\cos\vartheta-\cos\vartheta')\delta(\varphi-\varphi')$.

A.5. Solid Spherical Harmonics

A.5.1. Definition

$$\mathcal{Y}_{lm}(\mathbf{r}) = r^l Y_{lm}(\vartheta,\varphi) \tag{A.40}$$

where $Y_{lm}(\vartheta,\varphi)$ is the spherical harmonic (see Section A.3).

A.5.2. Table

Solid spherical harmonics for small l are given in Table A.4.

A.6. Rotations and Wigner D-Functions

A.6.1. Rotations and Euler Angles

We use the right-handed coordinate systems. An arbitrary rotation ω of the system of coordinates $S\{x,y,z\} \xrightarrow{\omega} S'\{x',y',z'\}$ can be executed by three successive rotations:

1. Rotation through the angle α around the z-axis $(0 \leq \alpha < 2\pi)$

2. Rotation through the angle β around the new y-axis $(0 \leq \beta \leq \pi)$

3. Rotation through the angle γ around the new z-axis $(0 \leq \gamma < 2\pi)$

The angles α, β, and γ are called the *Euler angles*. An angle of rotation about an axis is positive if, looking along the direction of the axis, the rotation is clockwise. In other words, the rotation through the positive angle about the directed axis leads to the movement of the axis in the positive direction in accordance with the right-handed screw.

A.6.2. Definition of D-Functions

Let $|jm\rangle$ be a state vector of a physical system with fixed angular momentum j and its projection m on the z-axis of the coordinate system S. Let $|\widetilde{jm}\rangle$ be another state vector (of the same physical system) that is characterized by the same j and the same projection m, but on the z-axis of another (rotated) coordinate system $\tilde{S}$. Let $D(\alpha,\beta,\gamma)$ be an operator that transforms the state vector $|jm\rangle$ into the state vector $|\widetilde{jm}\rangle$:

$$|\widetilde{jm}\rangle = D(\alpha,\beta,\gamma)\,|jm\rangle \tag{A.41}$$

The Euler angles α, β, and γ characterize the rotation $S \xrightarrow{\omega} \tilde{S}$. Then the state vector $|\widetilde{jm}\rangle$ is expanded in terms of the state vectors $|jm\rangle$ $(m = -j,\dots,+j)$ as follows:

$$|\widetilde{jm}\rangle = \sum_{m'} D^j_{m'm}(\alpha,\beta,\gamma)\,|jm'\rangle \tag{A.42}$$

where $D^j_{m'm}(\alpha,\beta,\gamma)$ is called the *Wigner D-function*. It is defined as the matrix element of the rotation operator in the representation of the state vectors $|jm\rangle$:

$$\langle jm\,|D(\alpha,\beta,\gamma)|\,j'm'\rangle = \delta_{jj'}\,D^j_{mm'}(\alpha,\beta,\gamma) \tag{A.43}$$

A.6.3.   Explicit Form

$$D^J_{mm'}(\alpha,\beta,\gamma) = \exp(-im\alpha)\, d^J_{mm'}(\beta)\, \exp(-im'\gamma) \qquad \text{(A.44)}$$

where $d^j_{mm'}(\beta)$ is a real function that can be expressed in terms of polynomials in $\cos(\beta/2)$ and $\sin(\beta/2)$ (see Table A.5 on p. 210).

A.6.4.   Useful Symmetries

$$D^J_{MM'}(\alpha,\beta,\gamma) = (-1)^{M'-M} D^{J*}_{-M-M'}(\alpha,\beta,\gamma) = (-1)^{M'-M} D^J_{M'M}(\gamma,\beta,\alpha)$$

$$= D^{J*}_{-M'-M}(\gamma,\beta,\alpha) = D^{J*}_{M'M}(-\gamma,-\beta,-\alpha) \qquad \text{(A.45)}$$

A.6.5.   Formula of Addition

Consider two successive rotations of the coordinate system $S \xrightarrow{\omega_1} S' \xrightarrow{\omega_2} S''$, where the first rotation, ω_1, is characterized by the Euler angles α_1, β_1, and γ_1 and the second rotation ω_2 is characterized by the Euler angles α_2, β_2, and γ_2 in the intermediate system S'. Let the resulting rotation $S \to S''$ be characterized by the Euler angles α, β, and γ. Then

$$\sum_{m''=-j}^{j} D^j_{mm''}(\alpha_2,\beta_2,\gamma_2)\, D^j_{m''m'}(\alpha_1,\beta_1,\gamma_1) = D^j_{mm'}(\alpha,\beta,\gamma) \qquad \text{(A.46)}$$

A.6.6.   Unitarity

$$\sum_{m=-j}^{j} D^j_{mm'}(\alpha,\beta,\gamma)\, D^{j*}_{mm''}(\alpha,\beta,\gamma) = \delta_{m'm''} \qquad \text{(A.47)}$$

A.6.7.   Selected Formulas

$$D^{j_1}_{m_1 m_1'}(\alpha,\beta,\gamma)\, D^{j_2}_{m_2 m_2'}(\alpha,\beta,\gamma)$$

$$= \sum_{j=|j_1-j_2|}^{j_1+j_2} \sum_{mm'} (j_1 m_1, j_2 m_2 \,|\, j m)\, (j_1 m_1', j_2 m_2' \,|\, j m')\, D^j_{mm'}(\alpha,\beta,\gamma) \qquad \text{(A.48)}$$

$$\sum_{m_1' m_2'} D^{j_1}_{m_1 m_1'} (\alpha, \beta, \gamma) D^{j_2}_{m_2 m_2'} (\alpha, \beta, \gamma) \, (j_1 m_1', j_2 m_2' \,|\, j m')$$

$$= (j_1 m_1, j_2 m_2 \,|\, j m) D^{j}_{mm'} (\alpha, \beta, \gamma) \qquad (A.49)$$

$$\int_0^{2\pi} d\alpha \int_0^{\pi} d\beta \sin\beta \int_0^{2\pi} d\gamma D^{j_2*}_{m_2 m_2'} (\alpha, \beta, \gamma) \, D^{j_1}_{m_1 m_1'} (\alpha, \beta, \gamma)$$

$$= \frac{8\pi^2}{2j_1 + 1} \, \delta_{j_1 j_2} \, \delta_{m_1 m_2} \, \delta_{m_1' m_2'}$$

$$(j_1 + j_2 = \text{integer}) \qquad (A.50)$$

A.6.8.   Connection with Spherical Harmonics

$$D^{l}_{m0} (\varphi, \vartheta, \psi) = (-1)^m \sqrt{\frac{4\pi}{2l+1}} \, Y_{l-m} (\vartheta, \varphi) = \sqrt{\frac{4\pi}{2l+1}} \, Y^{*}_{lm} (\vartheta, \varphi) \qquad (A.51)$$

$$D^{l}_{0m} (\varphi, \vartheta, \psi) = \sqrt{\frac{4\pi}{2l+1}} \, Y_{l-m} (\vartheta, \psi) = (-1)^m \sqrt{\frac{4\pi}{2l+1}} \, Y^{*}_{lm} (\vartheta, \psi) \qquad (A.52)$$

A.6.9.   Particular Arguments

$$D^{j}_{mm'} (0,0,0) = \delta_{mm'} \qquad (A.53)$$

$$D^{j}_{mm'} (\alpha, 0, \gamma) = \delta_{mm'} e^{-im(\alpha+\gamma)} \qquad (A.54)$$

$$D^{j}_{mm'} (0, \beta, 0) = d^{j}_{mm'} (\beta) \qquad (A.55)$$

$$D^{j}_{mm'} (0, \pm 2n\pi, 0) = \delta_{mm'} (-1)^{2nj} \qquad (A.56)$$

$$D^{j}_{mm'} (0, \pm(2n+1)\pi, 0) = \delta_{-mm'} (-1)^{\pm(2n+1)j+m} \qquad (A.57)$$

A.6.10.   Connection with Legendre Polynomials

$$D^{l}_{00} (\varphi, \vartheta, \psi) = P_l (\cos\vartheta) \qquad (A.58)$$

A.6.11. Tables

The functions $d^j_{mm'}(\vartheta)$ are given in Table A.5 on p. 210 and numerical values of $d^J_{MM'}(\frac{\pi}{2})$ for small J in Table A.6 on p. 211.

A.7. Irreducible Tensor Operators

A.7.1. Definition

One calls the irreducible tensor operator of the rank k a set of $2k+1$ operators T_{kq} $(q = k, k-1, \ldots, -k)$, which transform under the rotation of the coordinate frame $S \xrightarrow{\omega} \tilde{S}$ by the law

$$\widetilde{T_{k'q'}} = \delta_{kk'} \sum_q D^k_{qq'}(\omega) T_{kq} \tag{A.59}$$

where $D^k_{qq'}(\omega)$ is the Wigner D-function (see Section A.6).

A.7.2. Irreducible Tensor Product

The irreducible tensor product, or simply the tensor product, of two irreducible tensor operators, U_{k_1} and V_{k_2}, with the ranks k_1 and k_2, respectively, is a new irreducible tensor operator of the rank k:

$$|k_1 - k_2| \le k \le k_1 + k_2$$

the components of which are defined by the relation:

$$\{U_{k_1} \otimes V_{k_2}\}_{kq} = \sum_{q_1 q_2} (k_1 q_1, k_2 q_2 | kq) U_{k_1 q_1} V_{k_2 q_2} \tag{A.60}$$

A particular case of the tensor product is the scalar product of two irreducible tensors:

$$(U_k V_k) = (-1)^k \sqrt{2k+1} \{U_k \otimes V_k\}_{00} = \sum_q (-1)^q U_{kq} V_{k-q} \tag{A.61}$$

A.7.3. The Wigner–Eckart Theorem

Let T_{kq} be a component of an irreducible tensor operator. Then

$$\langle \alpha jm | T_{kq} | \alpha' j'm' \rangle = \frac{1}{\sqrt{2j+1}} (j'm', kq | jm) \langle j \| T_k \| j' \rangle \tag{A.62}$$

The theorem factorizes the dependence of the matrix elements on the projections and serves as a definition of the *reduced matrix element* $\langle j \| T_k \| j' \rangle$.

Table A.5. Functions $d^j_{mm'}(\vartheta)$

$$d^{1/2}_{mm'}(\vartheta)$$

$m' \rightarrow$		
m	$1/2$	$-1/2$
$\downarrow$		
$1/2$	$\cos\frac{\vartheta}{2}$	$-\sin\frac{\vartheta}{2}$
$-1/2$	$\sin\frac{\vartheta}{2}$	$\cos\frac{\vartheta}{2}$

$$d^1_{mm'}(\vartheta)$$

$m' \rightarrow$			
m	1	0	-1
$\downarrow$			
1	$\frac{1}{2}(1+\cos\vartheta)$	$-\frac{1}{\sqrt{2}}\sin\vartheta$	$\frac{1}{2}(1-\cos\vartheta)$
0	$\frac{1}{\sqrt{2}}\sin\vartheta$	$\cos\vartheta$	$-\frac{1}{\sqrt{2}}\sin\vartheta$
-1	$\frac{1}{2}(1-\cos\vartheta)$	$-\frac{1}{\sqrt{2}}\sin\vartheta$	$\frac{1}{2}(1+\cos\vartheta)$

$$d^{3/2}_{mm'}(\vartheta)$$

$m' \rightarrow$				
m	$\frac{3}{2}$	$1/2$	$-1/2$	$-3/2$
$\downarrow$				
$3/2$	$\cos^3\frac{\vartheta}{2}$	$-\sqrt{3}\sin\frac{\vartheta}{2}\cos^2\frac{\vartheta}{2}$	$\sqrt{3}\sin^2\frac{\vartheta}{2}\cos\frac{\vartheta}{2}$	$-\sin^3\frac{\vartheta}{2}$
$1/2$	$\sqrt{3}\sin\frac{\vartheta}{2}\cos^2\frac{\vartheta}{2}$	$\cos\frac{\vartheta}{2}\left(3\cos^2\frac{\vartheta}{2}-2\right)$	$\sin\frac{\vartheta}{2}\left(3\sin^2\frac{\vartheta}{2}-2\right)$	$\sqrt{3}\sin^2\frac{\vartheta}{2}\cos\frac{\vartheta}{2}$
$-1/2$	$\sqrt{3}\sin^2\frac{\vartheta}{2}\cos\frac{\vartheta}{2}$	$-\sin\frac{\vartheta}{2}\left(3\sin^2\frac{\vartheta}{2}-2\right)$	$\cos\frac{\vartheta}{2}\left(3\cos^2\frac{\vartheta}{2}-2\right)$	$-\sqrt{3}\sin\frac{\vartheta}{2}\cos^2\frac{\vartheta}{2}$
$-3/2$	$\sin^3\frac{\vartheta}{2}$	$\sqrt{3}\sin^2\frac{\vartheta}{2}\cos\frac{\vartheta}{2}$	$\sqrt{3}\sin\frac{\vartheta}{2}\cos^2\frac{\vartheta}{2}$	$\cos^3\frac{\vartheta}{2}$

$$d^2_{mm'}(\vartheta)$$

$m' \rightarrow$			
m	2	1	0
$\downarrow$			
2	$\frac{1}{4}(1+\cos\vartheta)^2$	$-\frac{1}{2}\sin\vartheta(1+\cos\vartheta)$	$\frac{1}{2}\sqrt{\frac{3}{2}}\sin^2\vartheta$
1	$\frac{1}{2}\sin\vartheta(1+\cos\vartheta)$	$\frac{1}{2}(2\cos^2\vartheta+\cos\vartheta-1)$	$-\sqrt{\frac{3}{2}}\sin\vartheta\cos\vartheta$
0	$\frac{1}{2}\sqrt{\frac{3}{2}}\sin^2\vartheta$	$\sqrt{\frac{3}{2}}\sin\vartheta\cos\vartheta$	$\frac{1}{2}(3\cos^2\vartheta-1)$
-1	$\frac{1}{2}\sin\vartheta(1-\cos\vartheta)$	$-\frac{1}{2}(2\cos^2\vartheta-\cos\vartheta-1)$	$\sqrt{\frac{3}{2}}\sin\vartheta\cos\vartheta$
-2	$\frac{1}{4}(1-\cos\vartheta)^2$	$\frac{1}{2}\sin\vartheta(1-\cos\vartheta)$	$\frac{1}{2}\sqrt{\frac{3}{2}}\sin^2\vartheta$

$$d^2_{mm'}(\vartheta)\ (continued)$$

$m' \rightarrow$		
m	-1	-2
$\downarrow$		
2	$-\frac{1}{2}\sin\vartheta(1-\cos\vartheta)$	$\frac{1}{4}(1-\cos\vartheta)^2$
1	$-\frac{1}{2}(2\cos^2\vartheta-\cos\vartheta-1)$	$-\frac{1}{2}\sin\vartheta(1-\cos\vartheta)$
0	$-\sqrt{\frac{3}{2}}\sin\vartheta\cos\vartheta$	$\frac{1}{2}\sqrt{\frac{3}{2}}\sin^2\vartheta$
-1	$\frac{1}{2}(2\cos^2\vartheta+\cos\vartheta-1)$	$-\frac{1}{2}\sin\vartheta(1+\cos\vartheta)$
-2	$\frac{1}{2}\sin\vartheta(1+\cos\vartheta)$	$\frac{1}{4}(1+\cos\vartheta)^2$

Table A.6. Numerical Values of $d^J_{MM'}(\frac{\pi}{2})$ for $J = 1/2, 1, 3/2, 2$

$d^{1/2}_{mm'}(\frac{\pi}{2})$

$m' \rightarrow$		
$m \downarrow$	$1/2$	$-1/2$
$1/2$	$\frac{1}{\sqrt{2}}$	$-\frac{1}{\sqrt{2}}$
$-1/2$	$\frac{1}{\sqrt{2}}$	$\frac{1}{\sqrt{2}}$

$d^1_{mm'}(\frac{\pi}{2})$

$m' \rightarrow$			
$m \downarrow$	1	0	-1
1	$\frac{1}{2}$	$-\frac{1}{\sqrt{2}}$	$\frac{1}{2}$
0	$\frac{1}{\sqrt{2}}$	0	$-\frac{1}{\sqrt{2}}$
-1	$\frac{1}{2}$	$\frac{1}{\sqrt{2}}$	$\frac{1}{2}$

$d^{3/2}_{mm'}(\frac{\pi}{2})$

$m' \rightarrow$				
$m \downarrow$	$3/2$	$1/2$	$-1/2$	$-3/2$
$3/2$	$\frac{1}{2\sqrt{2}}$	$-\frac{\sqrt{3}}{2\sqrt{2}}$	$\frac{\sqrt{3}}{2\sqrt{2}}$	$-\frac{1}{2\sqrt{2}}$
$1/2$	$\frac{\sqrt{3}}{2\sqrt{2}}$	$-\frac{1}{2\sqrt{2}}$	$-\frac{1}{2\sqrt{2}}$	$\frac{\sqrt{3}}{2\sqrt{2}}$
$-1/2$	$\frac{\sqrt{3}}{2\sqrt{2}}$	$\frac{1}{2\sqrt{2}}$	$-\frac{1}{2\sqrt{2}}$	$-\frac{\sqrt{3}}{2\sqrt{2}}$
$-3/2$	$\frac{1}{2\sqrt{2}}$	$\frac{\sqrt{3}}{2\sqrt{2}}$	$\frac{\sqrt{3}}{2\sqrt{2}}$	$\frac{1}{2\sqrt{2}}$

$d^2_{mm'}(\frac{\pi}{2})$

$m' \rightarrow$					
$m \downarrow$	2	1	0	-1	-2
2	$\frac{1}{4}$	$-\frac{1}{2}$	$\frac{\sqrt{3}}{2\sqrt{2}}$	$-\frac{1}{2}$	$\frac{1}{4}$
1	$\frac{1}{2}$	$-\frac{1}{2}$	0	$\frac{1}{2}$	$-\frac{1}{2}$
0	$\frac{\sqrt{3}}{2\sqrt{2}}$	0	$-\frac{1}{2}$	0	$\frac{\sqrt{3}}{2\sqrt{2}}$
-1	$\frac{1}{2}$	$\frac{1}{2}$	0	$-\frac{1}{2}$	$-\frac{1}{2}$
-2	$\frac{1}{4}$	$\frac{1}{2}$	$\frac{\sqrt{3}}{2\sqrt{2}}$	$\frac{1}{2}$	$\frac{1}{4}$

A.7.4.   Algebra of Tensor Operators

1. If irreducible tensor operators U_{k_1} and V_{k_2} operate in a common subspace, then

$$\langle \alpha j \| \{U_{k_1} \otimes V_{k_2}\}_k \| \alpha' j' \rangle = (-1)^{j+j'+k} \sqrt{2k+1}$$
$$\times \sum_{\alpha'' j''} \begin{Bmatrix} k_1 & k_2 & k \\ j' & j & j'' \end{Bmatrix} \langle \alpha j \| U_{k_1} \| \alpha'' j'' \rangle \langle \alpha'' j'' \| V_{k_2} \| \alpha' j' \rangle \qquad \text{(A.63)}$$

2. If U operates in a subspace of functions relating to the angular momentum j_1 and V operates in a subspace of functions relating to the angular momentum j_2 (for example, orbital angular momenta of a particle and its spin), then

$$\langle \alpha j_1 j_2 j \| \{U_{k_1} \otimes V_{k_2}\}_k \| \alpha' j_1' j_2' j' \rangle = \sqrt{(2j+1)(2j'+1)(2k+1)}$$
$$\times \begin{Bmatrix} j_1 & j_2 & j \\ j_1' & j_2' & j' \\ k_1 & k_2 & k \end{Bmatrix} \sum_{\alpha''} \langle \alpha j_1 \| U_{k_1} \| \alpha'' j_1' \rangle \langle \alpha'' j_2 \| V_{k_2} \| \alpha' j_2' \rangle \qquad \text{(A.64)}$$

The sum over α'' is included to account for a case when both operators, U and V, act on the quantum number α.

3. Replacing the operator V by unity, one obtains from Eq. (A.64):

$$\langle \alpha j_1 j_2 j \| U_k \| \alpha' j_1' j_2' j' \rangle = \delta_{j_2 j_2'} (-1)^{j_1+j_2+j'+k} \sqrt{(2j+1)(2j'+1)}$$

$$\times \left\{ \begin{matrix} j_2 & j_1 & j \\ k & j' & j_1' \end{matrix} \right\} \langle \alpha j_1 \| U_k \| \alpha' j_1' \rangle \qquad (A.65)$$

4. Replacing the operator U by unity, one obtains from Eq. (A.64):

$$\langle \alpha j_1 j_2 j \| V_k \| \alpha' j_1' j_2' j' \rangle = \delta_{j_1 j_1'} (-1)^{j_1'+j_2+j+k} \sqrt{(2j+1)(2j'+1)}$$

$$\times \left\{ \begin{matrix} j_1 & j & j_2 \\ k & j_2' & j' \end{matrix} \right\} \langle \alpha j_2 \| V_k \| \alpha' j_2' \rangle \qquad (A.66)$$

A.7.5.   Examples

Unity operator:

$$\langle j \| I \| j' \rangle = \delta_{jj'} \sqrt{2j+1} \qquad (A.67)$$

Spherical harmonic:

$$\langle l \| Y_k \| l' \rangle = \sqrt{\frac{(2l'+1)(2k+1)}{4\pi}} \; (l'0,k0|l0) \qquad (A.68)$$

Angular momentum:

$$\langle j \| \mathbf{j} \| j' \rangle = \delta_{jj'} \sqrt{j(j+1)(2j+1)} \qquad (A.69)$$

Operator of unity vector:

$$\langle l \| \mathbf{n} \| l' \rangle = \sqrt{2l+1} \; (l0,10|l'0) \qquad (A.70)$$

A.7.6.   Reduced Matrix Elements in the LS-Coupling Approximation

The following expressions are written for an arbitrary spin s of the outgoing particle. For an electron, one must put $s = \frac{1}{2}$. The reduced matrix element of the decay operator V in the case when the total spin and the total orbital angular momentum are conserved during the decay and the operator V does not act on spin variables is

$$\langle \alpha_f (L_f S_f) J_f, (ls)j : J \| V \| \alpha_i (L_i S_i) J_i \rangle$$

$$= \hat{J}_f \hat{j} \hat{S}_i \hat{J}_i \left\{ \begin{matrix} L_f & S_f & J_f \\ l & s & j \\ L_i & S_i & J_i \end{matrix} \right\} \langle \alpha_f L_f, l : L \| V \| \alpha_i L_i \rangle \delta_{LL_i} \delta_{JJ_i} \qquad (A.71)$$

Reduction of the dipole matrix element for excitation:

$$\langle \alpha(LS)J \| D \| \alpha_i(L_iS_i)J_i \rangle$$

$$= \delta_{SS_i} \hat{J} \hat{J}_i (-1)^{L+S+J_i+1} \begin{Bmatrix} S & L & J \\ 1 & J_i & L_i \end{Bmatrix} \langle \alpha L \| D \| \alpha_i L_i \rangle \qquad (A.72)$$

Reduction of the dipole matrix element for ionization:

$$\langle \alpha_f(L_fS_f)J_f, (ls)j : J \| D \| \alpha_i(L_iS_i)J_i \rangle = \sum_{LS} \delta_{SS_i} \hat{J}_f \hat{j} \hat{J} \hat{J}_i \hat{S} \hat{L} (-1)^{L+S+J_i+1}$$

$$\times \begin{Bmatrix} S & L & J \\ 1 & J_i & L_i \end{Bmatrix} \begin{Bmatrix} L_f & S_f & J_f \\ l & s & j \\ L & S & J \end{Bmatrix} \langle \alpha_f L_f, l : L \| D \| \alpha_i L_i \rangle \qquad (A.73)$$

where

$$D = \sum_n \mathbf{r}_n$$

Reduction of the plane-wave Born multipole amplitude for excitation:

$$\langle \alpha(LS)J \| T_\lambda^B \| \alpha_i(L_iS_i)J_i \rangle$$

$$= \delta_{SS_i} \hat{J} \hat{J}_i (-1)^{L+S+J_i+\lambda} \begin{Bmatrix} S & L & J \\ \lambda & J_i & L_i \end{Bmatrix} \langle \alpha L \| T_\lambda^B \| \alpha_i L_i \rangle \qquad (A.74)$$

Reduction of the plane-wave Born multipole amplitude for ionization:

$$\langle \alpha_f(L_fS_f)J_f, (ls)j : J \| T_\lambda^B \| \alpha_i(L_iS_i)J_i \rangle = \sum_{LS} \delta_{SS_i} \hat{J}_f \hat{j} \hat{J} \hat{J}_i \hat{S} \hat{L} (-1)^{L+S+J_i+\lambda}$$

$$\times \begin{Bmatrix} S & L & J \\ \lambda & J_i & L_i \end{Bmatrix} \begin{Bmatrix} L_f & S_f & J_f \\ l & s & j \\ L & S & J \end{Bmatrix} \langle \alpha_f L_f, l : L \| T_\lambda^B \| \alpha_i L_i \rangle \qquad (A.75)$$

where $T_\lambda^B = \sum_n j_\lambda (Qr_n) Y_\lambda (\vartheta_n, \varphi_n)$.

In practice, a unity operator which operates in the spin subspace is implied sometimes in the reduced matrix elements at the right side of Eqs. (A.71)–(A.75). In this case the right side of Eqs. (A.71)–(A.75) should be multiplied by $\hat{S}^{-1}$ in accordance with Eq. (A.67).

A.7.7. Transformations of Scattering Amplitudes

Consider three coupling schemes of the total atomic angular momentum $\mathbf{J}$ with angular momenta of the scattering particle: jj-coupling $[(\mathbf{J} + (\mathbf{l} + \mathbf{s})\mathbf{j} = \tilde{\mathbf{J}}]$,

jl-coupling (called also jK-coupling) $[(\mathbf{J}+\mathbf{l})\mathbf{K}+\mathbf{s}=\tilde{\mathbf{J}}]$, and the channel spin representation $[(\mathbf{J}+\mathbf{s})\mathbf{S}+\mathbf{l}=\tilde{\mathbf{J}}]$.

Transformation from jj-coupling to jl-coupling

$$\left\langle \alpha J, (ls)j : \tilde{J} \| T \| \alpha_0 J_0, (l_0 s)j_0 : \tilde{J} \right\rangle$$

$$= (-1)^{2\tilde{J}+2s+l_0+J_0+l+J} \hat{j}\hat{j_0} \sum_{KK_0} \hat{K}\hat{K_0} \begin{Bmatrix} J & l & K \\ s & \tilde{J} & j \end{Bmatrix} \begin{Bmatrix} J_0 & l_0 & K_0 \\ s & \tilde{J} & j_0 \end{Bmatrix}$$

$$\times \left\langle \alpha, (Jl)K, s : \tilde{J} \| T \| \alpha_0, (J_0 l_0)K_0, s : \tilde{J} \right\rangle \qquad (A.76)$$

Transformation from the jj-coupling to the channel spin representation

$$\left\langle \alpha J, (ls)j : \tilde{J} \| T \| \alpha_0 J_0, (l_0 s)j_0 : \tilde{J} \right\rangle$$

$$= (-1)^{2\tilde{J}+j_0+J_0+j+J} \hat{j}\hat{j_0} \sum_{SS_0} \hat{S}\hat{S_0} \begin{Bmatrix} J & s & S \\ l & \tilde{J} & j \end{Bmatrix} \begin{Bmatrix} J_0 & s & S_0 \\ l_0 & \tilde{J} & j_0 \end{Bmatrix}$$

$$\times \left\langle \alpha, (Js)S, l : \tilde{J} \| T \| \alpha_0, (J_0 s)S_0, l_0 : \tilde{J} \right\rangle \qquad (A.77)$$

Transformation from the channel spin representation to the jl-coupling

$$\left\langle \alpha, (Js)S, l : \tilde{J} \| T \| \alpha_0, (J_0 s)S_0, l_0 : \tilde{J} \right\rangle$$

$$= (-1)^{2s+l_0+S_0+l+S} \hat{S}\hat{S_0} \sum_{KK_0} (-1)^{K+K_0} \hat{K}\hat{K_0} \begin{Bmatrix} s & J & S \\ l & \tilde{J} & K \end{Bmatrix} \begin{Bmatrix} s & J_0 & S_0 \\ l_0 & \tilde{J} & K_0 \end{Bmatrix}$$

$$\times \left\langle \alpha, (Jl)K, s : \tilde{J} \| T \| \alpha_0, (J_0 l_0)K_0, s : \tilde{J} \right\rangle \qquad (A.78)$$

The above equations are the recouplings of three angular momenta with the help of $6j$-symbols [see Eqs. (A.96)–(A.98)]. Reverse transformations are obtained in the similar way.

The amplitudes in uncoupled representation $|\alpha JM, \mathbf{p}\mu\rangle$ can be related to the three above representations with the use of Clebsch–Gordan coefficients, the Wigner–Eckart theorem (A.62) for the transition operator T (which is a scalar in the total space of variables atom + particle), and Eq. (A.25) [in which (ϑ, φ) shows the direction of the linear momentum $\mathbf{p}$ of the particle]. For example, transformations (direct and reverse) between the jj-coupling representation and the uncoupled representation are of the form

$$\left\langle \alpha JM, \mathbf{p}\mu \mid T \mid \alpha_0 J_0 M_0, \mathbf{p}_0 \mu_0 \right\rangle = \sum_{l_0 j_0 l j \tilde{J}} \tilde{J}^{-1} \left\langle \alpha J, lj : \tilde{J} \| T \| \alpha_0 J_0, l_0 j_0 : \tilde{J} \right\rangle$$

$$\times \Bigg[\sum_{m_{l_0} m_{j_0} m_l m_j \tilde{M}} (JM, jm_j \mid \tilde{J}\tilde{M}) (J_0 M_0, j_0 m_{j_0} \mid \tilde{J}\tilde{M}) (lm_l, s\mu \mid jm_j)$$

$$\times (l_0 m_{l_0}, s\mu_0 \mid j_0 m_{j_0}) Y^*_{l_0 m_{l_0}}(\vartheta_0, \varphi_0) Y_{lm_l}(\vartheta, \varphi) \Bigg] \qquad (A.79)$$

$$\langle \alpha J, lj : \tilde{J}\tilde{M} \,|\, T \,|\, \alpha_0 J_0, l_0 j_0 : \tilde{J}\tilde{M} \rangle = \hat{\tilde{J}}^{-1} \langle \alpha J, lj : \tilde{J} \,\|\, T \,\|\, \alpha_0 J_0, l_0 j_0 : \tilde{J} \rangle$$

$$= \sum_{MM_0\mu\mu_0} \int d\Omega_0 \int d\Omega \langle \alpha JM, \mathbf{p}\mu \,|\, T \,|\, \alpha_0 J_0 M_0, \mathbf{p}_0\mu_0 \rangle$$

$$\times \left[\sum_{m_{l_0} m_{j_0} m_l m_j} (JM, jm_j \,|\, \tilde{J}\tilde{M}) \, (J_0 M_0, j_0 m_{j_0} \,|\, \tilde{J}\tilde{M}) \, (lm_l, s\mu \,|\, jm_j) \right.$$

$$\left. \times (l_0 m_{l_0}, s_0 \mu_0 \,|\, j_0 m_{j_0}) \, Y_{l_0 m_{l_0}} (\vartheta_0, \varphi_0) Y^*_{l m_l} (\vartheta, \varphi) \right] \qquad \text{(A.80)}$$

Another type of transformation accounts for internal atomic couplings of angular momenta. For example, if atom is characterized by the total orbital angular momentum L and total spin S, one can transform from jj-coupling $(\mathbf{L}+\mathbf{S}=\mathbf{J}, \mathbf{l}+\mathbf{s}=\mathbf{j}, \mathbf{J}+\mathbf{j}=\tilde{\mathbf{J}})$ to LS-coupling $(\mathbf{L}+\mathbf{l}=\tilde{\mathbf{L}}, \mathbf{S}+\mathbf{s}=\tilde{\mathbf{S}}, \tilde{\mathbf{L}}+\tilde{\mathbf{S}}=\tilde{\mathbf{J}})$:

$$\langle \alpha(LS)J, (ls)j : \tilde{J} \,\|\, T \,\|\, \alpha_0 (L_0 S_0) J_0, (l_0 s_0) j_0 : \tilde{J} \rangle$$

$$= \sum_{\tilde{S}\tilde{L}\tilde{S}_0\tilde{L}_0} \hat{J}_0 \hat{j}_0 \hat{J} \hat{j} \hat{\tilde{L}}_0 \hat{\tilde{S}}_0 \hat{\tilde{L}} \hat{\tilde{S}} \left\{ \begin{matrix} L & S & J \\ l & s & j \\ \tilde{L} & \tilde{S} & \tilde{J} \end{matrix} \right\} \left\{ \begin{matrix} L_0 & S_0 & J_0 \\ l_0 & s_0 & j_0 \\ \tilde{L}_0 & \tilde{S}_0 & \tilde{J} \end{matrix} \right\}$$

$$\times \langle \alpha, (Ll)\tilde{L}, (Ss)\tilde{S} : \tilde{J} \,\|\, T \,\|\, \alpha_0, (L_0 l_0)\tilde{L}_0, (S_0 s)\tilde{S}_0 : \tilde{J} \rangle \qquad \text{(A.81)}$$

Equation (A.81) is a recoupling of four angular momenta with the help of the $9j$-symbols [see Eq. (A.117)].

A.8.   Clebsch–Gordan Coefficients

The Clebsch–Gordan coefficients form the matrix of unitary transformation from the representation $|\, j_1 m_1 j_2 m_2 \rangle$ to the representation $|\, j_1 j_2 jm \rangle$, and the reverse:

$$(j_1 m_1, j_2 m_2 \,|\, jm) = \langle j_1 m_1 j_2 m_2 \,|\, j_1 j_2 jm \rangle = \langle j_1 j_2 jm \,|\, j_1 m_1 j_2 m_2 \rangle \qquad \text{(A.82)}$$

We use the standard phase convention when the Clebsch–Gordan coefficients are real.

A.8.1.   Restrictions on the Arguments

The Clebsch–Gordan coefficients are zero if at least one of the following rules is not fulfilled:

1. j_1, j_2, j are integer or half-integer nonnegative numbers

2. m_1, m_2, m are integer or half-integer positive or negative numbers

3. $|m_1| \leq j_1, |m_2| \leq j_2, |m| \leq j$

4. $j_1 + m_1, j_2 + m_2, j + m, j_1 + j_2 + j$ are integer nonnegative numbers

5. The triangle inequality: $|j_1 - j_2| \leq j \leq j_1 + j_2$

6. $m_1 + m_2 = m$

A.8.2. Relation to 3jm-Symbol

$$(a\alpha, b\beta | c\gamma) = (-1)^{a-b+\gamma} \hat{c} \begin{pmatrix} a & b & c \\ \alpha & \beta & -\gamma \end{pmatrix} \tag{A.83}$$

$$\begin{pmatrix} a & b & c \\ \alpha & \beta & \gamma \end{pmatrix} = (-1)^{c+\gamma+2a} \hat{c}^{-1} (a-\alpha, b-\beta | c\gamma) \tag{A.84}$$

A.8.3. Main Symmetry Relations

$$(a\alpha, b\beta | c\gamma) = (-1)^{a+b-c} (a-\alpha, b-\beta | c-\gamma) = (-1)^{a+b-c} (b\beta, a\alpha | c\gamma)$$
$$= (-1)^{a-\alpha} \frac{\hat{c}}{\hat{b}} (a\alpha, c-\gamma | b-\beta) = (-1)^{b+\beta} \frac{\hat{c}}{\hat{a}} (c-\gamma, b\beta | a-\alpha) \tag{A.85}$$

A.8.4. Formulas of Summation

We present only the relations most often used. If some practical expression shows another order of the angular momenta, one should use the symmetry relations (A.85) to rearrange the coupling.

$$\sum_\alpha (a\alpha, b0 | a\alpha) = \hat{a}^2 \, \delta_{b0} \tag{A.86}$$

$$\sum_{\alpha\beta} (a\alpha, b\beta | c\gamma)(a\alpha, b\beta | c'\gamma') = \delta_{cc'} \, \delta_{\gamma\gamma'} \tag{A.87}$$

$$\sum_{c\gamma} (a\alpha, b\beta | c\gamma)(a\alpha', b\beta' | c\gamma) = \delta_{\alpha\alpha'} \, \delta_{\beta\beta'} \tag{A.88}$$

$$\sum_{\alpha\beta\delta} (a\alpha,b\beta|c\gamma)\,(d\delta,b\beta|e\varepsilon)\,(a\alpha,f\varphi|d\delta)$$

$$= (-1)^{b+c+d+f}\,\hat{c}\hat{d}\,(c\gamma,f\varphi|e\varepsilon)\begin{Bmatrix} a\,b\,c \\ e\,f\,d \end{Bmatrix} \tag{A.89}$$

$$\sum_{\alpha\beta\delta} (-1)^{a-\alpha}\,(a\alpha,b\beta|c\gamma)\,(d\delta,b\beta|e\varepsilon)\,(d\delta,a-\alpha|f\varphi)$$

$$= (-1)^{b+c+d+f}\,\hat{c}\hat{f}\,(c\gamma,f\varphi|e\varepsilon)\begin{Bmatrix} a\,b\,c \\ e\,f\,d \end{Bmatrix} \tag{A.90}$$

$$\sum_{\beta\gamma\varepsilon\varphi} (b\beta,c\gamma|a\alpha)\,(e\varepsilon,f\varphi|d\delta)\,(e\varepsilon,b\beta|g\eta)\,(f\varphi,c\gamma|j\mu)$$

$$= (-1)^{a-b+c+d+e-f}\,\hat{a}\hat{g}\sum_{s\sigma}\hat{s}^2\,(a\alpha,s\sigma|j\mu)\,(g\eta,s\sigma|d\delta)\begin{Bmatrix} b\,c\,a \\ j\,s\,f \end{Bmatrix}\begin{Bmatrix} b\,e\,g \\ d\,s\,f \end{Bmatrix}$$

$$= \hat{a}\hat{d}\hat{g}\hat{j}\sum_{t\tau}(g\eta,j\mu\,|t\tau)\,(d\delta,a\alpha|t\tau)\begin{Bmatrix} c\,b\,a \\ f\,e\,d \\ j\,g\,t \end{Bmatrix} \tag{A.91}$$

A.8.5. Zero Values of Arguments

$$(a\alpha,00|c\gamma) = \delta_{ac}\,\delta_{\alpha\gamma} \tag{A.92}$$

$$(a\alpha,b\beta|00) = (-1)^{a-\alpha}\,\hat{a}^{-1}\,\delta_{ab}\,\delta_{\alpha,-\beta} \tag{A.93}$$

$$(a0,b0|c0) = 0 \qquad (a+b+c=2n+1) \tag{A.94}$$

$$(a0,b0|c0) = \frac{(-1)^{n-c}\,\hat{c}n!}{(n-a)!\,(n-b)!\,(n-c)!}$$

$$\times \left[\frac{(2n-2a)!\,(2n-2b)!\,(2n-2c)!}{(2n+1)!}\right]^{1/2}$$

$$(a+b+c=2n) \tag{A.95}$$

A.8.6. Tables

We give Clebsch–Gordan coefficients $(a\alpha,b\beta|c\alpha+\beta)$ for $c=1/2,1,2,3$ and $a,b\le 3$ in terms of simple fractions in Table A.7 on pp. 218–220.

Table A.7. Clebsch–Gordan Coefficients

$$Z = (a\alpha, b\beta \,|\, 1/2\, \alpha+\beta)$$

a	α	b	β	Z	a	α	b	β	Z
1/2	1/2	0	0	1	5/2	3/2	2	−1	$-2/\sqrt{3\cdot5}$
1	1	1/2	−1/2	$\sqrt{2}/\sqrt{3}$	5/2	3/2	2	−2	$1/\sqrt{3\cdot5}$
1	0	1/2	1/2	$-1/\sqrt{3}$	5/2	1/2	2	0	$1/\sqrt{5}$
3/2	3/2	1	−1	$1/\sqrt{2}$	5/2	1/2	2	−1	$-\sqrt{2}/\sqrt{3\cdot5}$
3/2	1/2	1	0	$-1/\sqrt{3}$	3	3	5/2	−5/2	$\sqrt{2}/\sqrt{7}$
3/2	1/2	1	−1	$1/\sqrt{2\cdot3}$	3	2	5/2	−3/2	$-\sqrt{5}/\sqrt{3\cdot7}$
2	2	3/2	−3/2	$\sqrt{2}/\sqrt{5}$	3	2	5/2	−5/2	$1/\sqrt{3\cdot7}$
2	1	3/2	−1/2	$-\sqrt{3}/\sqrt{2\cdot5}$	3	1	5/2	−1/2	$2/\sqrt{3\cdot7}$
2	1	3/2	−3/2	$1/\sqrt{2\cdot5}$	3	1	5/2	−3/2	$-\sqrt{2}/\sqrt{3\cdot7}$
2	0	3/2	1/2	$1/\sqrt{5}$	3	0	5/2	1/2	$-1/\sqrt{7}$
5/2	5/2	2	−2	$1/\sqrt{3}$					

$$Z = (a\alpha, b\beta \,|\, 1\, \alpha+\beta)$$

a	α	b	β	Z	a	α	b	β	Z
1/2	1/2	1/2	1/2	1	5/2	3/2	3/2	−3/2	$1/\sqrt{5}$
1/2	1/2	1/2	−1/2	$1/\sqrt{2}$	5/2	1/2	3/2	1/2	$\sqrt{3}/(2\sqrt{5})$
1	1	0	0	1	5/2	1/2	3/2	−1/2	$-\sqrt{3}/\sqrt{2\cdot5}$
1	0	0	0	1	5/2	1/2	3/2	−3/2	$1/(2\sqrt{5})$
1	1	1	0	$1/\sqrt{2}$	5/2	5/2	5/2	−3/2	$1/\sqrt{7}$
1	1	1	−1	$1/\sqrt{2}$	5/2	5/2	5/2	−5/2	$\sqrt{5}/\sqrt{2\cdot7}$
1	0	1	0	0	5/2	3/2	5/2	−1/2	$-2\sqrt{2}/\sqrt{5\cdot7}$
3/2	3/2	1/2	−1/2	$\sqrt{3}/2$	5/2	3/2	5/2	−3/2	$-3/\sqrt{2\cdot5\cdot7}$
3/2	1/2	1/2	1/2	$-1/2$	5/2	1/2	5/2	1/2	$3/\sqrt{5\cdot7}$
3/2	1/2	1/2	−1/2	$1/\sqrt{2}$	5/2	1/2	5/2	−1/2	$1/\sqrt{2\cdot5\cdot7}$
3/2	3/2	3/2	−1/2	$\sqrt{3}/\sqrt{2\cdot5}$	3	3	2	−2	$\sqrt{3}/\sqrt{7}$
3/2	3/2	3/2	−3/2	$3/(2\sqrt{5})$	3	2	2	−1	$-\sqrt{2}/\sqrt{7}$
3/2	1/2	3/2	1/2	$-\sqrt{2}/\sqrt{5}$	3	2	2	−2	$1/\sqrt{7}$
3/2	1/2	3/2	−1/2	$-1/(2\sqrt{5})$	3	1	2	0	$\sqrt{2\cdot3}/\sqrt{5\cdot7}$
2	2	1	−1	$\sqrt{3}/\sqrt{5}$	3	1	2	−1	$-2\sqrt{2}/\sqrt{5\cdot7}$
2	1	1	0	$-\sqrt{3}/\sqrt{2\cdot5}$	3	1	2	−2	$1/\sqrt{5\cdot7}$
2	1	1	−1	$\sqrt{3}/\sqrt{2\cdot5}$	3	0	2	1	$-\sqrt{3}/\sqrt{5\cdot7}$
2	0	1	1	$1/\sqrt{2\cdot5}$	3	0	2	0	$3/\sqrt{5\cdot7}$
2	0	1	0	$-\sqrt{2}/\sqrt{5}$	3	3	3	−2	$\sqrt{3}/(2\sqrt{7})$
2	2	2	−1	$1/\sqrt{5}$	3	3	3	−3	$3/(2\sqrt{7})$
2	2	2	−2	$\sqrt{2}/\sqrt{5}$	3	2	3	−1	$-\sqrt{5}/(2\sqrt{7})$
2	1	2	0	$-\sqrt{3}/\sqrt{2\cdot5}$	3	2	3	−2	$-1/\sqrt{7}$
2	1	2	−1	$-1/\sqrt{2\cdot5}$	3	1	3	0	$\sqrt{3}/\sqrt{2\cdot7}$
2	0	2	0	0	3	1	3	−1	$1/(2\sqrt{7})$
5/2	5/2	3/2	−3/2	$1/\sqrt{2}$	3	0	3	0	0
5/2	3/2	3/2	−1/2	$-\sqrt{3}/\sqrt{2\cdot5}$					

Table A.7. *(continued)*

$$Z = (a\alpha, b\beta \mid 2\alpha + \beta)$$

a	α	b	β	Z	a	α	b	β	Z
1	1	1	1	1	5/2	1/2	3/2	1/2	$-5/(2\sqrt{3\cdot7})$
1	1	1	0	$1/\sqrt{2}$	5/2	1/2	3/2	−1/2	$-1/\sqrt{2\cdot7}$
1	1	1	−1	$1/\sqrt{2\cdot3}$	5/2	1/2	3/2	−3/2	$3/2(\sqrt{7})$
1	0	1	0	$\sqrt{2}/\sqrt{3}$	5/2	5/2	5/2	−1/2	$\sqrt{5}/2(\sqrt{7})$
3/2	3/2	1/2	1/2	1	5/2	5/2	5/2	−3/2	$\sqrt{5}/\sqrt{2\cdot7}$
3/2	3/2	1/2	−1/2	1/2	5/2	5/2	5/2	−5/2	$5/(2\sqrt{3\cdot7})$
3/2	1/2	1/2	1/2	$\sqrt{3}/2$	5/2	3/2	5/2	1/2	$-3/(2\sqrt{7})$
3/2	1/2	1/2	−1/2	$1/\sqrt{2}$	5/2	3/2	5/2	−1/2	$-1/\sqrt{7}$
3/2	3/2	3/2	1/2	$1/\sqrt{2}$	5/2	3/2	5/2	−3/2	$1/(2\sqrt{3\cdot7})$
3/2	3/2	3/2	−1/2	$1/\sqrt{2}$	5/2	1/2	5/2	1/2	0
3/2	3/2	3/2	−3/2	1/2	5/2	1/2	5/2	−1/2	$-2/\sqrt{3\cdot7}$
3/2	1/2	3/2	1/2	0	3	3	1	−1	$\sqrt{5}/\sqrt{7}$
3/2	1/2	3/2	−1/2	1/2	3	2	1	0	$-\sqrt{5}/\sqrt{3\cdot7}$
2	2	0	0	1	3	2	1	−1	$\sqrt{2\cdot5}/\sqrt{3\cdot7}$
2	1	0	0	1	3	1	1	1	$1/\sqrt{3\cdot7}$
2	0	0	0	1	3	1	1	0	$-2\sqrt{2}/\sqrt{3\cdot7}$
2	2	1	0	$\sqrt{2}/\sqrt{3}$	3	1	1	−1	$\sqrt{2}/\sqrt{7}$
2	2	1	−1	$1/\sqrt{3}$	3	0	1	1	$1/\sqrt{7}$
2	1	1	1	$-1/\sqrt{3}$	3	0	1	0	$-\sqrt{3}/\sqrt{7}$
2	1	1	0	$1/\sqrt{2\cdot3}$	3	3	2	−1	$\sqrt{5}/\sqrt{2\cdot7}$
2	1	1	−1	$1/\sqrt{2}$	3	3	2	−2	$\sqrt{5}/\sqrt{2\cdot7}$
2	0	1	1	$-1/\sqrt{2}$	3	2	2	0	$-\sqrt{5}/\sqrt{2\cdot7}$
2	0	1	0	0	3	2	2	−1	0
2	2	2	0	$\sqrt{2}/\sqrt{7}$	3	2	2	−2	$\sqrt{5}/\sqrt{2\cdot7}$
2	2	2	−1	$\sqrt{3}/\sqrt{7}$	3	1	2	1	$\sqrt{3}/\sqrt{2\cdot7}$
2	2	2	−2	$\sqrt{2}/\sqrt{7}$	3	1	2	0	$-1/\sqrt{7}$
2	1	2	1	$-\sqrt{3}/\sqrt{7}$	3	1	2	−1	$-1/\sqrt{7}$
2	1	2	0	$-1/\sqrt{2\cdot7}$	3	1	2	−2	$\sqrt{3}/\sqrt{2\cdot7}$
2	1	2	−1	$1/\sqrt{2\cdot7}$	3	0	2	2	$-1/\sqrt{2\cdot7}$
2	0	2	0	$-\sqrt{2}/\sqrt{7}$	3	0	2	1	$\sqrt{2}/\sqrt{7}$
5/2	5/2	1/2	−1/2	$\sqrt{5}/\sqrt{2\cdot3}$	3	0	2	0	0
5/2	3/2	1/2	1/2	$-1/\sqrt{2\cdot3}$	3	3	3	−1	$\sqrt{5}/\sqrt{2\cdot3\cdot7}$
5/2	3/2	1/2	−1/2	$\sqrt{2}/\sqrt{3}$	3	3	3	−2	$5/(2\sqrt{3\cdot7})$
5/2	1/2	1/2	1/2	$-1/\sqrt{3}$	3	3	3	−3	$5/(2\sqrt{3\cdot7})$
5/2	1/2	1/2	−1/2	$1/\sqrt{2}$	3	2	3	0	$-\sqrt{5}/\sqrt{3\cdot7}$
5/2	5/2	3/2	−1/2	$\sqrt{2\cdot5}/\sqrt{3\cdot7}$	3	2	3	−1	$-\sqrt{5}/(2\sqrt{7})$
5/2	5/2	3/2	−3/2	$\sqrt{5}/\sqrt{2\cdot7}$	3	2	3	−2	0
5/2	3/2	3/2	1/2	$-2\sqrt{2}/\sqrt{3\cdot7}$	3	1	3	1	$\sqrt{2}/\sqrt{7}$
5/2	3/2	3/2	−1/2	$1/\sqrt{2\cdot3\cdot7}$	3	1	3	0	$1/\sqrt{2\cdot3\cdot7}$
5/2	3/2	3/2	−3/2	$\sqrt{3}/\sqrt{7}$	3	1	3	−1	$-\sqrt{3}/(2\sqrt{7})$
5/2	1/2	3/2	3/2	$1/\sqrt{7}$	3	0	3	0	$2/\sqrt{3\cdot7}$

Table A.7. *(continued)*

$$Z = (a\alpha, b\beta \mid 3\alpha + \beta)$$

a	α	b	β	Z	a	α	b	β	Z
3/2	3/2	3/2	3/2	1	5/2	3/2	5/2	-1/2	$1/\sqrt{2\cdot3\cdot5}$
3/2	3/2	3/2	1/2	$1/\sqrt{2}$	5/2	3/2	5/2	-3/2	$7/(2\cdot3\sqrt{5})$
3/2	3/2	3/2	-1/2	$1/\sqrt{5}$	5/2	1/2	5/2	1/2	$-2/\sqrt{3\cdot5}$
3/2	3/2	3/2	-3/2	$1/(2\sqrt{5})$	5/2	1/2	5/2	-1/2	$-2/(3\sqrt{5})$
3/2	1/2	3/2	1/2	$\sqrt{3}/\sqrt{5}$	3	3	0	0	1
3/2	1/2	3/2	-1/2	$3/(2\sqrt{5})$	3	2	0	0	1
2	2	1	1	1	3	1	0	0	1
2	2	1	0	$1/\sqrt{3}$	3	0	0	0	1
2	2	1	-1	$1/\sqrt{3\cdot5}$	3	3	1	0	$\sqrt{3}/2$
2	1	1	1	$\sqrt{2}/\sqrt{3}$	3	3	1	-1	1/2
2	1	1	0	$2\sqrt{2}/\sqrt{3\cdot5}$	3	2	1	1	-1/2
2	1	1	-1	$1/\sqrt{5}$	3	2	1	0	$1/\sqrt{3}$
2	0	1	1	$\sqrt{2}/\sqrt{5}$	3	2	1	-1	$\sqrt{5}/(2\sqrt{3})$
2	0	1	0	$\sqrt{3}/\sqrt{5}$	3	1	1	1	$-\sqrt{5}/(2\sqrt{3})$
2	2	2	1	$1/\sqrt{2}$	3	1	1	0	$1/(2\sqrt{3})$
2	2	2	0	$1/\sqrt{2}$	3	1	1	-1	$1/\sqrt{2}$
2	2	2	-1	$\sqrt{3}/\sqrt{2\cdot5}$	3	0	1	1	$-1/\sqrt{2}$
2	2	2	-2	$1/\sqrt{2\cdot5}$	3	0	1	0	0
2	1	2	1	0	3	3	2	0	$\sqrt{5}/(2\sqrt{3})$
2	1	2	0	$1/\sqrt{5}$	3	3	2	-1	$\sqrt{5}/(2\sqrt{3})$
2	1	2	-1	$\sqrt{2}/\sqrt{5}$	3	3	2	-2	$1/\sqrt{2\cdot3}$
2	0	2	0	0	3	2	2	1	$-\sqrt{5}/(2\sqrt{3})$
5/2	5/2	1/2	1/2	1	3	2	2	0	0
5/2	5/2	1/2	-1/2	$1/\sqrt{2\cdot3}$	3	2	2	-1	1/2
5/2	3/2	1/2	1/2	$\sqrt{5}/\sqrt{2\cdot3}$	3	2	2	-2	$1/\sqrt{3}$
5/2	3/2	1/2	-1/2	$1/\sqrt{3}$	3	1	2	2	$1/\sqrt{2\cdot3}$
5/2	1/2	1/2	1/2	$\sqrt{2}/\sqrt{3}$	3	1	2	1	-1/2
5/2	1/2	1/2	-1/2	$1/\sqrt{2}$	3	1	2	0	$-\sqrt{3}/(2\sqrt{5})$
5/2	5/2	3/2	1/2	$\sqrt{5}/(2\sqrt{2})$	3	1	2	-1	$1/\sqrt{2\cdot3\cdot5}$
5/2	5/2	3/2	-1/2	$\sqrt{5}/(2\sqrt{3})$	3	1	2	-2	$\sqrt{2}/\sqrt{5}$
5/2	5/2	3/2	-3/2	$1/(2\sqrt{2})$	3	0	2	2	$1/\sqrt{3}$
5/2	3/2	3/2	3/2	$-\sqrt{3}/(2\sqrt{2})$	3	0	2	1	$-1/\sqrt{2\cdot3\cdot5}$
5/2	3/2	3/2	1/2	$1/(2\sqrt{3})$	3	0	2	0	$-2/\sqrt{3\cdot5}$
5/2	3/2	3/2	-1/2	$7/(2\sqrt{2\cdot3\cdot5})$	3	3	3	0	$1/\sqrt{2\cdot3}$
5/2	3/2	3/2	-3/2	$\sqrt{3}/\sqrt{2\cdot5}$	3	3	3	-1	$1/\sqrt{3}$
5/2	1/2	3/2	3/2	$-1/\sqrt{2}$	3	3	3	-2	$1/\sqrt{3}$
5/2	1/2	3/2	1/2	$-1/(2\sqrt{3\cdot5})$	3	3	3	-3	$1/\sqrt{2\cdot3}$
5/2	1/2	3/2	-1/2	$1/\sqrt{5}$	3	2	3	1	$-1/\sqrt{3}$
5/2	1/2	3/2	-3/2	$3/(2\sqrt{5})$	3	2	3	0	$-1/\sqrt{2\cdot3}$
5/2	5/2	5/2	1/2	$\sqrt{5}/(3\sqrt{2})$	3	2	3	-1	0
5/2	5/2	5/2	-1/2	$\sqrt{5}/(2\sqrt{3})$	3	2	3	-2	$1/\sqrt{2\cdot3}$
5/2	5/2	5/2	-3/2	$1/\sqrt{3}$	3	1	3	1	0
5/2	5/2	5/2	-5/2	$\sqrt{5}/(2\cdot3)$	3	1	3	0	$-1/\sqrt{2\cdot3}$
5/2	3/2	5/2	3/2	$-2/3$	3	1	3	-1	$-1/\sqrt{2\cdot3}$
5/2	3/2	5/2	1/2	$-1/(2\sqrt{3})$	3	0	3	0	0

A.9. $6j$-Symbols

A.9.1.   Recoupling Coefficients

$$\langle (j_1 j_2) j_{12} j_3 jm \,|\, j_1, (j_2 j_3) j_{23} j'm' \rangle$$
$$= \delta_{jj'} \delta_{mm'} (-1)^{j_1+j_2+j_3+j} \hat{j}_{12} \hat{j}_{23} \begin{Bmatrix} j_1 & j_2 & j_{12} \\ j_3 & j & j_{23} \end{Bmatrix} \tag{A.96}$$

$$\langle (j_1 j_2) j_{12} j_3 jm \,|\, (j_1 j_3) j_{13} j_2 j'm' \rangle$$
$$= \delta_{jj'} \delta_{mm'} (-1)^{j_2+j_3+j_{12}+j_{13}} \hat{j}_{12} \hat{j}_{13} \begin{Bmatrix} j_2 & j_1 & j_{12} \\ j_3 & j & j_{13} \end{Bmatrix} \tag{A.97}$$

$$\langle j_1, (j_2 j_3) j_{23} jm \,|\, (j_1 j_3) j_{13} j_2 j'm' \rangle$$
$$= \delta_{jj'} \delta_{mm'} (-1)^{j_1+j+j_{23}} \hat{j}_{13} \hat{j}_{23} \begin{Bmatrix} j_1 & j_3 & j_{13} \\ j_2 & j & j_{23} \end{Bmatrix} \tag{A.98}$$

A.9.2.   Triangle Rules

The $6j$-symbol $\begin{Bmatrix} a & b & c \\ d & e & f \end{Bmatrix}$ vanishes when at least one of the following triangle rules is not satisfied:

$$a+b+c=0, \quad a+e+f=0, \quad d+b+f=0, \quad d+e+c=0 \tag{A.99}$$

A.9.3.   Relation to Other Coefficients

$$W(abed;cf) = (-1)^{a+b+d+e} \begin{Bmatrix} a & b & c \\ d & e & f \end{Bmatrix} \tag{A.100}$$

$$U(abed;cf) = (-1)^{a+b+d+e} \hat{c}\hat{f} \begin{Bmatrix} a & b & c \\ d & e & f \end{Bmatrix} \tag{A.101}$$

A.9.4. Classical Symmetry Properties

The value of 6j-symbol does not change under:

1. arbitrary permutation of columns:

$$\begin{Bmatrix} a\,b\,c \\ d\,e\,f \end{Bmatrix} = \begin{Bmatrix} b\,a\,c \\ e\,d\,f \end{Bmatrix} = \begin{Bmatrix} c\,a\,b \\ f\,d\,e \end{Bmatrix} = \begin{Bmatrix} a\,c\,b \\ d\,f\,e \end{Bmatrix}$$

$$= \begin{Bmatrix} b\,c\,a \\ e\,f\,d \end{Bmatrix} = \begin{Bmatrix} c\,b\,a \\ f\,e\,d \end{Bmatrix} \tag{A.102}$$

2. permutation of two arbitrary elements from the first line with the two corresponding elements from the second line, for example,

$$\begin{Bmatrix} a\,b\,c \\ d\,e\,f \end{Bmatrix} = \begin{Bmatrix} d\,e\,c \\ a\,b\,f \end{Bmatrix} = \begin{Bmatrix} d\,b\,f \\ a\,e\,c \end{Bmatrix} \dots \tag{A.103}$$

A.9.5. 6j-Symbol with Zero Element

$$\begin{Bmatrix} a\,b\,c \\ d\,e\,0 \end{Bmatrix} = \begin{Bmatrix} a\,c\,b \\ d\,0\,e \end{Bmatrix} = \begin{Bmatrix} c\,a\,b \\ 0\,d\,e \end{Bmatrix} = \begin{Bmatrix} 0\,d\,b \\ c\,a\,e \end{Bmatrix}$$

$$= \begin{Bmatrix} d\,0\,b \\ a\,c\,e \end{Bmatrix} = \begin{Bmatrix} d\,b\,0 \\ a\,e\,c \end{Bmatrix} = (-1)^{b+c+e} \frac{1}{\hat{a}\hat{b}} \delta_{bd}\,\delta_{ae} \tag{A.104}$$

A.9.6. 6j-Symbol with Angular Momentum $\frac{1}{2}$

In Eqs. (A.105)–(A.109), $s = a+b+c$.

$$\begin{Bmatrix} a & b & c \\ \frac{1}{2}\,c-\frac{1}{2}\,b+\frac{1}{2} \end{Bmatrix} = (-1)^s \left[\frac{(s-2b)(s-2c+1)}{(2b+1)(2b+2)2c(2c+1)} \right]^{1/2} \tag{A.105}$$

$$\begin{Bmatrix} a & b & c \\ \frac{1}{2}\,c-\frac{1}{2}\,b-\frac{1}{2} \end{Bmatrix} = (-1)^s \left[\frac{(s+1)(s-2a)}{2b(2b+1)2c(2c+1)} \right]^{1/2} \tag{A.106}$$

A.9.7.   6j-Symbol with Angular Momentum 1

$$\begin{Bmatrix} a & b & c \\ 1 & c-1 & b-1 \end{Bmatrix}$$

$$= (-1)^s \left[\frac{s(s+1)(s-2a-1)(s-2a)}{(2b-1)2b(2b+1)(2c-1)2c(2c+1)} \right]^{1/2} \tag{A.107}$$

$$\begin{Bmatrix} a & b & c \\ 1 & c-1 & b \end{Bmatrix}$$

$$= (-1)^s \left[\frac{2(s+1)(s-2a)(s-2b)(s-2c+1)}{2b(2b+1)(2b+2)(2c-1)2c(2c+1)} \right]^{1/2} \tag{A.108}$$

$$\begin{Bmatrix} a & b & c \\ 1 & c-1 & b+1 \end{Bmatrix}$$

$$= (-1)^s \left[\frac{(s-2b)(s-2b-1)(s-2c+1)(s-2c+2)}{(2b+1)(2b+2)(2b+3)(2c-1)2c(2c+1)} \right]^{1/2} \tag{A.109}$$

$$\begin{Bmatrix} a & b & c \\ 1 & c & b \end{Bmatrix} = (-1)^s \frac{2[a(a+1)-b(b+1)-c(c+1)]}{[2b(2b+1)(2b+2)2c(2c+1)(2c+2)]^{1/2}} \tag{A.110}$$

A.9.8.   6j-Symbol with Two Pairs of Equal Momenta

$$\begin{Bmatrix} a & a & c \\ b & b & f \end{Bmatrix} = (-1)^{a+b+c+f} \left[\frac{(2a-c)!(2b-c)!}{(2a+c+1)!(2b+c+1)!} \right]^{1/2} V_c(a,f,b) \tag{A.111}$$

where $V_c(a,f,b)$ is a polynomial of the degree c in the variable $x = f(f+1) - a(a+1) - b(b+1)$. V_c can be found by the recurrence relation

$$V_{c+1} = \frac{2c+1}{c+1} V_1 V_c - c(2c+1)V_c$$

$$- \frac{c}{c+1} [4a(a+1)+1-c^2][4b(b+1)+1-c^2]V_{c-1} \tag{A.112}$$

First polynomials:

$$V_0(a,f,b) = 1, \quad V_1(a,f,b) = -2x,$$

$$V_2(a,f,b) = 6x^2 + 6x - 8ab(a+1)(b+1) \tag{A.113}$$

Some particular values of f:

$$V_c(a, a-b, b) = \frac{(2b)!\,(2a+c+1)!}{(2a+1)!\,(2b-c)!}, \quad (a \geq b) \tag{A.114}$$

$$V_c(a, b-a, b) = \frac{(2a)!\,(2b+c+1)!}{(2b+1)!\,(2a-c)!}, \quad (a \leq b) \tag{A.115}$$

$$V_c(a, a+b, b) = (-1)^c \frac{(2a)!\,(2b)!}{(2a-c)!\,(2b-c)!} \tag{A.116}$$

A.9.9. Formulas of Summation

The formulas of summation, including the condition of unitarity, are given in Section A.11.

A.9.10. Tables

We give $6j$-symbols $\begin{Bmatrix} a & b & c \\ d & e & f \end{Bmatrix}$ in terms of simple fractions for all indices up to 3 in Table A.8 on pp. 226–228. The $6j$-symbols are collected into the following groups: $6j$-symbols with all integer indices, $6j$-symbols with three integer indices, and $6j$-symbols with two integer indices. $6j$-symbols with zeroes are not presented because they can be found easily from Eq. (A.104). To reduce the $6j$-symbols with all indices not greater than 3 to those included in the tables, one may use the symmetry properties (A.102) and (A.103).

A.10. $9j$-Symbols

A.10.1. Recoupling Coefficients

$$\langle (j_1 j_2) j_{12} (j_3 j_4) j_{34} jm \,|\, (j_1 j_3) j_{13} (j_2 j_4) j_{24} j'm' \rangle$$

$$= \delta_{jj'} \delta_{mm'} \, \hat{j}_{12} \hat{j}_{13} \hat{j}_{24} \hat{j}_{34} \begin{Bmatrix} j_1 & j_2 & j_{12} \\ j_3 & j_4 & j_{34} \\ j_{13} & j_{24} & j \end{Bmatrix} \tag{A.117}$$

$$\langle (j_1 j_2) j_{12} (j_3 j_4) j_{34} jm \,|\, (j_1 j_4) j_{14} (j_2 j_3) j_{23} j' m' \rangle$$

$$= \delta_{jj'} \,\delta_{mm'} \,(-1)^{j_3 + j_4 - j_{34}} \,\hat{j}_{12} \hat{j}_{14} \hat{j}_{23} \hat{j}_{34} \begin{Bmatrix} j_1 & j_2 & j_{12} \\ j_4 & j_3 & j_{34} \\ j_{14} & j_{23} & j \end{Bmatrix} \qquad (\text{A.118})$$

$$\langle (j_1 j_3) j_{13} (j_2 j_4) j_{24} jm \,|\, (j_1 j_4) j_{14} (j_2 j_3) j_{23} j' m' \rangle$$

$$= \delta_{jj'} \,\delta_{mm'} \,(-1)^{j_3 - j_4 - j_{23} - j_{24}} \,\hat{j}_{13} \hat{j}_{14} \hat{j}_{24} \hat{j}_{23} \begin{Bmatrix} j_1 & j_3 & j_{13} \\ j_4 & j_2 & j_{24} \\ j_{14} & j_{23} & j \end{Bmatrix} \qquad (\text{A.119})$$

A.10.2. Triangle Rules

The $9j$-symbol $\begin{Bmatrix} a & b & c \\ d & e & f \\ g & h & j \end{Bmatrix}$ vanishes if at least one of the following triangle rules is not satisfied:

$$\mathbf{a} + \mathbf{b} + \mathbf{c} = 0, \ \mathbf{d} + \mathbf{e} + \mathbf{f} = 0, \ \mathbf{g} + \mathbf{h} + \mathbf{j} = 0$$
$$\mathbf{a} + \mathbf{d} + \mathbf{g} = 0, \ \mathbf{b} + \mathbf{e} + \mathbf{h} = 0, \ \mathbf{c} + \mathbf{f} + \mathbf{j} = 0 \qquad (\text{A.120})$$

A rule to memorize the triangles is: three elements from either one line or one column.

A.10.3. Symmetry Properties

The value of the $9j$-symbol does not change under cyclic permutation of lines, cyclic permutation of columns, or transposing against both main diagonals. Under noncyclic permutations of lines and columns, the $9j$-symbol is multiplied by the phase factor $(-1)^S$, where S is the sum of all nine elements of the $9j$-symbol. (The cyclic permutation consists of even numbers of permutations of adjacent lines or columns, while the noncyclic permutation consists of odd numbers of such permutations.) An important property follows from the permutation symmetry: a $9j$-symbol with two equal lines or two equal columns is zero if the sum of the three remaining elements is odd:

$$\begin{Bmatrix} a & b & c \\ a & b & c \\ g & h & j \end{Bmatrix} = 0 \quad (g + h + j = 2k + 1); \qquad \begin{Bmatrix} a & a & c \\ d & d & f \\ g & g & j \end{Bmatrix} = 0 \quad (c + f + j = 2k + 1)$$

$$k = 0, 1, \dots \qquad (\text{A.121})$$

Table A.8. $6j$-Symbols

a	b	c	d	e	f	$\left\{\begin{matrix} a\,b\,c \\ d\,e\,f \end{matrix}\right\}$	a	b	c	d	e	f	$\left\{\begin{matrix} a\,b\,c \\ d\,e\,f \end{matrix}\right\}$
1	1	1	1	1	1	$1/2\cdot3$	3	3	1	2	1	2	$-1/\sqrt{3\cdot5\cdot7}$
2	1	1	1	1	1	$1/2\cdot3$	3	3	1	2	2	1	$-2\sqrt{2}/3\sqrt{5\cdot7}$
2	1	1	2	1	1	$1/2\cdot3\cdot5$	3	3	1	2	2	2	$1/\sqrt{2\cdot5\cdot7}$
2	2	1	1	1	1	$-1/2\sqrt{5}$	3	3	1	3	2	1	$-1/3\cdot7$
2	2	1	1	1	2	$1/2\cdot3\sqrt{5}$	3	3	1	3	3	1	$11/2\cdot2\cdot3\cdot7$
2	2	1	2	1	1	$-1/2\cdot5$	3	3	2	1	1	2	$\sqrt{2}/5\sqrt{7}$
2	2	1	2	2	1	$1/2\cdot3$	3	3	2	1	1	3	$1/2\sqrt{2\cdot7}$
2	2	2	1	1	1	$\sqrt{7}/2\cdot5\sqrt{3}$	3	3	2	2	1	2	$\sqrt{3}/5\sqrt{7}$
2	2	2	2	1	1	$\sqrt{7}/2\cdot5\sqrt{3}$	3	3	2	2	1	3	$-1/2\sqrt{2\cdot5\cdot7}$
2	2	2	2	2	1	$-1/2\cdot5$	3	3	2	2	2	1	$2\sqrt{2\cdot3}/5\cdot7$
2	2	2	2	2	2	$-3/2\cdot5\cdot7$	3	3	2	2	2	2	$-\sqrt{3}/5\cdot7\sqrt{2}$
3	2	1	1	1	2	$1/3\sqrt{5}$	3	3	2	2	2	3	$-11/2\cdot5\cdot7\sqrt{2\cdot3}$
3	2	1	1	2	1	$1/5$	3	3	2	3	1	2	$2/5\cdot7$
3	2	1	2	1	2	$1/5\sqrt{3\cdot7}$	3	3	2	3	2	1	$\sqrt{3}/7\sqrt{2\cdot5}$
3	2	1	2	2	1	$1/3\cdot5$	3	3	2	3	2	2	$-1/2\cdot5$
3	2	1	3	2	1	$1/3\cdot5\cdot7$	3	3	2	3	3	1	$-3/2\cdot2\cdot7$
3	2	2	1	2	1	$-\sqrt{2}/5\sqrt{3}$	3	3	2	3	3	2	$19/2\cdot2\cdot3\cdot5\cdot7$
3	2	2	1	2	2	0	3	3	3	2	2	1	$-\sqrt{3}/7\sqrt{5}$
3	2	2	2	2	1	$-\sqrt{2}/5\sqrt{7}$	3	3	3	2	2	2	$-\sqrt{3}/2\cdot7\sqrt{5}$
3	2	2	2	2	2	$2\cdot2/5\cdot7$	3	3	3	3	2	1	$-1/7\sqrt{2}$
3	2	2	3	2	1	$-1/5\cdot7$	3	3	3	3	2	2	$2/7\sqrt{3\cdot5}$
3	2	2	3	2	2	$1/2\cdot7$	3	3	3	3	3	1	$1/2\cdot7$
3	3	1	1	1	2	$-\sqrt{2}/3\sqrt{7}$	3	3	3	3	3	2	$1/2\cdot3\cdot7$
3	3	1	1	1	3	$1/2\cdot3\sqrt{2\cdot7}$	3	3	3	3	3	3	$-1/2\cdot7$

A.10.4. $9j$-Symbol with Zero Element

$$\left\{\begin{matrix} 0\,c\,c \\ g\,e\,b \\ g\,d\,a \end{matrix}\right\} = \left\{\begin{matrix} c\,0\,c \\ d\,g\,a \\ e\,g\,b \end{matrix}\right\} = \left\{\begin{matrix} g\,g\,0 \\ e\,d\,c \\ b\,a\,c \end{matrix}\right\} = \left\{\begin{matrix} g\,b\,e \\ 0\,c\,c \\ g\,a\,d \end{matrix}\right\} = \left\{\begin{matrix} a\,g\,d \\ c\,0\,c \\ b\,g\,e \end{matrix}\right\} = \left\{\begin{matrix} b\,a\,c \\ g\,g\,0 \\ e\,d\,c \end{matrix}\right\}$$

$$= \left\{\begin{matrix} c\,e\,d \\ c\,b\,a \\ 0\,g\,g \end{matrix}\right\} = \left\{\begin{matrix} d\,c\,e \\ a\,c\,b \\ g\,0\,g \end{matrix}\right\} = \left\{\begin{matrix} a\,b\,c \\ d\,e\,c \\ g\,g\,0 \end{matrix}\right\} = (-1)^{b+d+c+g}\,\frac{1}{\hat{c}\hat{g}}\left\{\begin{matrix} a\,b\,c \\ e\,d\,g \end{matrix}\right\} \qquad \text{(A.122)}$$

A case of a $9j$-symbol with two zero elements:

$$\left\{\begin{matrix} a\,b\,c \\ d\,0\,f \\ g\,h\,0 \end{matrix}\right\} = \delta_{df}\,\delta_{bh}\,\delta_{cf}\,\delta_{gh}\,(-1)^{a-b-c}\,\frac{1}{(\hat{b}\hat{c})^2} \qquad \text{(A.123)}$$

Table A.8. (*continued*)

a	b	c	d	e	f	$\begin{Bmatrix} a\,b\,c \\ d\,e\,f \end{Bmatrix}$
1	1	1	1/2	1/2	1/2	$-1/3$
1	1	1	3/2	1/2	1/2	$-1/2\cdot3$
1	1	1	3/2	3/2	1/2	$\sqrt5/2\cdot3\sqrt2$
1	1	1	3/2	3/2	3/2	$-1/3\sqrt{2\cdot5}$
1	1	1	5/2	3/2	3/2	$-1/2\sqrt{2\cdot5}$
1	1	1	5/2	5/2	3/2	$\sqrt7/2\cdot3\sqrt5$
1	1	1	5/2	5/2	5/2	$-1/3\sqrt{5\cdot7}$
2	1	1	1/2	3/2	1/2	$1/2\sqrt3$
2	1	1	1/2	3/2	3/2	$-1/2\sqrt{2\cdot3}$
2	1	1	3/2	3/2	1/2	$1/2\sqrt{2\cdot3\cdot5}$
2	1	1	3/2	3/2	3/2	$-\sqrt2/5\sqrt3$
2	1	1	3/2	5/2	1/2	$-1/2\sqrt5$
2	1	1	3/2	5/2	3/2	$\sqrt7/2\cdot5\sqrt2$
2	1	1	3/2	5/2	5/2	$-\sqrt7/3\cdot5\sqrt2$
2	1	1	5/2	3/2	3/2	$-1/2\cdot5\sqrt{2\cdot3}$
2	1	1	5/2	5/2	3/2	$1/2\cdot5$
2	1	1	5/2	5/2	5/2	$-2\cdot2\sqrt2/3\cdot5\sqrt7$
2	2	1	1/2	1/2	3/2	$-1/2\sqrt5$
2	2	1	1/2	1/2	5/2	$-1/3\sqrt5$
2	2	1	3/2	1/2	3/2	$-1/2\sqrt{2\cdot5}$
2	2	1	3/2	1/2	5/2	$\sqrt7/2\cdot3\sqrt5$
2	2	1	3/2	3/2	1/2	$-3/2\cdot5\sqrt2$
2	2	1	3/2	3/2	3/2	$1/5\sqrt2$
2	2	1	3/2	3/2	5/2	$-1/2\cdot3\cdot5\sqrt2$
2	2	1	5/2	3/2	1/2	$-1/2\cdot5\sqrt3$
2	2	1	5/2	3/2	3/2	$\sqrt7/2\cdot5\sqrt{2\cdot3}$
2	2	1	5/2	3/2	5/2	$-1/5\sqrt2$
2	2	1	5/2	5/2	1/2	$\sqrt7/3\cdot5$
2	2	1	5/2	5/2	3/2	$-11/2\cdot3\cdot5\sqrt7$
2	2	1	5/2	5/2	5/2	$1/5\sqrt7$
2	2	2	3/2	3/2	1/2	$\sqrt7/2\cdot5\sqrt2$
2	2	2	3/2	3/2	3/2	0
2	2	2	5/2	3/2	1/2	$1/2\cdot5$
2	2	2	5/2	3/2	3/2	$-1/2\sqrt{2\cdot7}$
2	2	2	5/2	5/2	1/2	$-\sqrt2/5\sqrt3$
2	2	2	5/2	5/2	3/2	$1/7\sqrt{2\cdot3}$
2	2	2	5/2	5/2	5/2	$1/7\sqrt{2\cdot3}$
3	2	1	1/2	1/2	5/2	$1/3\sqrt2$
3	2	1	1/2	3/2	3/2	$1/2\sqrt5$
3	2	1	1/2	3/2	5/2	$-1/3\sqrt5$
3	2	1	3/2	1/2	5/2	$1/3\sqrt{2\cdot7}$
3	2	1	3/2	3/2	3/2	$1/2\cdot5$
3	2	1	3/2	3/2	5/2	$-2\cdot2\sqrt2/3\cdot5\sqrt7$
3	2	1	3/2	5/2	1/2	$1/\sqrt{2\cdot3\cdot5}$
3	2	1	3/2	5/2	3/2	$-\sqrt2/5\sqrt3$
3	2	1	3/2	5/2	5/2	$\sqrt3/5\sqrt7$
3	2	1	5/2	3/2	3/2	$1/2\cdot5\sqrt{3\cdot7}$
3	2	1	5/2	3/2	5/2	$-1/5\sqrt{2\cdot7}$
3	2	1	5/2	5/2	1/2	$1/3\sqrt{2\cdot5\cdot7}$
3	2	1	5/2	5/2	3/2	$-2\cdot2\cdot2/3\cdot5\cdot7$
3	2	1	5/2	5/2	5/2	$3\sqrt3/5\cdot7\sqrt2$
3	2	2	1/2	3/2	3/2	$-1/5\sqrt2$
3	2	2	1/2	5/2	3/2	$-1/5\sqrt2$
3	2	2	1/2	5/2	5/2	$1/5\sqrt2$
3	2	2	3/2	3/2	3/2	$-1/5\sqrt2$
3	2	2	3/2	5/2	1/2	$-1/\sqrt{5\cdot7}$
3	2	2	3/2	5/2	3/2	$1/5\sqrt7$
3	2	2	3/2	5/2	5/2	$\sqrt2/5\cdot7$
3	2	2	5/2	3/2	3/2	$-3/5\cdot7\sqrt2$
3	2	2	5/2	5/2	1/2	$-1/\sqrt{2\cdot3\cdot5\cdot7}$
3	2	2	5/2	5/2	3/2	$13/2\cdot5\cdot7\sqrt3$
3	2	2	5/2	5/2	5/2	$-3\cdot3/2\cdot5\cdot7\sqrt2$
3	3	1	1/2	1/2	5/2	$-\sqrt2/3\sqrt7$
3	3	1	3/2	1/2	5/2	$-\sqrt5/2\cdot3\sqrt7$
3	3	1	3/2	3/2	3/2	$-1/\sqrt{5\cdot7}$
3	3	1	3/2	3/2	5/2	$\sqrt7/2\cdot2\cdot3\sqrt5$
3	3	1	5/2	3/2	3/2	$-1/\sqrt{2\cdot3\cdot5\cdot7}$
3	3	1	5/2	3/2	5/2	$\sqrt3/2\sqrt{2\cdot5\cdot7}$
3	3	1	5/2	5/2	1/2	$-\sqrt{2\cdot5}/3\cdot7$
3	3	1	5/2	5/2	3/2	$17/2\cdot3\cdot7\sqrt{2\cdot5}$
3	3	1	5/2	5/2	5/2	$-\sqrt2/7\sqrt5$
3	3	2	3/2	1/2	5/2	$\sqrt3/2\sqrt{5\cdot7}$
3	3	2	3/2	3/2	3/2	$\sqrt3/5\sqrt7$
3	3	2	3/2	3/2	5/2	$\sqrt3/2\cdot2\cdot5\sqrt7$
3	3	2	5/2	1/2	5/2	$1/\sqrt{3\cdot5\cdot7}$
3	3	2	5/2	3/2	3/2	$3\sqrt3/5\cdot7\sqrt2$
3	3	2	5/2	3/2	5/2	$-17/2\cdot5\cdot7\sqrt{2\cdot3}$
3	3	2	5/2	5/2	1/2	$1/7$
3	3	2	5/2	5/2	3/2	$-11/2\cdot2\cdot5\cdot7$
3	3	2	5/2	5/2	5/2	$-1/2\cdot3\cdot5\cdot7$
3	3	3	3/2	3/2	3/2	$-\sqrt3/7\sqrt{2\cdot5}$
3	3	3	5/2	3/2	3/2	$-3\sqrt3/2\cdot7\sqrt{2\cdot5}$
3	3	3	5/2	5/2	1/2	$-\sqrt5/7\sqrt{2\cdot3}$
3	3	3	5/2	5/2	3/2	$1/2\cdot7\sqrt{2\cdot3\cdot5}$
3	3	3	5/2	5/2	5/2	$17/2\cdot3\cdot7\sqrt{2\cdot3\cdot5}$

Table A.8. *(continued)*

a	b	c	d	e	f	$\begin{Bmatrix} a\,b\,c \\ d\,e\,f \end{Bmatrix}$	a	b	c	d	e	f	$\begin{Bmatrix} a\,b\,c \\ d\,e\,f \end{Bmatrix}$
1/2	1/2	1	1/2	1/2	1	$1/2\cdot3$	3/2	1/2	2	3/2	1/2	2	$1/2\cdot2\cdot5$
3/2	1/2	1	1/2	1/2	1	$-1/3$	3/2	3/2	2	3/2	1/2	2	$-1/2\cdot5$
3/2	1/2	1	3/2	1/2	1	$-1/2\cdot2\cdot3$	3/2	3/2	2	3/2	3/2	2	$3/2\cdot2\cdot5$
3/2	3/2	1	1/2	1/2	1	$\sqrt{5}/2\cdot3\sqrt{2}$	5/2	1/2	2	3/2	1/2	2	$-1/5$
3/2	3/2	1	3/2	1/2	1	$1/2\cdot3$	5/2	1/2	2	5/2	1/2	2	$-1/2\cdot3\cdot5$
3/2	3/2	1	3/2	3/2	1	$-11/2\cdot2\cdot3\cdot5$	5/2	3/2	2	1/2	3/2	2	$3/2\cdot2\cdot5$
5/2	3/2	1	1/2	3/2	1	$-1/2\cdot2$	5/2	3/2	2	3/2	1/2	2	$\sqrt{7}/2\cdot5\sqrt{2}$
5/2	3/2	1	3/2	3/2	1	$-1/2\cdot5$	5/2	3/2	2	3/2	3/2	2	$-1/2\cdot5$
5/2	3/2	1	5/2	3/2	1	$-1/2\cdot2\cdot3\cdot5$	5/2	3/2	2	5/2	1/2	2	$1/3\cdot5$
5/2	5/2	1	3/2	3/2	1	$\sqrt{7}/2\cdot5\sqrt{2}$	5/2	3/2	2	5/2	3/2	2	$-47/2\cdot2\cdot3\cdot5\cdot7$
5/2	5/2	1	5/2	3/2	1	$1/3\cdot5$	5/2	5/2	2	3/2	1/2	2	$-\sqrt{2}/5\sqrt{3}$
5/2	5/2	1	5/2	5/2	1	$-31/2\cdot3\cdot5\cdot7$	5/2	5/2	2	3/2	3/2	2	$-1/2\cdot5\sqrt{3\cdot7}$
3/2	1/2	2	1/2	3/2	1	$1/2\sqrt{2\cdot5}$	5/2	5/2	2	5/2	1/2	2	$-1/2\cdot5$
3/2	1/2	2	3/2	1/2	1	$1/2\cdot2$	5/2	5/2	2	5/2	3/2	2	$2\cdot2/5\cdot7$
3/2	3/2	2	3/2	1/2	1	$-1/2\sqrt{5}$	5/2	5/2	2	5/2	5/2	2	$-1/2\cdot2\cdot3\cdot5$
3/2	3/2	2	3/2	3/2	1	$-1/2\cdot2\cdot5$	3/2	3/2	3	3/2	3/2	2	$1/2\cdot2\cdot5$
5/2	1/2	2	1/2	3/2	1	$-1/\sqrt{3\cdot5}$	5/2	1/2	3	3/2	3/2	2	$1/2\sqrt{2\cdot5\cdot7}$
5/2	1/2	2	1/2	5/2	1	$\sqrt{7}/2\cdot3\sqrt{5}$	5/2	1/2	3	5/2	1/2	2	$1/2\cdot3$
5/2	1/2	2	3/2	3/2	1	$-1/2\sqrt{2\cdot3\cdot5}$	5/2	3/2	3	1/2	5/2	2	$-1/\sqrt{3\cdot5\cdot7}$
5/2	3/2	2	1/2	3/2	1	$\sqrt{7}/2\cdot2\sqrt{3\cdot5}$	5/2	3/2	3	3/2	3/2	2	$-\sqrt{2}/5\sqrt{7}$
5/2	3/2	2	1/2	5/2	1	$1/3\sqrt{5}$	5/2	3/2	3	3/2	5/2	2	$11/2\cdot2\cdot5\sqrt{3\cdot7}$
5/2	3/2	2	3/2	3/2	1	$\sqrt{7}/2\cdot5\sqrt{3}$	5/2	3/2	3	5/2	1/2	2	$-2\sqrt{2}/3\sqrt{5\cdot7}$
5/2	3/2	2	3/2	5/2	1	$-13/2\cdot3\cdot5\sqrt{2\cdot7}$	5/2	3/2	3	5/2	3/2	2	$23/2\cdot3\cdot5\cdot7$
5/2	3/2	2	5/2	3/2	1	$1/2\cdot2\cdot5$	5/2	5/2	3	3/2	3/2	2	$3\cdot3/2\cdot5\cdot7$
5/2	5/2	2	3/2	3/2	1	$-\sqrt{7}/2\cdot5\sqrt{3}$	5/2	5/2	3	5/2	1/2	2	$\sqrt{3}/2\sqrt{5\cdot7}$
5/2	5/2	2	5/2	3/2	1	$-\sqrt{2}/5\sqrt{7}$	5/2	5/2	3	5/2	3/2	2	$-\sqrt{3}/5\cdot7\sqrt{2}$
5/2	5/2	2	5/2	5/2	1	$23/2\cdot3\cdot5\cdot7$	5/2	5/2	3	5/2	5/2	2	$-29/2\cdot2\cdot3\cdot5\cdot7$
3/2	3/2	3	3/2	3/2	1	$3/2\cdot2\cdot5$	3/2	3/2	3	3/2	3/2	3	$1/2\cdot2\cdot5\cdot7$
5/2	1/2	3	1/2	5/2	1	$\sqrt{5}/2\cdot3\sqrt{7}$	5/2	1/2	3	5/2	1/2	3	$1/2\cdot3\cdot7$
5/2	1/2	3	3/2	3/2	1	$1/2\sqrt{2\cdot3}$	5/2	3/2	3	3/2	3/2	3	$-1/5\cdot7$
5/2	3/2	3	1/2	5/2	1	$-\sqrt{2}/3\sqrt{7}$	5/2	3/2	3	5/2	1/2	3	$-1/3\cdot7$
5/2	3/2	3	3/2	3/2	1	$-\sqrt{2}/5\sqrt{3}$	5/2	3/2	3	5/2	3/2	3	$71/2\cdot2\cdot2\cdot3\cdot5\cdot7$
5/2	3/2	3	3/2	5/2	1	$1/2\cdot3\cdot5\sqrt{2\cdot7}$	5/2	5/2	3	3/2	3/2	3	$3\cdot3/2\cdot2\cdot5\cdot7$
5/2	3/2	3	5/2	3/2	1	$-1/2\cdot5$	5/2	5/2	3	5/2	1/2	3	$1/2\cdot7$
5/2	5/2	3	3/2	3/2	1	$1/2\cdot5$	5/2	5/2	3	5/2	3/2	3	$-1/2\cdot5$
5/2	5/2	3	5/2	3/2	1	$\sqrt{3}/5\sqrt{7}$	5/2	5/2	3	5/2	5/2	3	$79/2\cdot2\cdot3\cdot5\cdot7$
5/2	5/2	3	5/2	5/2	1	$-11/2\cdot3\cdot5\cdot7$							

A case of $9j$-symbols with three zero elements:

$$\begin{Bmatrix} a\,b\,c \\ d\,e\,f \\ 0\,0\,0 \end{Bmatrix} = \delta_{ad}\,\delta_{be}\,\delta_{cf}\,\frac{1}{\hat{a}\hat{b}\hat{c}},$$

$$\begin{Bmatrix} 0\,b\,c \\ d\,0\,f \\ g\,h\,0 \end{Bmatrix} = \delta_{bc}\,\delta_{bd}\,\delta_{bf}\,\delta_{bg}\,\delta_{bh}\,(-1)^{2b}\,\frac{1}{\hat{b}^4} \tag{A.124}$$

A.10.5. Formulas of Summation

The formulas of summation, including the condition of unitarity, are given in Section A.11.

A.11. Sums of Products of the 3nj-Symbols

In this section we use the $3j$-symbol $\{abc\}$ defined by

$$\{abc\} = \begin{cases} 1 \text{ if } a+b+c \text{ is an integer and } |a-b| \le c \le a+b \\ 0 \text{ otherwise} \end{cases}$$

A.11.1. One 3nj-Symbol

$$\sum_X \hat{X}^2\{abX\} = (\hat{a}\hat{b})^2 \tag{A.125}$$

$$\sum_X (-1)^{2a+X}\hat{X}^2\{aaX\} = (\hat{a})^2 \tag{A.126}$$

$$\sum_X \hat{X}^2 \begin{Bmatrix} a\,b\,X \\ a\,b\,c \end{Bmatrix} = (-1)^{2c}\{abc\} \tag{A.127}$$

$$\sum_X (-1)^{a+b+X}\hat{X}^2 \begin{Bmatrix} a\,b\,X \\ b\,a\,c \end{Bmatrix} = \hat{a}\hat{b}\,\delta_{c0} \tag{A.128}$$

$$\sum_X \hat{X}^2 \left\{ \begin{matrix} a\,b\,e \\ c\,d\,f \\ e\,f\,X \end{matrix} \right\} = (\hat{b})^{-2}\{abe\}\{bdf\}\delta_{bc} \tag{A.129}$$

$$\sum_X (-1)^{a+b+c+d-X} \hat{X}^2 \left\{ \begin{matrix} a\,b\,e \\ c\,d\,f \\ f\,e\,X \end{matrix} \right\} = (\hat{a})^{-2}\{dbe\}\{acf\}\delta_{ad} \tag{A.130}$$

A.11.2. Two 3nj-Symbols

$$\sum_X \hat{X}^2 \left\{ \begin{matrix} a\,b\,X \\ c\,d\,p \end{matrix} \right\} \left\{ \begin{matrix} c\,d\,X \\ a\,b\,q \end{matrix} \right\} = (\hat{p})^{-2}\{adp\}\{bcp\}\delta_{pq} \tag{A.131}$$

$$\sum_X (-1)^{p+q+X} \hat{X}^2 \left\{ \begin{matrix} a\,b\,X \\ c\,d\,p \end{matrix} \right\} \left\{ \begin{matrix} c\,d\,X \\ b\,a\,q \end{matrix} \right\} = \left\{ \begin{matrix} c\,a\,q \\ d\,b\,p \end{matrix} \right\} \tag{A.132}$$

$$\sum_X \hat{X}^2 \left\{ \begin{matrix} a\,f\,X \\ d\,q\,e \\ p\,c\,b \end{matrix} \right\} \left\{ \begin{matrix} a\,f\,X \\ e\,b\,s \end{matrix} \right\} = (-1)^{2s} \left\{ \begin{matrix} a\,b\,s \\ c\,d\,p \end{matrix} \right\} \left\{ \begin{matrix} c\,d\,s \\ e\,f\,q \end{matrix} \right\} \tag{A.133}$$

$$\sum_X (-1)^{a+b+c+d+e+f+p+q+X} \hat{X}^2 \left\{ \begin{matrix} a\,f\,X \\ d\,q\,e \\ p\,c\,b \end{matrix} \right\} \left\{ \begin{matrix} a\,f\,X \\ b\,e\,s \end{matrix} \right\}$$
$$= (-1)^{2s} \left\{ \begin{matrix} p\,q\,s \\ e\,a\,d \end{matrix} \right\} \left\{ \begin{matrix} p\,q\,s \\ f\,b\,c \end{matrix} \right\} \tag{A.134}$$

$$\sum_{XY} (\hat{X}\hat{Y})^2 \left\{ \begin{matrix} a\,b\,X \\ c\,d\,Y \\ e\,f\,j \end{matrix} \right\} \left\{ \begin{matrix} a\,b\,X \\ c\,d\,Y \\ g\,h\,j \end{matrix} \right\} = (\hat{e}\hat{f})^{-2}\delta_{eg}\delta_{fh}\{ace\}\{bdh\}\{gfj\} \tag{A.135}$$

$$\sum_{XY} (-1)^Y (\hat{X}\hat{Y})^2 \left\{ \begin{matrix} a\,b\,X \\ c\,d\,Y \\ e\,f\,j \end{matrix} \right\} \left\{ \begin{matrix} a\,b\,X \\ d\,c\,Y \\ g\,h\,j \end{matrix} \right\} = (-1)^{2b+f+h} \left\{ \begin{matrix} a\,d\,g \\ c\,b\,h \\ e\,f\,j \end{matrix} \right\} \tag{A.136}$$

$$\sum_{XYZ} (\hat{X}\hat{Y}\hat{Z})^2 \begin{Bmatrix} X & Y & Z \\ a & b & c \end{Bmatrix}^2 = (\hat{a}\hat{b}\hat{c})^2 \tag{A.137}$$

$$\sum_{XYZ} (\hat{X}\hat{Y}\hat{Z})^2 \begin{Bmatrix} X & Y & Z \\ a & b & c \\ b & c & a \end{Bmatrix} \begin{Bmatrix} X & Y & Z \\ c & a & b \end{Bmatrix} = \{abc\} \tag{A.138}$$

$$\sum_{XYZ} (\hat{X}\hat{Y}\hat{Z})^2 \begin{Bmatrix} X & Y & Z \\ a & b & c \\ d & e & f \end{Bmatrix}^2 = \{abc\}\{def\} \tag{A.139}$$

A.11.3.   Three 3nj-Symbols

$$\sum_X (-1)^{a+b+c+d+e+f+p+q+r+X}\hat{X}^2 \begin{Bmatrix} a & b & X \\ c & d & p \end{Bmatrix} \begin{Bmatrix} c & d & X \\ e & f & q \end{Bmatrix} \begin{Bmatrix} e & f & X \\ b & a & r \end{Bmatrix}$$
$$= \begin{Bmatrix} p & q & r \\ e & a & d \end{Bmatrix} \begin{Bmatrix} p & q & r \\ f & b & c \end{Bmatrix} \tag{A.140}$$

$$\sum_X (-1)^{2X}\hat{X}^2 \begin{Bmatrix} a & b & X \\ c & d & p \end{Bmatrix} \begin{Bmatrix} c & d & X \\ e & f & q \end{Bmatrix} \begin{Bmatrix} e & f & X \\ a & b & r \end{Bmatrix} = \begin{Bmatrix} a & f & r \\ d & q & e \\ p & c & b \end{Bmatrix} \tag{A.141}$$

$$\sum_{XY} (\hat{X}\hat{Y})^2 \begin{Bmatrix} a & b & X \\ c & d & p \end{Bmatrix} \begin{Bmatrix} c & d & X \\ a & b & Y \end{Bmatrix} \begin{Bmatrix} a & b & q \\ c & d & Y \end{Bmatrix} = \begin{Bmatrix} a & b & q \\ c & d & p \end{Bmatrix} \tag{A.142}$$

$$\sum_{XY} (-1)^{X+Y+p}(\hat{X}\hat{Y})^2 \begin{Bmatrix} a & b & X \\ c & d & p \end{Bmatrix} \begin{Bmatrix} a & c & Y \\ d & b & X \end{Bmatrix} \begin{Bmatrix} a & d & q \\ b & c & Y \end{Bmatrix} = (\hat{p})^{-2}\delta_{pq}\{adp\}\{bcp\} \tag{A.143}$$

$$\sum_{XY} (-1)^{X+Y+2a-s-t}(\hat{X}\hat{Y})^2 \begin{Bmatrix} a & b & p \\ c & d & X \\ q & Y & g \end{Bmatrix} \begin{Bmatrix} c & d & X \\ p & g & s \end{Bmatrix} \begin{Bmatrix} b & d & Y \\ q & g & t \end{Bmatrix} = \begin{Bmatrix} a & b & p \\ c & g & s \\ q & t & d \end{Bmatrix} \tag{A.144}$$

$$\sum_{XY}(\hat{X}\hat{Y})^2 \begin{Bmatrix} a\,b\,p \\ c\,d\,X \\ q\,Y\,g \end{Bmatrix} \begin{Bmatrix} c\,d\,X \\ g\,p\,s \end{Bmatrix} \begin{Bmatrix} b\,d\,Y \\ g\,q\,t \end{Bmatrix} = (-1)^{2s}(\hat{s})^{-2}\delta_{st}\{dgs\}\begin{Bmatrix} a\,b\,p \\ s\,c\,q \end{Bmatrix}$$

(A.145)

$$\sum_{XY}(\hat{X}\hat{Y})^2 \begin{Bmatrix} a\,b\,X \\ c\,d\,Y \\ e\,f\,g \end{Bmatrix} \begin{Bmatrix} a\,b\,X \\ Y\,g\,h \end{Bmatrix} \begin{Bmatrix} c\,d\,Y \\ b\,h\,j \end{Bmatrix} = (-1)^{2h}(\hat{f})^{-2}\delta_{fj}\{bdf\}\begin{Bmatrix} e\,j\,g \\ h\,a\,c \end{Bmatrix}$$

(A.146)

$$\sum_{XY}(\hat{X}\hat{Y})^2 \begin{Bmatrix} a\,f\,X \\ d\,q\,e \\ p\,c\,b \end{Bmatrix} \begin{Bmatrix} a\,f\,X \\ h\,r\,e \\ Y\,g\,b \end{Bmatrix} \begin{Bmatrix} a\,h\,Y \\ g\,b\,s \end{Bmatrix} = \begin{Bmatrix} a\,b\,s \\ c\,d\,p \end{Bmatrix} \begin{Bmatrix} c\,d\,s \\ e\,f\,q \end{Bmatrix} \begin{Bmatrix} e\,f\,s \\ g\,h\,r \end{Bmatrix}$$

(A.147)

$$\sum_{XYZ}(\hat{X}\hat{Y}\hat{Z})^2 \begin{Bmatrix} a\,b\,p \\ c\,d\,X \\ q\,Y\,Z \end{Bmatrix} \begin{Bmatrix} p\,X\,Z \\ d\,e\,c \end{Bmatrix} \begin{Bmatrix} q\,Y\,Z \\ d\,e\,b \end{Bmatrix} = (-1)^{2e}\hat{d}^2 \begin{Bmatrix} a\,b\,p \\ e\,c\,q \end{Bmatrix} \quad \text{(A.148)}$$

$$\sum_{XYZ}(-1)^{X+Y-e-h}(\hat{X}\hat{Y}\hat{Z})^2 \begin{Bmatrix} a\,b\,X \\ c\,d\,Z \\ e\,f\,g \end{Bmatrix} \begin{Bmatrix} g\,b\,Y \\ c\,d\,Z \\ h\,j\,a \end{Bmatrix} \begin{Bmatrix} a\,b\,X \\ g\,Z\,Y \end{Bmatrix}$$
$$= (\hat{f})^{-2}\delta_{fj}\{bdf\}\begin{Bmatrix} e\,f\,g \\ h\,c\,a \end{Bmatrix}$$

(A.149)

$$\sum_{XYZ}(-1)^{X+f-a-l}(\hat{X}\hat{Y}\hat{Z})^2 \begin{Bmatrix} a\,b\,p \\ c\,d\,X \\ q\,Y\,Z \end{Bmatrix} \begin{Bmatrix} b\,d\,Y \\ c\,f\,Z \\ l\,k\,q \end{Bmatrix} \begin{Bmatrix} p\,X\,Z \\ c\,f\,d \end{Bmatrix}$$
$$= (\hat{k})^{-2}\delta_{pk}\{dfp\}\begin{Bmatrix} a\,b\,p \\ l\,q\,c \end{Bmatrix}$$

(A.150)

$$\sum_{XYZ}(\hat{X}\hat{Y}\hat{Z})^2 \begin{Bmatrix} a\,b\,p \\ c\,d\,X \\ q\,Y\,Z \end{Bmatrix} \begin{Bmatrix} c\,d\,X \\ e\,f\,Z \\ g\,h\,p \end{Bmatrix} \begin{Bmatrix} b\,d\,Y \\ e\,f\,Z \\ j\,k\,q \end{Bmatrix} = (\hat{h})^{-2}\delta_{hk}\{dfh\}\begin{Bmatrix} a\,b\,p \\ c\,e\,g \\ q\,j\,h \end{Bmatrix}$$

(A.151)

A.11.4.   Four 3nj-Symbols

$$\sum_{XY}(\hat{X}\hat{Y})^2 \begin{Bmatrix} a\,b\,c \\ d\,X\,Y \end{Bmatrix} \begin{Bmatrix} a\,e\,f \\ g\,X\,Y \end{Bmatrix} \begin{Bmatrix} c\,g\,p \\ f\,d\,X \end{Bmatrix} \begin{Bmatrix} b\,g\,q \\ e\,d\,Y \end{Bmatrix}$$

$$= (-1)^{2(c+e)} \begin{Bmatrix} a\,b\,c \\ g\,p\,q \end{Bmatrix} \begin{Bmatrix} a\,e\,f \\ d\,p\,q \end{Bmatrix} \tag{A.152}$$

$$\sum_{XY}(-1)^{X+Y}(\hat{X}\hat{Y})^2 \begin{Bmatrix} a\,b\,c \\ d\,X\,Y \end{Bmatrix} \begin{Bmatrix} a\,e\,f \\ g\,X\,Y \end{Bmatrix} \begin{Bmatrix} c\,f\,p \\ g\,d\,X \end{Bmatrix} \begin{Bmatrix} b\,e\,q \\ g\,d\,Y \end{Bmatrix}$$

$$= (-1)^{-a+b+c-d+e+f-g-p}(\hat{p})^{-2}\delta_{pq}\{dgp\} \begin{Bmatrix} a\,b\,c \\ p\,f\,e \end{Bmatrix} \tag{A.153}$$

$$\sum_{XY}(-1)^Y(\hat{X}\hat{Y})^2 \begin{Bmatrix} a\,b\,X \\ c\,d\,Y \\ p\,q\,r \end{Bmatrix} \begin{Bmatrix} X\,Y\,r \\ j\,h\,g \end{Bmatrix} \begin{Bmatrix} a\,b\,X \\ h\,g\,e \end{Bmatrix} \begin{Bmatrix} c\,d\,Y \\ j\,g\,f \end{Bmatrix}$$

$$= (-1)^{a-b+d+e-j-p+r} \begin{Bmatrix} e\,b\,h \\ f\,d\,j \\ p\,q\,r \end{Bmatrix} \begin{Bmatrix} a\,c\,p \\ f\,e\,g \end{Bmatrix} \tag{A.154}$$

$$\sum_{XYZ}(-1)^{X+Y+Z+a+b+c+d+e+f+p}(\hat{X}\hat{Y}\hat{Z})^2$$

$$\times \begin{Bmatrix} a\,b\,c \\ X\,Y\,Z \end{Bmatrix} \begin{Bmatrix} a\,e\,f \\ p\,Z\,Y \end{Bmatrix} \begin{Bmatrix} b\,f\,d \\ p\,X\,Z \end{Bmatrix} \begin{Bmatrix} c\,d\,e \\ p\,Y\,X \end{Bmatrix}$$

$$= (\hat{p})^2 \begin{Bmatrix} a\,b\,c \\ d\,e\,f \end{Bmatrix} \tag{A.155}$$

$$\sum_{XYZ}(\hat{X}\hat{Y}\hat{Z})^2 \begin{Bmatrix} X\,Y\,Z \\ d\,e\,f \\ a\,b\,c \end{Bmatrix} \begin{Bmatrix} X\,Y\,Z \\ f\,c\,g \end{Bmatrix} \begin{Bmatrix} X\,a\,d \\ b'\,g\,c \end{Bmatrix} \begin{Bmatrix} Y\,b\,e \\ d'\,f\,g \end{Bmatrix}$$

$$= (-1)^{2(b+d)}(\hat{b}\hat{d})^{-2}\delta_{bb'}\delta_{dd'}\{abc\}\{def\}\{bdg\} \tag{A.156}$$

$$\sum_{XYZ}(-1)^{2Z}(\hat{X}\hat{Y}\hat{Z})^2 \begin{Bmatrix} a\,b\,X \\ c\,d\,Y \\ e'\,f'\,g \end{Bmatrix} \begin{Bmatrix} a\,b\,X \\ f\,Z\,d \end{Bmatrix} \begin{Bmatrix} c\,d\,Y \\ Z\,e\,a \end{Bmatrix} \begin{Bmatrix} e\,f\,g \\ X\,Y\,Z \end{Bmatrix}$$

$$= (\hat{e}\hat{f})^{-2}\delta_{ee'}\delta_{ff'}\{ace\}\{bdf\}\{egf\} \tag{A.157}$$

$$\sum_{XYZ}(-1)^Z(\hat{X}\hat{Y}\hat{Z})^2\begin{Bmatrix}a & b & c' \\ d & e & f \\ X & Y & g\end{Bmatrix}\begin{Bmatrix}l & e & j \\ a & b & c \\ Z & Y & h\end{Bmatrix}\begin{Bmatrix}a & d & X \\ k & Z & l\end{Bmatrix}\begin{Bmatrix}X & Y & g \\ h & k & Z\end{Bmatrix}$$

$$=(-1)^{2c+b-d+g+j}(\hat{c})^{-2}\delta_{cc'}\{abc\}\begin{Bmatrix}k & j & f \\ c & g & h\end{Bmatrix}\begin{Bmatrix}k & j & f \\ e & d & l\end{Bmatrix} \qquad (A.158)$$

$$\sum_{XYZ}(-1)^{2Y-Z}(\hat{X}\hat{Y}\hat{Z})^2\begin{Bmatrix}a & b & X \\ c & d & Y \\ e & f & Z\end{Bmatrix}\begin{Bmatrix}g & h & X \\ k & l & Y \\ f & e & Z\end{Bmatrix}\begin{Bmatrix}a & b & X \\ g & h & j\end{Bmatrix}\begin{Bmatrix}c & d & Y \\ k & l & j'\end{Bmatrix}$$

$$=(-1)^{-b+c-h+k}(\hat{j})^{-2}\delta_{jj'}\begin{Bmatrix}a & c & e \\ l & h & j\end{Bmatrix}\begin{Bmatrix}b & d & f \\ k & g & j\end{Bmatrix} \qquad (A.159)$$

$$\sum_{XYZ}(-1)^{Y-a-b-c-f-h-p+q}(\hat{X}\hat{Y}\hat{Z})^2\begin{Bmatrix}a & b & X \\ g & c & q \\ p & Z & Y\end{Bmatrix}\begin{Bmatrix}b & d & f' \\ c & h & j \\ Z & Y & p\end{Bmatrix}\begin{Bmatrix}a & b & X \\ d & e & f\end{Bmatrix}\begin{Bmatrix}d & h & Y \\ q & X & e\end{Bmatrix}$$

$$=(-1)^{2j}(\hat{f})^{-2}\delta_{ff'}\{bdf\}\begin{Bmatrix}e & g & j \\ p & f & a\end{Bmatrix}\begin{Bmatrix}e & g & j \\ c & h & q\end{Bmatrix} \qquad (A.160)$$

$$\sum_{XYZ}(\hat{X}\hat{Y}\hat{Z})^2\begin{Bmatrix}a & b & X \\ c & d & Y \\ t & s & r\end{Bmatrix}\begin{Bmatrix}a & b & X \\ h & j & q \\ e & f & Z\end{Bmatrix}\begin{Bmatrix}k & l & p \\ c & d & Y \\ e & f & Z\end{Bmatrix}\begin{Bmatrix}p & q & r \\ X & Y & Z\end{Bmatrix}$$

$$=\begin{Bmatrix}k & l & p \\ h & j & q \\ t & s & r\end{Bmatrix}\begin{Bmatrix}k & h & t \\ a & c & e\end{Bmatrix}\begin{Bmatrix}l & j & s \\ b & d & f\end{Bmatrix} \qquad (A.161)$$

A.11.5. *Rearrangement of Angular Momenta during Summation*

Sometimes a summation may be completed after angular momenta are exchanged between different $3nj$-symbols. Numerous corresponding formulas can follow from the symmetry properties of the $3nj$-symbols of a higher order. Some useful expressions are presented below. Note that when handling sums, sometimes it is necessary to take a step back and even to increase a number of summation indices to achieve further progress (i.e., to use equations of Section A11 from right to left):

$$\sum_{\gamma}(a\alpha,b\beta|c\gamma)(d\delta,e\varepsilon|c\gamma)$$

$$= (-1)^{a+b+e+d}\frac{\hat{c}^2}{\hat{a}\hat{d}}\sum_{t\tau}\hat{t}^2\,(t\tau,e\varepsilon|a\alpha)(t\tau,b\beta|d\delta)\begin{Bmatrix}a\,b\,c\\d\,e\,t\end{Bmatrix} \qquad (A.162)$$

$$\sum_{X}\hat{X}^2\begin{Bmatrix}a\,b\,X\\c\,d\,p\end{Bmatrix}\begin{Bmatrix}c\,d\,X\\e\,f\,q\end{Bmatrix}\begin{Bmatrix}e\,f\,X\\g\,h\,r\end{Bmatrix}\begin{Bmatrix}g\,h\,X\\a\,b\,s\end{Bmatrix} = \sum_{X}\hat{X}^2\begin{Bmatrix}a\,f\,X\\d\,q\,e\\p\,c\,b\end{Bmatrix}\begin{Bmatrix}a\,f\,X\\h\,r\,e\\s\,g\,b\end{Bmatrix}$$

$$= (-1)^{q+s+a+f-p-r-e-b}\sum_{Y}\hat{Y}^2\begin{Bmatrix}p\,r\,Y\\d\,e\,q\\a\,h\,s\end{Bmatrix}\begin{Bmatrix}p\,r\,Y\\c\,f\,q\\b\,g\,s\end{Bmatrix} \qquad (A.163)$$

$$\sum_{X}(-1)^{a+b+c+d+e+f+g+h+p+q+r+s-X}\hat{X}^2\begin{Bmatrix}a\,b\,X\\c\,d\,p\end{Bmatrix}\begin{Bmatrix}c\,d\,X\\e\,f\,q\end{Bmatrix}\begin{Bmatrix}e\,f\,X\\g\,h\,r\end{Bmatrix}\begin{Bmatrix}g\,h\,X\\b\,a\,s\end{Bmatrix}$$

$$= \sum_{Y}(-1)^{2Y+a+b+e+f}\hat{Y}^2\begin{Bmatrix}s\,h\,b\\g\,r\,f\\a\,e\,Y\end{Bmatrix}\begin{Bmatrix}b\,f\,Y\\q\,p\,c\end{Bmatrix}\begin{Bmatrix}a\,e\,Y\\q\,p\,d\end{Bmatrix} \qquad (A.164)$$

$$\sum_{XY}(-1)^{X+Y}(\hat{X}\hat{Y})^2\begin{Bmatrix}a\,b\,X\\c\,d\,Y\\p\,q\,s\end{Bmatrix}\begin{Bmatrix}e\,f\,X\\g\,h\,Y\\r\,t\,s\end{Bmatrix}\begin{Bmatrix}a\,b\,X\\f\,e\,k\end{Bmatrix}\begin{Bmatrix}c\,d\,Y\\h\,g\,l\end{Bmatrix}$$

$$= (-1)^{k+l+a+c+p-s-h-t-f}\sum_{Z}(-1)^{Z}\hat{Z}^2\begin{Bmatrix}c\,g\,l\\a\,e\,k\\p\,r\,Z\end{Bmatrix}\begin{Bmatrix}f\,b\,k\\h\,d\,l\\t\,q\,Z\end{Bmatrix}\begin{Bmatrix}p\,r\,Z\\t\,q\,s\end{Bmatrix} \qquad (A.165)$$

Bibliography

1. Balashov, V. V., Belyaev, V. B., Korenman, G. Ya., Smirnov, Yu. F., and Yudin, N. P. (1965). *Theoretical Practicum in Nuclear Physics*, Moscow University Press, Moscow (in Russian).

2. Balashov, V. V., Kabachnik, N. M., Korenman, G. Ya., Korotkikh, V. L., Leonova, S. V., Mileev, V. N., Popov, V. P., Senashenko, V. S., and Strakhova, S. I. (1979). *Theoretical Practicum in Atomic and Nuclear Physics, Part II*, Moscow University Press, Moscow (in Russian).

3. Balashov, V. V., Korenman, G. Ya., Smirnov, Yu. F., and Yudin, N. P. (1980). *Theoretical Practicum in Atomic and Nuclear Physics, Part I*, Moscow University Press, Moscow (in Russian).

4. Balashov, V. V., Grum-Grzhimailo, A. N., Dolinov, V. K., Korenman, G. Ya., Krementsova, Yu. N., Smirnov, Yu. F., and Yudin, N. P. (1984). *Theoretical Practicum in Nuclear and Atomic Physics*, Energoatomizdat, Moscow (in Russian).

5. Devons, S., and Goldfarb, L. J. B. (1957). in: *Handbuch der Physik*, Springer-Verlag, Berlin, S. Flügge, ed., **42**, 362.

6. Fano, U., and Macek, J. H. (1973). *Rev. Mod. Phys.* **45**, 553.

7. Kessler, J. (1985). *Polarized Electrons*, 2d ed., Springer-Verlag, Berlin.

8. Blum, K. (1996). *Density Matrix Theory and Applications*, 2d ed., Plenum Press, New York.

9. Andersen, N., Gallagher, J. W., and Hertel, I. V. (1988). *Phys. Rep.*, **165**, 1.

10. Andersen, N., Broad, J. T., Campbell, E. E. B., Gallagher, J. W., and Hertel, I. V. (1997). *Phys. Rep.*, **278**, 107.

11. Andersen, N., Bartschat, K., Broad, J. T., and Hertel, I. V. (1997). *Phys. Rep.*, **279**, 251.

12. Andersen, N., and Bartschat, K. (2000). *Polarization, Alignment, and Orientation in Atomic Collisions*, Springer-Verlag, New York.

13. Danos, M., and Fano, U. (1998). *Phys. Rep.* **304**, 155.

14. Beyer, H. J., Blum, K., and Hippler, R., eds., (1988). *Coherence in Atomic Collision Physics*, Plenum Press, New York.

15. Kleinpoppen, H., and Newell, W. R., eds., (1995). *Polarized Electron/Polarized Photon Physics*, Plenum Press, New York.

16. Becker, U., and Shirley, D. A., eds., (1996). *VUV and Soft X-ray Photoionization*, Plenum Press, New York.

17. Campbell, D. M., and Kleinpoppen, H., eds., (1996). *Selected Topics on Electron Physics*, Plenum Press, New York.

18. Burke, P. G., and Joachain, C. J., eds., (1997). *Photon and Electron Collisions with Atoms and Molecules*, Plenum Press, New York.

19. Balashov, V. V., Grum-Grzhimailo, and A. N., Romanovsky, E. A., eds., (1996). *Autoionization Phenomena in Atoms*, Moscow University Press, Moscow.

20. Varshalovich, D. A., Moskalev, A. N., and Khersonskii, V. K. (1988). *Quantum Theory of Angular Momentum*, World Scientific, Singapore.

21. Born, M., and Wolf, E. (1970). *Principles of Optics*, Pergamon Press, New York.

22. Ferguson, A. T. (1965). *Angular Correlation Method in Gamma-Ray Spectroscopy*, North-Holland, Amsterdam.

23. Bederson, B. (1969). *Comm. At. Mol. Phys.*, **1**, 41, 65; (1970). **2**, 160.

24. Kleinpoppen, H. (1999). *Rev. Mod. Phys.*, **72**, S226.

25. Van der Burgt, P. J. M., Westeveld, W. B., and Risley, J. S. (1989). *J. Phys. Chem. Ref. Data*, **18**, 1757.

26. Eminyan, M., MacAdam, K. B., Slevin, J., and Kleinpoppen, H. (1973). *Phys. Rev. Lett.*, **31**, 576.

27. Hertel, I. V., and Stoll, W. (1977). *Adv. At. Mol. Phys.*, **13**, 113.

28. Percival, I. C., and Seaton, M. J. (1958). *Phil. Trans. R. Soc. London*, **A251**, 113.

29. Fano, U. (1961). *Phys. Rev.*, **124A**, 1866.

30. McCarthy, I. V., and Weigold, E. (1978). *Endeavour*, New Series, ABC, **2**, 72.

31. Shore, B. W. (1967). *J. Opt. Soc. Am.*, **57**, 881.

Index

Series Publications

Below is a chronological listing of all the published volumes in the *Physics of Atoms and Molecules* series.

ELECTRON AND PHOTON INTERACTIONS WITH ATOMS
Edited by H. Kleinpopper and M. R. C. McDowell

ATOM–MOLECULE COLLISION THEORY: A Guide for the Experimentalist
Edited by Richard B. Bernstein

COHERENCE AND CORRELATION IN ATOMIC COLLISIONS
Edited by H. Kleinpoppen and J. F. Williams

VARIATIONAL METHODS IN ELECTRON–ATOM SCATTERING THEORY
R. K. Nesbet

DENSITY MATRIX THEORY AND APPLICATIONS
Karl Blum

INNER-SHELL AND X-RAYS PHYSICS OF ATOMS AND SOLIDS
Edited by Derek J. Fabian, Hans Kleinpoppen, and Lewis M. Watson

INTRODUCTION TO THE THEORY OF LASER–ATOM INTERACTIONS
Marvin H. Mittleman

ATOMS IN ASTROPHYSICS
Edited by P. G. Burke, W. B. Eissner, D. G. Hummer, and I. C. Percival

ELECTRON–ATOM AND ELECTRON–MOLECULE COLLISIONS
Edited by Juergen Hinze

ELECTRON–MOLECULE COLLISIONS
Edited by Isao Shimamura and Kazuo Takayanagi

ISOTOPE SHIFTS IN ATOMIC SPECTRA
W. H. King

AUTOIONIZATION: Recent Developments and Applications
Edited by Aaron Temkin

ATOMIC INNER-SHELL PHYSICS
Edited by Barnd Crasemann

COLLISIONS OF ELECTRONS WITH ATOMS AND MOLECULES
G. P. Drukarev

THEORY OF MULTIPHOTON PROCESSES
Farhad H. M. Faisal

PROGRESS IN ATOMIC SPECTROSCOPY, Parts A, B, C, and D
Edited by W. Hanle, H. Kleinpoppen, and H. J. Beyer

RECENT STUDIES IN ATOMIC AND MOLECULAR PROCESSES
Edited by Arthur W. Kingston

QUANTUM MECHANICS VERSUS LOCAL REALISM: The Einstein-Podolsky-Rosen Paradox
Edited by Franco Selleri

ZERO-RANGE POTENTIALS AND THEIR APPLICATIONS IN ATOMIC PHYSICS
Yu. N. Demkov and V. N. Ostrovskii

COHERENCE IN ATOMIC COLLISION PHYSICS
Edited by H. J. Beyer, K. Blum, and J. B. West

ELECTRON–MOLECULE SCATTERING AND PHOTOIONIZATION
Edited by P. G. Burke and J. B. West

ATOMIC SPECTRA AND COLLISIONS IN EXTERNAL FIELDS
Edited by K. T. Taylor, M. H. Nayfeh, and C. W. Clark

ATOMIC PHOTOEFFECT
M. Ya. Amusia

MOLECULAR PROCESSES IN SPACE
Edited by Tsutomu Watanabe, Isao Shimamura, Mikio Shimizu, and Yukikazu Itikawa

THE HANLE EFFECT AND LEVEL CROSSING SPECTROSCOPY
Edited by Giovanni Moruzzi and Franco Strumia

ATOMS AND LIGHT: INTERACTIONS
John N. Dodd

POLARIZATION BREMSSTRAHLUNG
Edited by V. N. Tsytovich and I. M. Ojringel

INTRODUCTION TO THE THEORY OF LASER–ATOM INTERACTIONS (Second Edition)
Marvin H. Mittleman

ELECTRON COLLISIONS WITH MOLECULES, CLUSTERS, AND SURFACES
Edited by H. Ehrhardt and L. A. Morgan

THEORY OF ELECTRON–ATOM COLLISIONS, Part 1: Potential Scattering
Philip G. Burke and Charles J. Joachain

POLARIZED ELECTRON/POLARIZED PHOTON PHYSICS
Edited by H. Kleinpoppen and W. R. Newell

INTRODUCTION TO THE THEORY OF X-RAY AND ELECTRONIC SPECTRA OF FREE
ATOMS
Romas Karazija

VUV AND SOFT X-RAY PHOTOIONIZATION
Edited by Uwe Becker and David A. Shirley

DENSITY MATRIX THEORY AND APPLICATIONS (Second Edition)
Karl Blum

SELECTED TOPICS ON ELECTRON PHYSICS
Edited by D. Murray Campbell and Hans Kleinpoppen

PHOTON AND ELECTRON COLLISIONS WITH ATOMS AND MOLECULES
Edited by Philip G. Burke and Charles J. Joachain

COINCIDENCE STUDIES OF ELECTRON AND PHOTON IMPACT IONIZATION
Edited by Colm T. Whelan and H. R. J. Walters

PRACTICAL SPECTROSCOPY OF HIGH-FREQUENCY DISCHARGES
Sergei A. Kazantsev, Vyacheslav K. Khutorshchikov, Günter H. Guthöhrlein, and Laurentius
Windholz

IMPACT SPECTROPOLARIMETRIC SENSING
S. A. Kazantsev, A. G. Petrashen, and N. M. Firstova

NEW DIRECTIONS IN ATOMIC PHYSICS
Edited by Colm T. Whelan, R. M. Dreizler, J. H. Macek, and H. R. J. Walters

ELECTRON MOMENTUM SPECTROSCOPY
Erich Weigold and Ian McCarthy

POLARIZATION AND CORRELATION PHENOMENA IN ATOMIC COLLISIONS: A Practical
Theory Course
Vsevolod V. Balashov, Alexei N. Grum-Grzhimailo, and Nikolai M. Kabachnik